AF769473

Preparation for

Introductory College Physics
A GUIDED STUDENT PRIMER

Jon A. Celesia
San Francisco State University

Brooks/Cole Publishing Company
ITP® An International Thomson Publishing Company

Pacific Grove • Albany • Belmont • Bonn • Boston • Cincinnati • Detroit • Johannesburg • London
Madrid • Melbourne • Mexico City • New York • Paris • Singapore • Tokyo • Toronto • Washington

Developmental Editor: *Keith Dodson*
Assistant Editor: *Linda Row*
Editorial Assistant: *Georgia Jurickovich*
Marketing Communications: *Margaret Parks*
Art Editor: *Lisa Torri*
Production Editor: *Mary Vezilich*
Printing and Binding: *West Publishing Co.*
Cover Design: Laurie Albrecht

For more information, contact:

BROOKS/COLE PUBLISHING
511 Forest Lodge Road
Pacific Grove, CA 93950
USA

International Thomson Publishing Europe
Berkshire House 168-173
High Holborn
London WC1V 7AA
England

Thomas Nelson Australia
102 Dodds Street
South Melbourne, 3205
Victoria, Australia

Nelson Canada
1120 Birchmount Road
Scarborough, Ontario
Canada M1K 5G4

International Thomson Editores
Seneca 53
Col. Polanco
11560 México, D. F., México

International Thomson Publishing Japan
Hirakawacho Kyowa Building, 3F 418
2-2-1 Hirakawacho
Chiyoda-ku, Tokyo 102
Japan

International Thomson Publishing Asia
221 Henderson Road
#05-10 Henderson Building
Singapore 0315

International Thomson Publishing GmbH
Königswinterer Strasse
53227 Bonn
Germany

Printed in the United States of America

5 4 3 2 1

ISBN 0-314-20933-6

Vol. 1 Contents

Hey You!! There are a few things that you should know before jumping into this tutorial-primer. So for your own sake, please skim the few pages under "You Should Know." For the moment, I'll give you a quick overview of an effective way to use this material to build your skills. First, skim over Part A and highlight techniques to use as you study. Discipline yourself to practice and develop these study habits. Return to Part A periodically to see what techniques you have forgotten about or could use more powerfully. Second, you can use Part B either as a reference to help you when you find a weakness in your math, or as a primer that you work through before the course really takes off. Third, Part I corresponds with the **Lea & Burke calculus-based text** ***Physics: The Nature of Things***. Since these topics are covered by all physics texts, you can still get value from this tutorial if you are using another text. You can even use this tutorial if you are in an algebra-based course. Simply skip over anything that gets too deep or uses calculus.

In Part I the page numbering system is: PART.CHAPTER.PAGE

TO THE STUDENT

Physics is many things. It has the reputation for being very difficult, time consuming, and even inaccessible. To *some* degree, this is true. I tell my students that the course is going to be like running an uphill race. The catch is that the surface that you're running on is a conveyor belt going the other way! "It's like running up the down escalator," as one student put it. "If you don't do *some* amount of work **daily**–at least a quick review, a power preview, or a problem or two–the stuff slips out of you memory like mercury!"

Your final grade usually depends on what level you reach by the end of the semester. And that's fair enough. But what may not be fair is that each individual starts the race at different places on the escalator! Some haven't even gotten to the escalator because their algebra and trig. skills are too weak, possibly from rust. Others are in their fourth semester of calculus (a nice place to be!). Such is life. You can't expect to go out for a football team and make it if you've never played. And it doesn't matter how hard you work if you can't get the job done. It's true in academia, industry, medicine, and most job environments.

Can't wait? ... There is a bright side. If you approach physics correctly, it can actually be fun, exciting, mind-expanding, and even addicting. Soon you'll be seeing the physics in everything around you! Sure you may lose a friend or two from your zeal, but hey, they won't be appreciating a gymnast or dancer they way you will be! (Nose snub.) After all, you're going to be learning the very rules by which the physical universe operates. You'll learn how our "common sense" is sometimes reliable and sometimes completely wrong. It took many men and women mucho tiempo thinking about the nature of things, battling conventional thought, and devising more and more precise experiments to delve into realms that are inaccessible to our senses. This knowledge is the fabric of any technologically based field (just about everything, these days). And this is what you are endeavoring to learn. ... There. Feel better?

So read on to see how this guide can help you run the race effectively. There will be frustration, and it's not easy, but what of value is obtained from an arm chair...unless there is a computer in front of it? So GO FOR IT!.......This knowledge is a gift to yourself, and, if used constructively, to others.

ABOUT THIS TUTORIAL

This tutorial-primer does not simply outline the text and work a few examples. Rather it is designed to help layer in the information and techniques in a way that expedites and deepens the learning experience. It certainly does not cover everything, but it attempts to clear up many common misconceptions and help you, the student, over many difficult spots.

This is the first printing of the first volume of this tutorial. Subsequent volumes will be published if students find it helpful. This volume is also being used by some schools as the basis for a Physics Preparation course for both algebra-based and calculus-based college physics. Any ideas that you have to improve the tutorial will be closely considered and greatly appreciated. (Feedback is probably best channeled through the publishers, although you may try me at jcelesia@stars.sfsu.edu or see me at my budding web site, http://www.physics.sfsu.edu/~celesia.)

Structure

The approach of layering in the material is reflected in the structure of the tutorial. Each chapter is addressed sequentially in Conceptual, Basic, Intermediate, and occasional Advanced sections. Ideally, this will help make the course accessible and enjoyable without it turning into a plug and chug exercise in applying algebra and trigonometry. (Why keep using 10 newtons, or 100 newtons, without having any idea what it physically feels like?)

Much of what is shared in this primer-tutorial is contained in the exercises. This reflects the philosophy that people learn better if they are actively engaged. The exercises vary greatly in length and are not all of the standard physics-problem style. Some require simple thought experiments, others require throwing an object across the room and noticing what happens, and others require a full-fledged mathematical analysis. All are designed to bring out the conceptual and mathematical subtleties that often trip unsuspecting prey. Every exercise has an instructive solution at the end of the chapter, and some have hints or solution plans to nudge the student over difficult spots. There is also a quiz section, followed by solutions, that incorporates true-false, multiple choice, and short answer questions to reinforce the distinctions that you should be making.

There are two reference sections provided: Part A - How To Study Physics, and Part B - Mathematic Land. Surprisingly, they cover tips on how to effectively study physics and a review of the basic math required for the course. You may want to review and train these skills–study, math, visualization, and problem solving–in a preparation course.

Effective Use

The most effective way to approach a course (lecture, office hours, T.A.s, text, study guide, tutors...) is to become *actively* involved. Hence, you shouldn't expect to get much out of this tutorial by merely reading it whenever you get around to it. Rather, the guide should actually be used as a *tutor* – a study companion of sorts. Here are some tips:

- Start by reading over **Parts A and B** ASAP. Try to implement the ideas in Part A immediately. Be sure to flip back to Part A throughout the course to see if you can pick up some useful tips. Use Part B to get you up to speed for the course. The sooner you do this, the better. If you have any weaknesses in your math background attend to them as soon as possible. After your initial diagnostic, use Part B as a reference.

- On the weekend, read the **Conceptual and Basic sections** of the chapter (or topics) to be covered in lecture the next week. Keep a **Study Journal** where you work the exercises and practice problems, maybe write out explanations to yourself, and definitely list questions that arise. (Also preview the material in the text.) It won't take you *that* long and it will certainly save you time in the long run!

- If you get stuck on an exercise, check the **Hint/Solution Plan section** to see if there is something there that will get you rolling. (I've tried to note the handful of exercises that have hints provided.)

- Bring your **questions to class**. Also bring them to your instructor's or T.A.'s office hours, and ask your **study partners**.

- Get deeper into the material in the **Intermediate and Advanced sections**. This should support your progress on your assigned problem set.

Final note: It may seem as though this just increases the amount that you have to do. Not so! The questions and subtleties raised in this tutorial are already part of a physics course. You can either tackle them randomly on your own, or knock 'em down one by one. The text and your instructor can only do so much in the limited space and time that they have. Learning physics takes an active role on your part and a hefty time investment. This guide will help show you how to spend your time studying the subject. It will be challenging, rewarding and empowering.

So Get Active...and Enjoy The Process!

Life Science students and everyone taking an algebra-based (i.e. no calculus) physics course, read on!

ALGEBRA-BASED/LIFE SCIENCE PHYSICS

If you are a life science major, or for whatever other reason are taking an algebra-based (without calculus) physics course, I have good news for you. You get to cross off some of the sections in this tutorial. It's not that much, but I think you'll be happy that I've told you which sections you can completely skip and which you can just skim.

Omit

Put a big red X through these sections of the tutorial. I suggest doing it now so that you don't start fretting over some material that you don't even need. [Be aware if your instructor tells you that she or he feels that one of these topics *is* important to you.]

PART B:

Section B.5.d
- exercise B.5, part d

Section B.6.b
- exercise B.7, parts c & d

Section B.6.d
- exercises B.10 & 11

Section B.6.e
- the trig identities listed as "Basic" and "Not So Basic"
- exercise B.12

Section B.8.e
- component method for the cross product; includes permutations and determinants.
- exercise B.17, parts b & c

Section B.8.f through B.10
- all (although you may enjoy the section on imaginary numbers)

CHAPTER 2:

exercise 2.26

any "differentiation" or "taking derivatives" of kinematics vectors

Skim

PART B:

Section B.7 should just be skimmed

CHAPTER 0:

You should not attempt to memorize this short summary of philosophical ideas about and historical development of physics. I hope it is, however, enjoyable and stimulating.

CHAPTER 1:

exercise 1.23 & 1.24 (some instructors *may* include this)

Special Note: "Unit Vectors" vs. "Component Vectors"

Any concept that includes direction is called a "vector" concept: the location of the moon relative to you (you have to point at it, so there is an associated direction); your velocity when you walk (you can't walk without having a direction!); and the force that you exert on a page as you turn it (you can't push or pull on something without doing it in some direction). There are, of course, plenty of non vector concepts (called "scalar" to distinguish the two types)–speed, time, and amount of money in your pocket, for example. As you can imagine, you will encounter plenty of both vector and non vector concepts in physics and these will be what the algebraic variables stand for. Soooooooooo..........

What I want to call to your attention here is that algebra-based physics texts often represent vectors in a slightly different manner than calculus-based texts. It's not a big deal, but you will notice the difference. Let me illustrate this with an example.

EXAMPLE: Say that you are traveling in a train at a speed of 100 mi/h and in the direction 30° north of east (see the picture below). How would you represent this in a picture, how would you represent this with an algebraic variable, and how would you communicate how much you were traveling eastward and how much northward (this latter representation is called "component" form).

Picture ("Graphical") Representation:

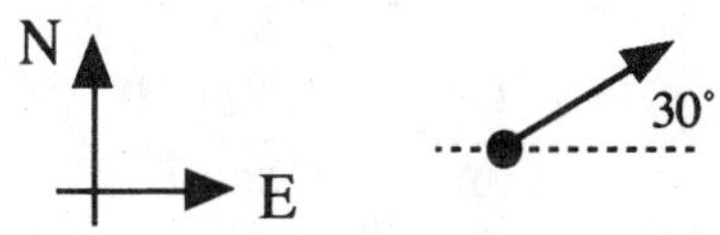

Vector Variable Representation:

Put an arrow over the variable, and if you are typing on a computer, make the letter bold. So velocity is written as "$\vec{\mathbf{v}}$."

Component Vector Representation:

Used by many algebra-based texts.

The actual velocity is broken down into the eastward velocity and the northward velocity. These "component" velocities are each written as vectors

$$\vec{\mathbf{v}}_{\text{actual}} = (\text{eastward velocity}) + (\text{northward velocity}) = \vec{\mathbf{v}}_{\text{eastward}} + \vec{\mathbf{v}}_{\text{northward}}$$

Unit Vector Representation:

Used by calculus-based texts and this tutorial.

The total velocity is again broken down into the eastward velocity and the northward velocity, but now we split the "component" velocities into speed and direction. The *unit vectors* are called "i-hat" and "j-hat" and are simply symbolic for each direction

$$\begin{aligned}\vec{\mathbf{v}}_{\text{actual}} &= (\text{eastward velocity}) + (\text{northward velocity})\\ &= (\text{eastward speed})(\text{east}) + (\text{northward speed})(\text{north})\\ &= v_{\text{eastward}}\,\hat{\mathbf{i}} + v_{\text{northward}}\,\hat{\mathbf{j}}\end{aligned}$$

Notice that algebra-based physics books (for life science folks) and calculus-based texts (for physical science folks) use the same arrow and variable representations, but the component representations are just a bit different. The unit vector notation is more powerful in complex calculations, but the component vector representation is a bit more transparent.

TO THE INSTRUCTOR

About This Project

As you may have read in the section above, this tutorial-primer can be, and is being, used as the basis for a preparatory physics course. (Preliminary versions have been very well received by students so far.) In the prep. course, I also have students buy the text that they will be using and provide them with worksheets on such topics as significant figures, ugly algebra, tricky trig, words to pictures, graphs to words, breaking processes into states and stages, and free body free for alls (soon available from the publisher?). I am currently looking into ways that we can integrate interactive lecture demonstrations, for example 1-D motion sensors, into the class. Contact me if you are interested in seeing the course schedule and collaborative learning worksheets. [Note: I really should say "we," are developing this course, as it is an NSF funded project, through MASTEP (Math and Science Teacher Education Program), that includes Barbara Uchida and Mohsen Janatpour from The College of San Mateo, and Annette Rapplyea and Peter Rowe from City College of San Francisco.]

This tutorial-primer may also be used by students entering their first semester courses. It is particularly well suited as a companion to the Lea & Burke text, *Physics: The Nature of Things*. I also use it for my algebra-based class, since the conceptual, basic, and intermediate level concepts and problems are much the same.

Obviously this volume is limited in scope. It serves the purpose of a prep. tutorial well, and it gives students a strong start in the introductory courses. If what I have written so far is well received by students and instructors, I will expand the tutorial to cover later chapters in volumes 2-4. As this is a very time consuming task, I am hesitant to proceed without further verification that it is well received.

One final note: Please consider this a work in progress. Every time I turn around I think of something that I could say more clearly, an exercise that would help students get over a conceptual hump, or a clearer method outline that will help structure the students' thinking and get them to focus on the physics. Already there are some pretty significant changes that will appear in future editions. Thus I will be very receptive to any suggestions that you might have.

So Much To Say, So Little Time

[The following addresses the use of this tutorial in your introductory physics courses.] As you well know, there is an enormous amount to communicate when teaching physics. There are a plethora of new concepts and the physical intuition to go along with them, the logical development and methods, and the coupling of problem solving and mathematical skills. On top of which, many students have very poor study skills. At every turn there is some kernel of wisdom or point of clarification that could benefit many of your students... and you have 2½ hours of class time a week to get it all across. To make matters worse, the students are in a state of information overload and sleep deprivation, and often have weak math and physics backgrounds coming into the class. The net effect is that a kernel passed on is often a kernel lost–your family, friends, and colleagues have heard all about it.

To assist all of us in our endeavor to truly *teach* physics, I have attempted to write a tutorial/study guide that both prepares the student for the lecture, and highlights those kernels of insight that are either mentioned briefly or are left in the "If I only had another 20 minutes" closet. If I have succeeded to some degree, this guide will make the question of how much time to spend on conceptual development, demonstrations, and problem solving more manageable. In fact, one of my goals is to free us up to use more class time for collaborative learning exercises and interactive lecture demonstrations. I am currently working on creating a very structured learning environment for my students. This will take the form of detailed study guidelines such as "On Monday we will begin working on Ch. X. To prepare yourself for the in class exercises, skim Ch. X of the text, work through the conceptual and basic parts of the tutorial, and try a few of the assigned problems." Please feel free to share any comments or suggestions that may improve this tutorial. (jcelesia@stars.sfsu.edu, or via the publisher.)

Using This Tutorial In Your Class

We all know that previewing material before a lecture greatly increases retention, yet few students do it. We also know that it generally takes a few readings before the material starts to make sense. This wall of apparent inaccessibility needs to be broken down before many students loosen up and liven up enough to engage in the class. The following plan is one attempt at doing just that.

- "Require" the students to read the Conceptual and Basic sections of the guide for the chapter (or topics) *before* your lecture. This should be quite accessible, and prepare them to benefit from class time. Recommend that they work and think through the exercises in those sections and bring to class any questions that arise. They should also do a 30 minute preview of the text material. In class, cover the material conceptually and elucidate with demonstrations. Have the students work a few simple problems, ideally in groups of 3.

- The students should then move on to their problem sets. Reading the text more carefully and working through the intermediate portion of this guide should support their efforts. Encourage them informally or formally to work in groups.

- In your next class, get deeper into the material and work or set up more difficult problems. Again, the ideal would be to have them break into small (but diverse) groups and get their hands and minds engaged in the process–they need to get it now!

I have used take home and in class quizzes to give a little more weight to the reading requirements. It may be enough to reflect the requirement in the class pace and interactions (ask individual students answerable questions?). Certainly it is the student's responsibility to keep up, but it is our responsibility to both convey the material as clearly as possible and enforce high standards.

About My Style

You will find that I am at times redundant, colloquial, and creative in grammatical construction. For the most part, this is purposeful. Based on student feedback, I believe that an informal tome ... ooh, nice Freudian typo, I think I'll leave it...an informal tone tends to allay the hypnotic state cast by hours, days, weeks, months, and years of dry – but very interesting! – reading. I simply want them to learn this material well, no matter what it takes. If my style gets in the way of, rather than supports, your students' comprehension, please let me know about it.

THANKS

My Students - For their probing questions that teach me what to say and how to say it.

Arinna Hollander - My wife, who has given up too many trips to see the whales and the wildflowers while I sit at the computer. Thank you for supporting, pushing, counseling, and waiting.

Jacqueline Johanson - My mom. Thanks for your endless support and love.

Dr. David Eakin - John Swett High School physics teacher, department head, life changer, and more. Thank you for changing my life (Viva la Grand Canyon!).

Keith Dodson - Assistant Editor at Brooks/Cole. For his support, wisdom, clarity, and timely pressure. Thanks for hanging in there.

Susan Lea & John Burke - San Francisco State Univ. Thank you both for the exceptional text, your commitment to teaching, and your guidance and support.

Mary Kay Horvath - Thank you for your invaluable comments as an editor and student.

Family & Friends - Maybe now I'll stop talking about this project when you ask me how I'm doing...maybe. Thanks for caring. [Dr. Jerry Fisher, Dr. Bob Rogers, Veronica Crane, Nola Burger, Dr. Mark & Kathy Detzer, Sid & Renee Hollander, Don Johanson, Kym & Jason Nesbit, Anthony Celesia, & in loving memory of my father, Jerry Celesia.]

PART A...

A. How To Study Physics . . .
(efficiently & effectively)

> "Active learning is FUN!
> Passive learning will put you to sleep, at best."

A.1. Read Me, Hear Me, Trust Me, Use Me

[My soap box, please.] Do you really want to learn physics? You need to decide soon, because it will affect how you approach the class. Consciously or not, you go into a class with a certain attitude and certain expectations. You might think that you know it all and don't need any help. (It's lonely at the top.) You might think that physics is much too difficult for you to do anything other than simply get through the course. You might think that *you* are the problem - maybe you weren't born with some inherent skills that the people getting A's were born with. You might think that your physics courses are "just hoops to jump through" or "weeder courses," i.e., not really pertinent to your career. Any one of the mind sets above, and certainly many more, will work against you. **Give yourself a chance to learn**.

My purpose in writing this study guide is to help you achieve your goal. It's easier for me to do that in person than in print, so I'll need your help. First, **consider me to be your trainer**, your guide, one of your study companions. My advice comes from years of studying, tutoring, and teaching physics. While you may not understand why I've recommended something, in fact you may think that it is a waste of time, I encourage you to give my advice a chance and see if it doesn't increase your facility with physics. Second, **actively APPLY the advice given to you**. (Now put your tongue on this here 9 volt battery.) Seriously, nodding your head is not even close to good enough in this game. You've got to **practice by doing**. You need to be able to meet the challenge of seeing a downright evil problem for the first time by methodically working through it. You've got to learn to lead. To do that, *you* will need to get clear on and quick with the material through intelligent questioning and drilling. Third, **lively-up yourself!** By that I mean get involved in the class and interested, if not excited, about the material. Simply wanting, really wanting, to learn something is, in my opinion, one of the best ways to increase your retention of the material, as well as enjoyment of the process. If you give it a chance, I know that you will find that physics is absolutely mind-blowing and exceedingly applicable to all science based fields.

A.1.a. Physics Made Easy?

> ***Everything should be made as simple as possible, but not simpler. - Albert Einstein***

There is no question, physics is a tough subject. It takes time, determination, and a "smart" approach. It is a good idea to assess your strengths and weaknesses going into the course. In this way, you can set realistic goals (like working 40 hours/wk, taking 3 science courses, and shooting for straight A's). Consider the following success factors: background (math, logical thinking, science); desire; effort (with solid concentration); hefty time investment; and a "smart" approach to studying. For example, going into a physics course with weak algebra and trigonometry skills is like entering a race with your shoe laces tied together. No one could win that way. Unrealistic goals are self-defeating and demoralizing. (On the other hand, wimpy goals, for example to "just get through the class," don't raise one's self-esteem either.) So, be realistic and determined. I'll do my best to help you sort out the methods, logic, concepts, and facts of physics. Together with your instructor, classmates, text, and any other resources that you enlist to help you, you should be in for a rewarding experience.

ADVICE: There's no way, nor any need, to memorize all of this in the first week of classes. **Use it as a reference.** Skim through this section (Part A) before and after studying each chapter. Continue doing this throughout the course and look for ideas and study techniques that you aren't fully utilizing. "Just do it!" You will fully appreciate this advice only after you have followed it for a few chapters. This is part of your aggressive, "active," "smart" approach to learning.

A.2. What Are Good Study Habits?

An unexpected freedom can be found through discipline. Good study habits → freedom. Check it out!

Read, reread, re-reread, drill, train, practice, work things out without looking at the solutions, ask yourself questions, work in groups, ask your instructor questions, work with classmates, read other resources, work in groups - in short, do whatever it takes and be an *active* learner. Need I say more? In my experience, yes. The following elements are fundamental to any effective and efficient study routine. Discipline yourself to implement the plan and you will reap the benefits.

1. **Pre-read** the material to be covered in class. Then reread as needed.
2. Keep a **study journal** - include anything that you aren't totally clear on, big or small, conceptual or mathematical. Include confusing concepts, words, phrases, logic, mathematics, plenty of doodles, or anything else. Also, use this to work the tutorial exercises and any other practice problems.
3. Get your **questions answered.** Then check off the questions in your journal.
4. Create **summaries.**
5. **Practice regularly**. An absolute minimum of 15 minutes a day. (It makes a bigger difference than most people realize.)
6. **Talk** about the material with others on a regular basis–an absolute must do!
7. Consciously implement **memorization techniques**. (But don't think memorizing every formula is the key to success in physics. It's often the road to ruin.)

You see that this will indeed take a conscious, active, disciplined effort on your part. You will gain more knowledge and a well-earned sense of accomplishment. In addition, you are likely to get a better grade, earn the respect of your instructor and classmates, win a

Nobel prize–just checking–, and reduce the amount of time spent stuck in paralyzed frustration. So many students nod their heads at this, but they don't *apply* it. Be different.

It may be worthwhile to go into some of the guidelines mentioned above in more detail, since I find that students, almost uniformly, do not apply these techniques powerfully. (Interestingly enough, it appears to me that anyone who is good in their field applies techniques along these lines.) If you think you are completely clear, see if you can't find something new here that will be of use to you. And as you know, **USE** is the key. **If you don't use it, you don't get it.**

A.2.a. Circling In (Hi. What's Your Major? Will You Marry Me?)

As you study, USE the image of spiraling in to help guide your efforts and track your progress.

Here we play the circle game, or better yet the spiral game. Your target, a detailed understanding with a clear view of the big picture, lies at the center of your circle. With each pass–first reading of the chapter and attempt of easy problems, lecture, rewriting lecture notes, second reading of the chapter and attempt of more problems, and so on, your circle gets smaller and smaller as you get closer and closer to your target. It is very helpful to know that you can't expect to retain much of the details on the first pass at the new material. (That's why waiting until lecture for you're first pass is a BAD idea!) Should a beginning juggler start with five bowling pins? Of course not. One or two bean bags are a good place to start.

Another analogy I like to use for this very efficient and effective study plan is that of a relationship. It goes something like this:

1. **HANDSHAKE** - hi, glad to meet you.
2. **GETTING ACQUAINTED** - what's your major? Where did you come from?
3. **BECOMING CLOSE FRIENDS** - I want to know *all* about you.

The handshake, i.e. first reading of the chapter, should absolutely be done *before* going to the lecture on the subject. Again, don't expect to understand it all on the first pass. That'll just demoralize you and slow down the process. Try to outline the chapter by picking out the most important points. You might even try a few easy problems. Then in lecture you won't be totally lost and you can get acquainted by layering in more details. I hope it is obvious to you how much more efficient this is than waiting until lecture to be introduced to a topic. You will move from getting acquainted to being old friends through working more and more problems, discussing the material with classmates, the instructor, T.A.s, rereading the text and lecture notes, and maybe even composing some explanatory notes of your own. By **spiraling in from the big picture to the fine details,** your progress will be steady, discernible, more efficient, less haphazard, and much less frustrating. Now that's smart!

A.2.b. The Power Preview

This is so important that it warrants a little more discussion. I like to quiz students at the beginning of a class on material that we are going to cover that day. Unfortunately, I find that most students have a hard time sorting out the main ideas from the fine details. Try this: read a page, or even a paragraph, and memorize every word the first time through. Silly, of course. But notice that you not only aren't able to memorize the material, but you also don't come away knowing what is being said. As I've said, the **broad ideas provide the scaffolding for the details.** You simply can't understand new material deeply on the first pass. (This is why cramming isn't very effective.)

Here are some suggestions for what you can do on that first pass. Keep in mind that you don't have to get it all from the first reading. More accurately, you *won't* get it all from the first reading.

1. **Do it *before* class.** (You've got many passes to make, so get started. Besides, the lecture will make more sense. If lecture doesn't make sense, you need to take some responsibility for that.)
2. **Create a chapter outline in your own words.** Do it quickly. This is only an initial step.
3. **Read the chapter summary first!** You've already got an outline there for you.
4. **Read the headings in the chapter.** You might also read the first and last paragraphs of each section. What are some of the main ideas? What methods are going to be used? What special cases are you going to be studying in detail? Set these ideas firmly in your mind.
5. **Look over many of the figures.** Skim-read the captions.
6. Get out your **Study Journal** and take notes.
7. Try some of the **easy, "get acquainted," problems** before lecture.
8. I almost forgot. **Read me!** Go as far as you can *before* class, but at least get an introduction to the concepts and basic relations to be discussed.

Remember, 15 minutes is much better than none. Just having seen the main ideas, words, and equations will help in lecture. Of course you won't have all of the details worked out, but with a clear chapter outline in mind, the details will fall into place more easily. Keep the following analogy in mind: if you were going to study the human body for the first time, a good place to start would be with the major bones, muscles, and organs, not with hair follicles.

A.2.c. The Study Journal (Huh?)

If you read or hear something and you don't understand it, what do you do? In the information-overload state of studentdom, it is impossible to remember everything that we are not yet clear on. Too often there are questions that we forget to ask and so don't get answered. These unanswered questions will continue to pop up as stumbling blocks in the way of our success.

The solution is simple and concrete: keep a "study journal" (I also call this a "questions journal") where you should keep your questions, worked tutorial exercises and all other practice problems in it. **When you read something that you don't fully understand, or when you get stuck in the course of solving a problem, make a note of it in your journal and MOVE ON!. Then cross off the question when you get it answered.** Your instructor (or friend, T.A., the text on rereading, etc.) may be able to clear up your confusion in a few minutes. That's better than not getting anywhere for hours as you curse Zeus, beat up your stuffed animals, snack on junk food, and organize your paper clips. Your time and mental health are valuable, and the study journal is a big step toward time-efficient learning and peace of mind.

A.2.d. Out Of The Closet (Working With Others)

Cooperation far outshines isolation or competition.

Have you ever had the experience where you were confused about something and as you explained your confusion to someone else you answered your own question? Why couldn't you solve it without someone there as a sounding board? I'm not sure either, but it's a common phenomenon. Again, we're looking at saving time and keeping the flow of learning moving forward. **You *must* work with others!** Well, okay, you can spin your wheels in the dungeon of your mind, spending 10 hours on a simple misunderstanding if you like. But I recommend forming a study group of three or four committed students and making regular visits to your instructor during office hours (have clear questions, general or specific, prepared in your study journal). Another benefit of working with others is that it gives you the opportunity to explain what you know. Teaching a subject is one of the best ways to deepen your own understanding of the material. It requires you to organize the information clearly in your own mind and tie up any loose ends. If you haven't done this on a regular basis, you may be amazed at how much time it saves and how much more engaged you become in the class. [**TIP**: If you don't know anyone in your class, meet people during your instructor's office hours. If you think you're on the same wavelength, ask them if they want to get together to study. If you're shy, and I have bouts of shyness too, get over it. You haven't got time for it.] – There truly *is* power in numbers!

A.2.e. Memory Techniques

[I'm NOT a psychologist ... but I play one on TV. So you may want to consult a resource written by an expert for the full scoop on memory and learning. Just check the library or your campus bookstore.] While physics instructors emphatically stress working through problems with physical intuition, logic, and mathematics, we also understand that some information needs to be firmly set in your memory banks. If you had to figure out which pedal was the gas and which the brake every time you got into your car, it would take you quite a while to get around town. This is true as well in problem solving–particularly on an exam. To my mind, there are 3 general categories of memorization: rote, repeated use, and mnemonics, i.e. memory aids.

Rote memorization is achieved by repetition, such as with flash cards. It is usually a slow process and tends to neglect the meaning of what is being memorized. Still, I suppose there is a place for it.

Memorization by repeated USE is ideal. By using the information in context over and over again, the material is set in our long term memory. This type of learning tends to include a deeper understanding of the meaning of the material. But this only occurs if we are paying attention to and questioning the meaning and subtleties of the material as we use it.

Mnemonics are surprisingly neglected by most students. These are clever memory techniques which make efficient use of the way our brains work. Much, if not all, of our memory is based on *association*. Unconsciously, an unrelated object or idea may "jog our memory" if it is somehow associated with what we are trying to remember. Consciously, we can associate a wild visual image, an interesting phrase, a purposely misplaced object such as a string around a finger, or some previous knowledge to that which we are trying to memorize.

My point here is that if you're going to memorize something, do it efficiently and effectively. Be conscious of the memorization technique that you're choosing to use. Then **get on with solving problems**, since that's where you'll put the meat on the bones of what it is you've memorized. Here are just a few ideas to help you on your way.

- **ACRONYMS** - Perhaps the most famous acronym is "Every Good Boy Does Fine." The first letters of the words give us the notes that correspond to the five lines on the music staff. Another one is "HOMES." I've capitalized every letter because each letter corresponds to one of the five Great Lakes (check it out). The order of the colors of a rainbow (white light spectra) in order of increasing frequency are: Red, Orange, Yellow, Green, Blue, & Violet. Let's see, how about "ROY Gets Between Villains"? Can you imagine a mechanic with the name Roy on his shirt breaking up a fight between two villains? Peace ensues, and a rainbow appears. The vivid mental image can help. Now you make one up. To distinguish the many possible "orbits" that an electron could have in an atom, we label the shells by n = 1,2,3, and so on, and the subshells by s, p, d, f, g, h, ..., where s is the first subshell, p is the second, and so on. Come up with at least two acronyms for the first five or six subshell letters. Don't read on until you've come up with two or more. I use "SPeeD Freaks Get Hurt." Okay, so some acronyms are better than others. How about "Some People Don't Forget Good Habits"? If you've actually tried to come up with some of your own, you've discovered something. Not only does the acronym itself aid your memory, but the *process* of coming up with an acronym engages you in concentrating on the material and so helps you remember.

- **YOU WILD AND CRAZY IMAGE!** - Visualization, using imagery, can be a wonderful tool when trying to memorize something. It is especially effective if the images you conjure up are vivid, exaggerated, and very unusual, since usual images are easily forgotten. A good friend of mine used this technique powerfully when she was studying the 300 herbs used in Chinese medicine. Not speaking Chinese, what chance did she have to remember 300 Chinese words and their meanings? For each herb, she and her study partner drew elaborate pictures which reminded them of the name, function, symptoms treated by it, and so on. For Dong Quai (don't cry), she may have used a person consoling someone crying. If the herb helped the urinary tract, the person crying may have wet their pants. The **ZANIER**, the better!!!!!!!

- **THE BRANDING IRON** - Often, even in trying to memorize something by rote, our concentration is not that intensely focused. This increases the time required to set the information into our grey matter. If you're going to memorize by rote, why not do it powerfully? Look at an equation and apply *searing* concentration. Imagine that the equation is being branded onto your brain. Visualize the formula being placed there with a branding iron and actually see the smoke come off of your brain as the information is burned into it. Do this with as much information as you like. There is plenty of room. You'll still need to set it into your long term memory and learn the actual meaning of the equation by further study and use.

- **SHORT & FREQUENT** - If it's the night before an exam and you haven't been an active student, I say "Cram away!" But we all know what a painful and ineffective process that is. The experts and experience tell us that many short study sessions are far better than a few long sessions. So make sure that you don't put off looking at the material because you haven't got enough time before the softball game. Those 15 minute sessions scattered throughout the day can powerfully reinforce what you've learned either during longer study sessions, in class, in your study group, or with the instructor during office hours. It keeps the material from fading away, and consequently reduces your stress and helps you maintain a positive mental attitude. Then you can say: "Every day, in every way, I'm being the best I can be." [Group hug.]

- **TALK IT UP!** - As a student, you're overloaded with new information that you need to absorb in a four month semester. You can't get a career's worth of repetition in that time, but you can make it a point to get as much repetition in as possible. By talking with others as frequently as possible, you will make a huge leap in your retention of the material. HUGE! This is highly researched and absolutely proven. So get yapping!

- **SEE IT FROM ALL ANGLES** - Think of something that you know very well. I bet you can explain and describe it from a number of different perspectives. Certainly, the more facets of something that we see, the more everything else falls into place and makes sense. So when you're studying something, try to see it and study it from a variety of perspectives. Get a conceptual understanding by visualizing physical processes, doing mini experiments, and talking with others. Get a mathematical understanding by looking at the defining relationships, checking the units, and working problems. Be mentally agile. Mix up your approach. In terms of memory, you **create a web of associations** that secures the information in your brain.

- **USE IT OR LOSE IT** - The amount you learn depends on the amount you practice! If you don't spend lots of time solving problems, you'll be like a person trying to learn how to drive a car by reading about it. Come time for your driving exam, you're not going to do so well no matter how much you've memorized and studied. Furthermore, you'll probably forget most of what you've learned soon after the final exam, and won't be able to apply it in later courses or in your career. **Problem solving is the mortar for the tower of knowledge in your head** ... or something like that.

There are other memory techniques out there. Keep in mind that the objective in physics is NOT to memorize the equations as you would Chinese herbs or the information in a zoology class. Science-related fields do use a fact base which incorporates many of the ideas that you will learn in your physics course. And you do need to be quite familiar with the equations so that you can readily think in terms of them. However, the essence of doing "real world" work is problem solving. You need to be able to start with basic principles, gather information, and be able to *construct* a solution.

A.3. Studying Physics

Many people find that studying physics is different from studying other subjects. You need to learn lots of new concepts that are often abstract but always precisely defined, and put these concepts together in a language, namely mathematics, that you probably aren't fluent in. If you aren't clear on the components of studying physics, you'll likely find it as difficult as cow tipping–which is especially tough if you don't know what it is you're trying to do. So, in addition to applying the guidelines given in section A.2, you'll probably have more success learning physics if you keep in mind the nature of the beast. To get good at physics, you need to **develop conceptual *and* mathematical understanding, and put it all together in your problem solving.**

A.3.a. See, Feel, & Think Physics (Feel The Force, Luke!)

Physics courses are NOT algebra classes. Learning physics is NOT a matter of finding the right equations, plugging in the knowns, and solving for the unknowns... Sorry. I get a little testy around this point. I'll try to be good. The thing is, physics courses can give you so much more than that. They can change the way you view the world as well as give you tools for analyzing biology, engineering, and home improvement questions. But you will only obtain this pearl, Grasshopper, if you learn to see, feel, and think physics.

"But how, Master?", you ask. First, make the following distinction: **you sense the world through sense organs, you observe the world with your mind.** Ever look at a picture and not been able to tell what it was you were looking at until someone told you? Then you know what I mean. Listening to a piece of classical music is another great example. You physically hear it, but do you notice all of the intricacies worked in by the composer? Do you notice how the various instruments play off of or with one another? What keys are being used and what emotions do they help conjure up? What about the tempo changes from one movement to another? All of this and more are what a trained ear observes as they physically hear the piece played. Without noticing such things, how can we really say that we are fully experiencing the music? Now watch something fall to the ground. In a visual sense, you see the object fall. But I bet there are things that you don't see (notice) in an observational sense. (Unless, of course, you've already taken the first semester course. In which case, you know what I mean!)

"How, Master, is it that I can see what it is I do not see, notice what it is I do not notice, and think what I have never thought?" Study, contemplate, practice, seek guidance from your best resources, ... and stop talking like that! There really is an element of **awareness training** in physics, though. Look at an object or a process. What do you observe? Can you see all of the forces (pushes and/or pulls) involved? How strong are these forces? In what direction, exactly, are they pushing or pulling? We could go on. In fact, we will do just that as you work through this course. Now do you understand why I say that physics is not simply about doing the math?

As you learn new concepts in physics, try to **take the dry definitions off the page and see them in the world around you.** Make the distinctions clear and vivid. Imagine that *you* are the object that is accelerating, *you* are the dancer jumping into the air, *you* are the horizontal beam mounted to the side of building and holding up the cafe sign, and *you* are the electron doing whatever it is the electron is doing. Physics is all around you. It is, after all, about the universe in which we live. Then start to think in terms of magnitudes, methods of measurement, relationships involving other concepts, and general principles that hold true in some or all situations.

In sum, develop the following:

Feel the physics.

- **Physical Intuition** – Get a sense for the physics going on. What's going to happen next? Get in a car and step on the accelerator. Step on the brake. Turn the steering wheel. Push a piano up some stairs. Get one of those kid's electricity kits. Look at the colors on a soap bubble. Listen to a siren as it approaches and then recedes. Take a hot mud bath. Inhale helium.

See the physics.

- **Observation Skills** – Know the conceptual definitions precisely and break down the world in terms of these. Clearly distinguish velocity from acceleration, momentum from energy, and electric force from magnetic force. Notice what you've taken for granted. Notice what you've never noticed. Observe the details. Actually do it. Look at something now. What do you notice?

Analyze the physics.

- **Logic & Math Skills** – Much of mathematics was created to describe the physical world. It is a quantitative language with self-consistent logic. Obviously, you need to carry out the mathematical "mechanics" flawlessly or your conclusions will be ... flawed. But you also need to be careful to use the equations only in cases where the equation is valid. That is, you have to know what the equation-sentence is saying, and what the physical parameters of your problem are. If the equations that you are using are approximate, you'd better know whether or not your approximate result is "good enough" for what you are doing.

Measure the physics.

- **Measurement Skills** – Data collection is essential to all of science and technology. You'll work on this in lab, but the three skills I've discussed above will certainly be put to good use. What units do you want to use to quantify your measurements? What are the errors inherent in your technique?

A.3.b. Conceptual & Mathematical Understanding

(Indulge me just a bit more. This is VERY important.) If you don't understand the concepts and the relationships between the concepts as you work the mathematics of a problem, you are like an illiterate person copying down letters, words, and sentences without understanding their meaning. (Isn't it true?) Just as one must learn to read before one writes, **one *must* learn to make the conceptual distinctions before one can set up the relationships to quantitatively solve physics problems.**

The novice often starts, in desperation, with the equations. That's okay for simple plug-in problems that are meant to get you acquainted with the equations and the variables that they relate. In this way, the equations can help develop your conceptual understanding. But if you don't gain a good conceptual grasp of the material, you will be stumped on the harder problems and most likely on the exam.

That you need to have solid math skills for physics should go without saying. Yet, students often enter their physics courses thinking that they'll somehow be able to get by with weak math skills. Don't do this to yourself. In fact, I highly recommend that you take as much math as you can before taking your physics courses. Of course there are time and schedule limitations, but the more math you know, the easier you'll find physics.

Unquestionably, to be competent at solving physics problems, one needs both conceptual understanding/intuition and mathematical competence. (More on this in just a bit.)

> **Warning Sign:** *If after a while you are still looking at the equations as letters and numbers to be manipulated, you have a BIG problem. You need to be able to read the equations like sentences. Just as the letters c-i-r-c-l-e conjure up the image O without effort, the mathematical symbols must conjure up physical concepts and relations if you hope to truly learn some physics. Just as with any new language, it takes training. But at least you now know what you're aiming for.*

A.3.c. The 15 Minute Commitment

Making a real commitment means that you will find a way to fulfill your commitment no matter what occurs. That said, I'd like you to consider making a very important commitment to yourself: not a day will go by that you don't **spend at least 15 minutes doing physics problems.** (Okay, you can stop when the semester is over.) Note: spending 30 min. on one day does not count for two days worth of problem solving!

The 15 min. practice session is one of the best ways to ensure progress. Do it no matter what. It will empower you in ways that you can't imagine until you do it for an extended period. Figuratively, **the dividends of daily practice are exponential.** A day missed is like running up that down escalator and stopping for a while–you lose ground and ultimately make more work for yourself. This occurs because new knowledge rapidly slips out of the short term memory banks. Regular review and study keeps it in place until it is secure.

A.3.d. So Here's The Plan. OK?

In order to stay on track, it will help if you have a rough checklist to monitor your progress. Here's my version of the ideal study regimen. Augment it as you will. Do what you can. Are there any improvements that you can make to this plan?

For each chapter (at a rate of about 1 chapter/wk), follow this 5 level study plan. You may even want to make a copy of it for each chapter and check off the levels as you do them. Oh, and don't forget that daily commitment.

- **LEVEL 1, HANDSHAKE** ("Power Preview" chapter): *Weekend.* TIME - 1-2 hr. In the text, read the chapter introduction and then the summary. Create a short outline of the key topics and results. Include the main concepts and check some of the units. (30 min.) Do only the simplest problems. (30 min.) Get a good sense of the big picture. In this tutorial, read and work through the Conceptual and Basic sections for the material to be covered. Again, keep your work and questions list in a study journal. (30-60 min.)

- **LEVEL 2, GET ACQUAINTED I**: *Night before each class.* Read sections to be covered more carefully (but don't get hung up and take too much time). Is it time to begin the Intermediate section of this guide? Continue writing questions down in your questions/study journal. Try a few more exercises/problems. Verbalize your understanding with your group and work problems with them. You're layering in some details now.

- **LEVEL 3, GET ACQUAINTED II**: *After lecture.* Get questions answered, meet again with your group, attend office hours and discussion sections, summarize lecture notes a.s.a.p., get deeper into the problem set (intermediate problems finished, start the harder problems and find out where you get stuck), and add more questions to your questions journal. Work through all of the Intermediate section of this guide.

- **LEVEL 4, BECOMING OLD FRIENDS**: *End of the week.* Get questions answered, finish and understand the problem set, and add any outstanding questions to your journal. You should be tackling some of the advanced problems by now.

- **LEVEL 5, BRING ON THE EXAM!**: *Reviewing for the exam.* Clarify the fine points. Be clear on the big picture. Streamline problem solving (get fast!). Set up lots of problems without working out the details. Solve some problems completely. Verbalize, verbalize, verbalize your understanding with others.

Yeah, it does take a hearty effort to do well. But why fool yourself into thinking you're ready to ace the exam when you haven't really done what it takes? Keep in mind that this is for you, and you will reap the many benefits of this investment.

A.3.e. Check 'em Out!

Often it helps to see something presented in a few different ways. If you really want to dive into physics, or if you just find that you still need some help, I recommend the following sources. Check with your instructor for other recommendations that may be high on her or his list.

CD's: More and more study aids are showing up as interactive computer guides. They can be fantastic learning tools! What you do NOT want is simply written words and static images (text or problem solver) on disk. What you do want is a full fledged, animated guide through the concepts and problem solving methods of physics. One of the difficulties students have is visualizing the physics. These can be a great help. I'm reviewing one called *Interactive Journey through Physics*, by Prentice Hall that may be available to purchase now. It looks pretty good, but I need to take a closer look. Check bookstores and computer stores, and ask your instructor.

THE WEB: There are some great resources on the web. Rather than list them all, I'll direct you first to the San Francisco State University pages [http://www.physics.sfsu.edu] where you can find some useful links. Check out Susan Lea's page for a link to the UC Berkeley physics tutorial. You can check out my page too. (Warning: wear a construction hat! I'm just getting started.) You might also see what Geoff Marcy, the discoverer (along with Paul Butler) of many planets, has on his page. Or search for "physics tutorials" and see what you get!

A STEP BACK: If you find that you're still having trouble catching on even with the text, tutorial, classmates, and assistance from your instructor, look in your library or check with your friends for the following:

Conceptual Physics by Paul Hewitt; HarperCollins - Actually, this book, or one like it, should be read by all physics students. As you might guess, it focuses on a conceptual understanding of physics with very little math. Some of the problems can stump even physics grad. students!

An algebra-based text, usually used for life science majors. If you are taking the calculus-based course, you may find it helpful to see the material without quite so much math. Some are certainly better than others, so ask your instructor for advice. Borrow one from a friend or the library.

A STEP UP: If you really want to get GOOOD, try these out.

An Introduction To Mechanics by Kleppner & Kolenkow; McGraw-Hill - A great bridge to upper division work. Very clear. Short. A bit mathematically sophisticated in spots, but that's you're goal.
Berkeley Physics Course, vol. 1-5; McGraw-Hill - Especially check out vol. 2, Electricity and Magnetism by Purcell. These are on par with the Kleppner & Kolenkow text above.
The Feynman Lectures On Physics, vol. 1-3, by Richard Feynman; Addison-Wesley - These are filled with the wonderful insights of a legendary physicist and teacher.

MATH FOR PHYSICS: Consider these if you need more assistance with developing your math skills. There's no way around it though, it'll take work on your part to stretch your skills.

Various review guides - Check your bookstore for review guides along the lines of Schaum's, etc. There are quite a few available.
Div, Grad, Curl, And All That by Schey; Norton - As the subtitle on the cover says, this is an informal text on vector calculus. It is done in the context of electricity and magnetism and is short, accessible, and wonderful. Some might think that this is a bit too advanced for this course, but I think it will help. So, before you get to electricity, you might want to check this one out. (It's a must for you physics majors!)

NOTEWORTHY: If you're interested, the following books will help you understand the structure and development of physics. These are not textbooks and contain very few equations and no problems. (Ask your instructor for more that might suit your interests, like *QED* ...)

The Character Of Physical Law by Feynman; MIT Press - Topics include: physical law, the relation of mathematics to physics, conservation principles, symmetry, past vs. future, probability and uncertainty, and the search for new laws.
The Evolution Of Physics by Einstein and Infeld; Simon and Schuster - This exceptional, highly accessible book takes us through the conceptual development of physics from Galileo through relativity and quantum mechanics. It probably won't directly help with your problem sets, but it's great stuff!
A Brief History of Time by Stephen Hawking; Bantam - This is a non expert's guide through some pretty wild physics.Topics include: space and time, elementary particles, black holes, and the origin and fate of the universe!

A.4. Building A Solid Conceptual Framework

Learning to think conceptually about the physics is probably the most neglected aspect of learning physics. It will most likely take a real effort on your part. We get so answer oriented that we are quick to plug in the numbers anywhere that looks remotely promising and then hope for the best. (Let's hope actual engineering isn't done that way!) It is essential that you know what it is that the equations are "saying." Make sure that you are developing your conceptual understanding of the material as you go. In this section I want to offer a few more little pointers that you may find useful.

A.4.a. Fuzzy Wuzzy & The Strict Definitions Blues Band

Fuzzy thinking will inevitably get you in trouble and waste your time. Granted, true learning is genuinely difficult. There are often an assortment of new ideas, relationships, and methods to keep straight. But that's the task. **Push yourself to think and speak precisely.** Velocity is NOT the same as acceleration, force is NOT the same as torque, momentum is NOT the same as kinetic energy, heat is NOT the same as temperature, and electric fields are NOT the same as magnetic fields. Sure, they're related. But far too often students are mixing up these ideas a day before the exam. It's not easy to keep everything straight at first. That's why you must keep a constant vigil to ward off fuzzy thinking...and talk, talk, talk about physics with others.

A.4.b. Visualize Analogies!

A force, i.e. a push or a pull, is one of the most concrete concepts in physics and not too difficult to get a handle on. (Until you start adding them up! More on that when we get to vectors.) There are, however, a lot of rather intangible concepts in physics (electric potential, magnetic flux, quantum states, wave-particle duality – you know, easy stuff). A good way to get a handle on these abstract ideas is to visualize concrete analogies. In this guide, I'll mention the analogies that I think are helpful, but feel free to lively-up your imagination.

One note of caution: **be careful to note the limitations of the analogy.** Analogies usually fall into the "it's kind of like this" category of explanations. So when you have an analogy in mind, also note the way in which the concept or process isn't exactly like your analogy.

A.4.c. Movie or Still Life?

There are three main ways to visualize a physical situation or process. It is very important that you know which you need to use or have chosen to use. This will effect the way you go about setting up and solving the problem you are working on. For example: with Newton's second law ($\sum F = ma$), we will often identify all of the forces on an object at an instant; for kinematics, we will often relate the positions and velocities at an earlier ("before," "initial") state to those at a later ("after," "final") state; and with conservation principles, we will compare the amount of some quantity at an earlier ("before") state to that at a later ("after") state and check to see if any was lost or gained between the states (during the "stage"). Believe me, you'll get plenty of practice.

A single "state."

1. **FREEZE FRAME** – Here you imagine the situation at an *instant* of time, as in a photograph. Keep in mind that the objects in your picture may not be at rest. This image works both in static and dynamic situations (In this snapshot we capture a "**state**" of the process…think instant = in "state".)

Multiple "states."

2. **SEQUENCE OF FREEZE FRAMES** – Here we compare the situation at two or more instants of time. This is an extremely valuable way to look at processes. Whenever you compare an earlier state to a later state, you imagine and draw a sequence of still lifes. This is particularly valuable when describing the motion (kinematics) or applying conservation principles, as noted above.

Single or multiple "stages."

3. **MOVIE** - here we imagine the process in motion. This allows us to visualize the process as it evolves in time. This really gives us a feel for what's going on. We also may need to know what's going on between the earlier and later states. For instance, is heat being added, friction at play, or the force changing throughout the process?

A.4.d. Visualizing Spatial Geometry

As you may have ascertained, it helps to be able to visualize physical processes. Many people have trouble translating a 2-D picture to a 3-D mental picture. If you find this difficult, you'll have to keep hammering away at it until you start to see in 3-D. An engineering drawing class or art class working on perspective helps. Check out a book from the library or just keep working with people until you begin to see in 3-D. Keep trying. It will prove to be an important skill in your physics courses. Try these:

A Table | A Flat ("Planar") Surface | 3-D Axes

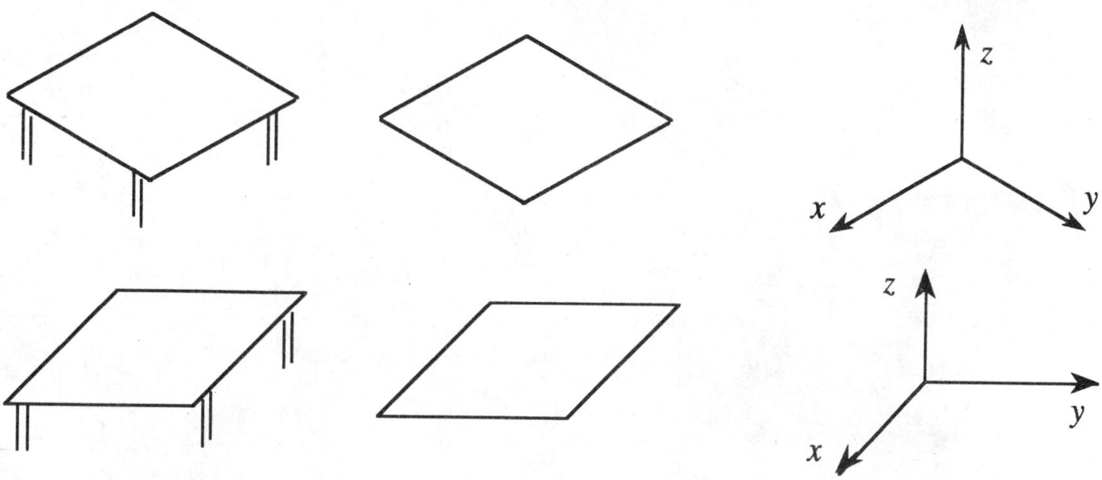

A.4.e. Concept Maps

Okay, so you're trying to sort out this overwhelming amount of material and you find that you can't see the forest for the trees. You've gone through it over and over, but by the time you get to the end, it's still a blur. Your "linear" approach may not be what you need now. Perhaps you should try some "mosaic" thinking. After all, the ideas in physics are not all related in a D follows C follows B follows A way. They are interconnected in a definite, albeit confusing at first, web. A "concept map," also known as a "mind map," is an **association map**, a sort of **mosaic flowchart**, which allows you to see the scaffolding of ideas within which you are working. This grand-scheme clarity will help you straighten out the pile of ideas and equations in your mind.

First, take out a clean sheet of paper. Next, decide what topic you want to sort out. You might select a big topic, such as "physics," a smaller topic, such as "kinematics" or "magnetism," or an even smaller topic, such as "gravitational free-fall" or "magnetic induction." Put that topic in the center of your page and circle it. This is the trunk of your tree. Now branch out with the next level of topics and circle them. You may have 5, 10, or any number of level two topics. From these, branch out to level 3 subtopics, and so on. It may help to write in the connecting ideas or equations along the branch lines. You may also like to draw connecting lines between subtopics to more fully show the interconnections. Have fun with it, it needn't be perfect–just accurate. Some people find this incredibly helpful, some don't. Be your own judge.

Here is the start of a concept map. (Don't spend time polishing up your map when you should be working on problems! This is a quick exercise to help you see the scaffolding.)

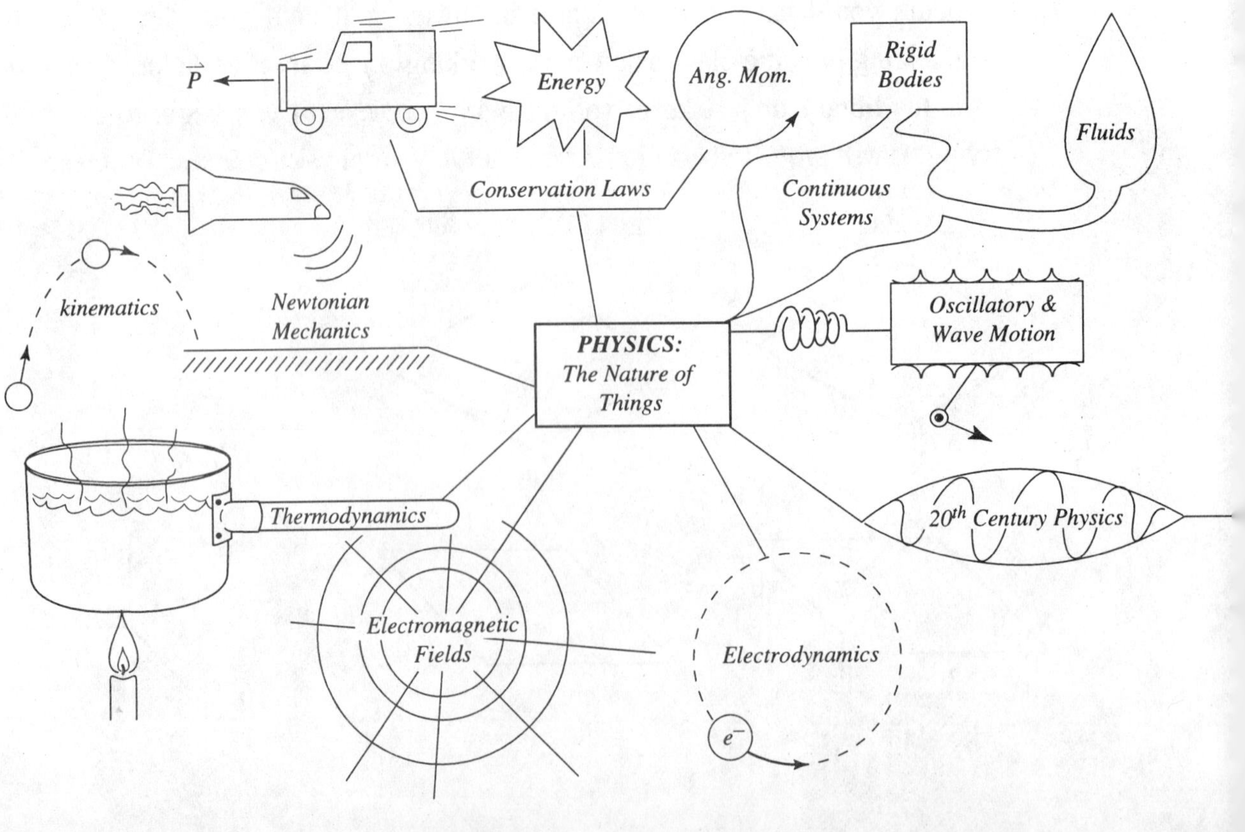

A.4.f. "But It's Not Exact!"

There is a time for precision and a time for rough approximations. There is a time to know that 1 meter is equivalent to 3.281 feet, and a time to think of 1 meter as 3 ft + a bit (about 3 inches). When you are trying to get a feel for the physics, work in rough numbers–throw around some numbers quickly and get a sense of the magnitudes involved. When you're working out a problem or a calculation to be used later, be as precise as your known data, equipment, and analysis allow.

It's ironic that we can easily lose our conceptual clarity as we strive for mathematical clarity. Still, it happens all too often. For example, a common metric unit for speed is meters/second. You'll use it over and over again. But how fast *is* 1 m/s? (It's the speed of a slow walk, or about 2.2 miles/hr. Better yet, start out thinking 2 mi/hr.) Can a car travel at 100 m/s? Not mine! These are the sorts of questions that will help bring the physics to life. Without these rough numerical checks, you may not have any idea what's really going on and/or what your numbers *mean*. [Note: see the "back of the envelope" calculations in the text.]

A.5. Mathematical Thinking

Some specific mathematical tools are dealt with in Part B. In this section I give you some general guidelines for working with the mathematics of a problem. You may find that you are slow at first. Don't give up. If you start applying these right away, you'll soon speed up and be on your way toward a much better understanding of what you are doing. If you don't, you'll never know how much these pointers could have helped.

Basically, you want to move in the direction of being able to *read* an equation as a very precise sentence. Learning to read equations is learning to notice the variables/concepts involved and their relationship to one another. You need to be able to spot proportionalities and inverse proportionalities, check the consistency of the units, be very clear on any special restrictions on when the equation is valid, and understand the basic logic (causality, equality, definition, or implication) of the equation.

A.5.a. Get To Know Your Equations

Here is a simple three step guide to assessing how well you know your equations.

1. <u>GENERAL APPLICABILITY</u> - Many equations apply in some situations and not others. You should classify the equations not only according to topic, but also according to the type of problems that the equation is valid for. You should also be clear under what conditions the equation does NOT apply.

2. PRECISE MEANING OF VARIABLES - You should know what each variable means conceptually and you should be able to see each variable in the problem. For example, if there is an angle in the equation, where exactly is it in the problem?

3. VERY PRECISE MEANING OF EACH VARIABLE - Push your clarity even further now. What don't you understand about this equation? If you are very clear on the meaning and use of an equation, you will think of it when needed and never use it when it doesn't apply. TEST: Can you "read" the equation as a sentence in English?

A.5.b. Always Check Proportionalities!

To "get to know" an equation, it helps a great deal to check which factors mathematically increase or decrease another factor. (Physically, changing one variable may change more than one of the other variables.) We must also be aware of which variables are constant in the process. For our purposes here, let's assume all but the two factors considered are held constant. We must further assume that all variables are positive (why?). Let's try and get a feel for the following:

$$A \cdot B = \frac{C \cdot D^2}{E \cdot \sqrt{F}}$$

We see that:

A is proportional to C............that is, ... as C increases ("↑"), A increases ("↑")
Similarly, ... as C decreases ("↓"), A decreases ("↓")
A is proportional to D^2..........that is, ... as D increases ("↑"), A increases ("↑")
A is inversely prop. to E........that is, ... as E increases ("↑"), A decreases ("↓")
A is inversely prop. to $\sqrt{F}$ .....that is, ... as F increases ("↑"), A decreases ("↓")

We also see that:

A is inversely prop. to B........that is, ... as B increases ("↑"), A decreases ("↓")

Notice that I've introduced the arrows "↑" and "↓", as shorthand for "increases" and "decreases," respectively. I will continue to do this from time to time throughout the tutorial. Further notice that I've chosen to write "as D↑, A ↑" rather than "as D^2 ↑, A ↑." Both statements are true, but when I'm just getting a "feel" for the relationship I may only be concerned with which factors make A bigger and which make A smaller.

So far we have investigated the mathematical relationship. The next question we need to ask is **"Does this make physical sense?"** In this case, who's to say? No particular physical meaning has been ascribed to the variables. In physics equations, we will always be looking at a relationship between physical quantities. We can then check our physical intuition with the mathematical relationship. For example, the ideal gas law says that $PV =$

$\mathcal{N}RT$. Since R is a known constant, if we keep the amount of gas, $\mathcal{N}$ and the volume, V, constant, an increase in temperature will cause an increase in pressure. This makes physical sense. I leave the remaining possibilities to you for now.

Let's take a look at the first equation with a slight variation. What do you notice?

$$A \cdot B = \frac{C \cdot D^2}{E \cdot \sqrt{F}} + G$$

Now there is only ONE *multiplicative* proportionality in the equation. What is it?......Yes, as $B \uparrow$, $A \downarrow$. We can still say that as $C \uparrow$, $A \uparrow$, as $E \uparrow$, $A \downarrow$, and even as $G \uparrow$, $A \uparrow$. It's just that now the relationships are not simply multiplicative. These are good things to notice when first meeting an equation.

A.5.c. Always Check The Units!

Before you buy a car, you'll look at it from all angles, inside and out. Similarly, in order to understand something fully, you need to study it from all angles. Checking the units of every equation that you encounter is a great way to familiarize yourself with the equation and the relationships encoded in it. It *will* accelerate your learning! If you repeatedly check the units of the same equation, you'll see that it takes you less and less time to run through it. Now you're *really* learning what the equation is saying!

Additionally, thinking in terms of the units involved is one approach to solving or at least getting a better handle on a problem. It will also tell you if an equation that you've derived makes any sense or not. Remember, just as you can't add apples and oranges, you can't add seconds to meters and get a number that has any physical meaning. (2 seconds + 3 meters = nonsense!)

Distinction: Dimensions vs. Units

This is as good a place as any to get this straight once and for all. Look at the two columns below. Notice that a **"dimension"** is a general physical concept while a **"unit"** is any chosen system of measurement of a physical quantity (or dimension). Systems of units are made-up by humans for convenience and convention, the dimensions of physical reality were constructed in the Acme Reality Factory (ARF), Chicago.

Dimension	Some corresponding units
time	seconds; minutes; hours; dog years
distance	meters; kilometers; inches; feet; miles
mass	kilograms; grams; slugs (really!)

If you're checking the "dimensions" of an equation, this technique is called "dimensional analysis." If you're checking units, you're additionally making sure that you haven't used different units for the same dimension, for example feet and meters for length, in the same equation. [See section B.1 & Ch. 1]

A.5.d. Always Check Numerical Values!

In physics, the number 2 could refer to many very different things. Are we talking about 2 hours, 2 seconds, 2 nanoseconds, 2 meters, 2 volts, or 2 objects? As you continue to try and understand what it is you are doing, you need to consider what the numbers mean. This not only increases your understanding of the process, but it provides a valuable way to check your calculations. If you are calculating the time it takes to drive from New York to San Francisco, and you come up with 2 nanoseconds = 2×10^{-9} sec., you'd better check your work! **Make it a habit to check the reasonableness of your numbers.** Also, try to **get a feel for the relative magnitudes of various quantities.** (How far is 100 meters?) In so doing, you will increase your physical intuition, avoid glaring math errors, and win points with your instructor!

A.5.e. Develop Fluency With The Mathematical Tools

Changing the form of an equation by using a math identity turns out to be a very important and powerful part of problem solving.

I recall thinking that certain seemingly esoteric trig and geometric relations, series expansions, approximations, and the like were not to be memorized but looked up, or maybe derived, as needed. To some degree I was wrong. Imagine if you had to look up the value of 2+3 every time you needed it. The fact of the matter is, the more you have math identities, relations, and methods at your mental finger tips, the better. Certainly, this comes from use. But it also comes from recognizing the importance and making a conscious effort to plant the ideas in your grey matter.

The point that I want to share with you is that it does pay to **know as much math as possible**. Pay attention, learn it well, and don't think you can get away without knowing the Pythagorean theorem, law of cosines, law of sines, double angle formulae, equation for an ellipse or hyperbola, logarithmic relations, and so on...and on...and on. Okay, you don't have to be a walking handbook of mathematics, but keep at it. Over the course of time and use those relations tend to feel at home in the old cerebral cortex. And like a carpenter who knows how to use every tool made, you will become much more skilled at constructing mathematical solutions.

A.5.f. Logic Lost

The last bit of advice that I'll give you on understanding the equations of physics is a simple, yet very helpful, point about logic. This really falls under "knowing what an equation means" and "reading an equation," but it's worthy of it's own little section. Let's

We often use the logic to set up the equations. Then we venture into Mathematic Land.

look at some common types of logical statements in physics. It would get a bit cumbersome to have different symbols for every different type of logical statement, but there are three symbols that are relatively common. Make sure that you see that **the logic of each of the cases below is quite different from the others!**

1. **"EQUALS,"** = : This is the basic symbol used for equations in math. In physics, this symbol may carry different physical meanings while retaining the same mathematical meaning of "quantitatively equal." Here are two important categories of equality:

   (a) CAUSE(S) = EFFECT. Causal logic simply tells us that some causes are responsible for some effect. (It is an enormous triumph of physics to be able to write equations that give a *quantitative* causal equality.) The ability to distinguish the causes from the effect will help you to avoid common mistakes (see fictitious forces, for example) when setting up an equation. In fact, one of the cornerstones of physics, Newton's 2nd law, says that forces cause a mass to accelerate ($\Sigma F = m\,a$).

   (b) BEFORE = AFTER. More accurately, this is "*earlier value of some property of the system = later value of the same property of the system,*" when the "before" and "after" times are depends on what's important in your analysis. This is called a conservation law (see Part II of the text Lea & Burke and Vol. 2 of this tutorial.). Here's a silly conservation law that's usually true: *your name now = your name in 10 minutes*. Here's one that's always true (as far as we know): *energy of universe at any time = energy of universe at any other time*. And here's one that's only true in special situations: *energy of a certain system of objects at some time = energy of that same system of objects at another time.* (Note the difference between this logic and causal logic.)

2. **"DEFINED AS," or "IDENTICAL WITH,"** ≡ : Both "=" and "≡" mathematically mean "quantitatively equals." However the 3-bar equality goes to the trouble of saying that the expression of the right equals the expression on the left because they refer to the same thing. Writing A ≡ B is the same as saying that A *is* B. For example, we will often define a concept quantitatively by showing what factors its magnitude (and possibly direction) depends on.

   - The force of friction on a solid object sliding over another solid object depends on the roughness of the surfaces, μ, and the amount that the objects are pressed against each other, n. Thus the equation for kinetic friction can be written: $f_k \equiv \mu\, n$. This should be read "f_k is $\mu\, n$." This is logically different than saying that the friction force *causes* an object to slow down: "$f_k = m\,a$.." In this last equation, I could replace f_k with $\mu\,n$., since they refer to the same physical thing.

- The torque on an object depends on the force on the object, F, and the "lever arm." This can be written: $\tau \equiv F\,d \sin\theta$. This should be read τ is $F\,d\sin\theta$.. More on this later!

Not everyone goes to the trouble of using the 3-bar equality sign for definitions, but you should be aware when an equality is in fact an identity. This will help clear up some common points of confusion. For instance, you may want to replace F_g with mg in an equation. You can do this because the two refer to the same thing.

3. **"IMPLIES,"** $\Rightarrow$: This is not an equality. A common mistake that will result in immediate loss of your physics scout badge is illustrated below. As you "read" the equation, decide if it makes logical sense.

$$v_f^2 = v_o^2 + 2a\Delta x = \Delta x = \frac{v_f^2 - v_o^2}{2a}$$

This should be written:

$$v_f^2 = v_o^2 + 2a\Delta x \quad \Rightarrow \quad \Delta x = \frac{v_f^2 - v_o^2}{2a}$$

How about "$V = IR = I = V/R$"? Don't laugh. I bet your instructor will see this type of mistake more than once this semester. But not from you .

In some cases, the logic doesn't jump out at you. Then you just see the relationship between variables as reflected in the proportionalities. **Frequently, an equation's logic is lost upon reorganizing it with algebraic manipulation. That's perfectly fine.** If you've done your algebra correctly, the equality holds no matter how you mix it up.

A.6. Problem Solving Strategy

[**The next four sections are pearls of wisdom. Make them yours, Grasshopper.**] I want to shift the way you look at and approach problem solving. First, we all get stuck. The question is whether you work through your confusion systematically or you get hopelessly entangled in the process. Second, working through a tough problem requires a clear, methodical approach. Sure, every problem is different to some degree, but the good news is that people that get good at physics, consciously or not, **learn some specific problem solving techniques that help them through the valley of despair**. And third, memorizing these guidelines won't do you much good. You must train yourself to **USE** them as you solve problems. Keep these guidelines and problem solution "template" in mind as you work problems. With time, the process will become natural – and that's a great achievement. Once again, this will take discipline on your part, and again, it will ultimately make your experience a positive and rewarding one with maximized efficiency. [This stuff may seem abstract, but if you train yourself to use it, you will see just how powerful it is.]

A.6.a. Be A Detective, Not An Answer-Person!

Most students look for the answer before working through the clues.

Problem solving is a process, not a yes/no, true/false, spit out the answer to win double jeopardy kind of thing. If I ask you what the capitol of China is, you'll answer "Beijing." If I ask you what color the sky is on a clear day at noon, you'll say "blue." That's fine. A simple question about facts warrants a simple factual answer. Now if I ask you what strength magnetic field will make an electron, traveling perpendicular to the magnetic field with a velocity of one tenth the speed of light, move in a circle of radius 1 cm, you'll respond "0.02 tesla, of course." . . . NOT! You're going to need to sort out the information and appropriate relations before you even begin to approach a numerical result. (Actually, the problem stated above is pretty straightforward.)

As obvious as this point is, I bet you'll easily be able to catch yourself in the **answer-person frame of mind; namely, you read a problem and then you immediately go for the answer.** Except in matter of fact or basic problems, this "get the answer as quickly as you can" approach will tend to lead you into confusion, frustration, and too much time spent spinning your wheels. For more involved problems, it is much better to use a **detective's approach: meticulously sort out what you know, what you want, and the connections between them. From there you methodically piece together the solution.** You may not know how the plot is going to work out until the very end; a detective usually doesn't either. But isn't that what makes mysteries challenging and fun? Be at peace with your not knowing, Grasshopper. The answer will come if you keep searching!

A.6.b. Formulate Questions!

If you're not asking questions, you're probably not learning.

[As far as I can tell, this applies to most, if not all, aspects of life!] Of all that I've written and will write, I think this is the most important piece of advice that I can offer: **guide yourself through a problem with probing questions.** Countless times I've had students come to me with problems that they were "totally stumped" by. I sat next to them and just asked questions. As they answered the questions, the solution to the problem was unveiled. Why couldn't they solve the problem on their own? They weren't leading themselves through with questions.

Granted, asking the "right" questions is easier said than done. But if you aren't even asking well formulated questions, what hope is there? You do have a built-in radar for what issues to probe with your questioning. **Your mental block, your confusion, the thing**

that irritates or scares you the most, turns out to be your best guide. If there is a word or phrase that you don't really understand, ask yourself "what does that really mean?" and seek out the answer. If there is a bit of math that you're fuzzy on, ask yourself "what do I understand? Is there a rule or identity that I am forgetting? Am I missing something conceptually or logically?" "Have I got the picture right?" "What does '20° north of east' mean?" "How do you get the components of a vector? I never did fully understand that." "Why is this or that true?" "How can you have negative energy?" "What is the sound of one hand clapping?" Then read, think, or ask someone.

Instructors love to hear well formulated questions. It lets us know you're in there getting dirt under your fingernails. Even if it's a simple question that you've researched but can't get anywhere with, ask it. The only stupid thing in that situation is not asking for help. You may just have an unconscious preconception that's getting in your way. Whatever it is, get it straightened out!

"But I'm Stuck, You @#!*$@**¤!"

It's tough to do something when you don't know what to do. What would Madame Curie do? What would Sherlock Holmes do? They would investigate the matter. Stare in the face of confusion, Luke (or is it Lucia?), and **keep questioning**!

If you're fuzzy, get clear!

1. **GO INTO THE FUZZY** - As I've said, a good place to begin investigating is wherever you're fuzzy. Conceptual definitions, pictures, graphs, math tools, and limitations on an equation's applicability, for example. Maybe it's simply the wording of the problem that is throwing you.

If it's so obvious, prove it!

2. **GO INTO THE "OBVIOUS"** - Sometimes what you don't know is cloaked by what you *think* you know. (Einstein looked into physicist's preconceptions about space and time and came up with a surprising and extremely powerful new theory.) The solution to uncovering such hidden sources of confusion is to **explain, clearly and thoroughly, what you "know."** This is MUCH more potent if you do it on paper (or at least verbally). Surely you've argued a point passionately only to end up uncovering your own erroneous assumptions. If you keep the argument in your head, you'll tend to wash over the "obvious truths." We tend to do this on paper as well. This is why we must lay out the ideas and logic without any gaps. Challenge everything that you assert. If you can back the assertion with solid logic, it's probably okay. [After a while, you can enjoy having your assumptions trashed. It keeps one open minded and restores a sense of wonder.]

A.6.c. Use "Top Down Thinking" To Think Like An Expert!

Start with the big, conceptual picture and work toward a detailed mathematical analysis.

It appears to me that most, if not all, efficient, effective, powerful workers approach their **work in layers**. (You're probably doing this to some degree already. Now try to push your use of this approach further and/or just apply what you've been doing to physics.) For example, I have a great book on writing called *A Writer's Workbook*, by Norman E. Stafford. What do you think the first part is on? Of course, Part 1 is "Planning and Writing." The most effective way to begin a paper or book such as this is to lay out the big pieces first. Things change in the process, but you need a framework within which to begin. Once you have an outline, you can begin to fill in the details.

If you were building a house, I hope you wouldn't put the carpet in before you've laid the foundation! Similarly, to learn a martial arts move or set of movements, you might want to follow the advice of Peter Ralston, World Champion in the 1978 full-contact Martial Arts Tournament held in Taiwan, and "chunk it out," then refine (layer in the details).

Why then do most students attempt to tackle difficult problems by grasping at equations and cranking out numbers before they have even thought about what's going on? (Probably because they are looking for an easy way out. Well, I'm glad you don't do that.) In the next section I'll give you a problem solving routine to follow, but for now try to get this general scheme of **working from the big picture to the fine details**. When you're laying out the big picture, don't hesitate to say the "obvious." Describe what's going on using the pertinent concepts. Make diagrams and label them. Tabulate what's known (given explicitly or implicitly) and what's unknown (wanted or possibly needed). Then, and only then, start thinking about the physical principles that apply and the mathematical relationships between the variables. Now that's exemplary detective work, my dear Watson!

In the course of solving a complex problem, it can be very helpful to know where you are in the process. If you get stuck, isolate the difficulty rather than assume that you're completely lost. For example, when you enter "Mathematic Land" you can let go of trying to follow every step conceptually and just turn the mathematical crank quickly, but carefully. A problem here may just mean that you need to review the properties of the trig. functions or how to get the components of vectors. Then, if your objective is to find a numerical result, you'll need to pop back into concept mode and see if the number you obtain makes physical sense.

This is the approach of an expert. You need to make this your approach. This "top down," hierarchical, outline view, layering in from big ideas to fine details approach will increase the rate at which you learn and work. It's up to you to train yourself to think and work in this way. It takes a conscious and determined effort. I have tried to help by laying out the structure of this tutorial to reflect this approach. I hope you will come to appreciate, and even revel in, this approach.

A.6.d. An Effective Problem Solving Routine

There is a time to think conceptually (usually 1st), a time to think in terms of the physical principles (usually 2nd), a time to think in terms of the equations and relations (usually 3rd), and a time to think mathematically (often 4th). While I've placed these in order, you will ultimately bounce between these types of thinking as you work through a problem. Nevertheless, the following provides an outline for structuring your solutions to physics problems. [A very nice, in some ways more detailed, problem solving template is in the text Interlude following chapter 3.]

PROBLEM SOLVING TEMPLATE

Mentally check off each step as you work through problems.

1. [MODEL]

   Read the problem. Read it quickly the first time to get a general idea. Then take note what special assumptions can be made to simplify the problem. We generally model real physical systems with ideal models that provide sufficient accuracy for our purposes.

   Visualize the process. Get a feel for what's going on. (Where are the forces or torques? What's moving? What about energy? Where's the electric current going?) Never skip this step! As you imagine the process in your mind's eye, sketch what's going on.

   Draw pictures - Make them big enough and detailed enough to be clear. If the diagram gets cluttered, draw another one isolating, and perhaps enlarging, the section that needs closer inspection. (There are some problems that don't require or warrant a diagram.) Include all states and stages of interest.

2. [SETUP]

   Sort out the problem. Read the problem for the second, third, or even fourth time to extract all of the information provided. After this stage, you probably won't need to read the problem again.

   (a) **Gather information** - You may want to make a table like the one below. The way I've divided the categories may prove valuable in sorting out what you have and what you need. In particular, information that is "hidden" in the problem statement falls under the implicit column. Often the "trick" or missing link in a problem is "hidden" information that you haven't extracted from the problem statement or from a conceptual analysis of the physics.

KNOWN		
GIVEN	IMPLICIT	WANT (UNKNOWN)

If you've drawn the diagrams and gathered the information, you won't need to reread the problem statement except to comb it for more information when you get stuck. You can glance up at your work and easily remind yourself what you have and what you need. This keeps you from spinning your wheels and saves time.

(b) **Classify the problem.** What principles apply? Is it a kinematics, F = ma, conservation of momentum, or conservation of energy problem? Is it a projectile problem? Is it a statics problem? A rotational motion problem? The problem may very well combine many "types" of problems. Break it down into sub-problems. Write down you're assessment of the problem. Be succinct.

(c) **Construct equations appropriate to the problem.** You might briefly explain what each equation is and why it is applicable. These will come from applicable

- principles/laws
- defining relations
- physical conditions or "constraints" in the problem
- wherever else you can obtain a pertinent equation

3. [SOLVE]
Enter Mathematic Land. Take the information and pertinent equations that you've gathered and turn the mathematical crank. Check the number of unknowns and the number of equations. Draw clear triangles. Create differential expressions. Check Part B of this guide or any other reference for help when needed.

Cycle through the steps as needed. Make sure you aren't missing something.

4. [ANALYZE]
Goal accomplished? Did you answer all parts of the question? What do your results tell you? Are you surprised by your results?
Reality check! Are your numbers physically reasonable or absurd?

<u>IN SUM</u>: [Model] → [Setup] → [Solve] → [Analyze]

So memorize and religiously follow these steps…don't skip any…I'm warning you, you'll waste time if you do…and I'll tell on you. Here is Sherlock's method:

This problem solving outline allows you to focus on layering in the details and to diagnose your confusion.

1. *Quick Read*: quickly get a sense of what's going on in the problem.
2. *Comb Through*: while you carefully read word-by-word and phrase-by-phrase,

<u>Visualize & Sketch</u>	<u>Gather & Sort Data</u>
(include states & stages)	(Given, Hidden, Want)

3. *Choose Your Method(s)*: Which method(s) applies? Which is easiest to use? Kinematics, force analysis, momentum analysis, and/or energy analysis?
4. *Set It Up*: Refer to your sketch(es) and data as you set up the analysis and construct appropriate equations.
5. *Solve*: Now you can turn the mathematical crank and eventually put in the numbers.
6. *Analyze*: Reasonable results? Check extreme cases. Interesting generalizations?

<u>STUCK? USE THE ROAD MAP ANALOGY. ASK YOURSELF:</u>

1. Where am I? (What do I want to know?)
2. Where do I want to go? (What do I need to solve for ?)
3. What are the pertinent avenues (relations/principles) to get me from 1 to 2?
4. What haven't I answered completely in 1-3 above? Is there any implicit information that I'm not using? (For example, "constant velocity" implies that the acceleration is 0 and the net force is 0.)

* To get past mental blocks, ASK an answerable question. (See A.6.b above.)
* WRITE IT DOWN! Work the problem on paper, not in your head. So many times people get stuck in their head. When they finally start sorting it out on paper, the problem gets solved. This is very common!
* Catch yourself when your mind goes blank or wanders! Then get back to formulating and answering questions.

There, I've said it. Phew! I'll put away my soap box now. It's on your shoulders. I encourage you to train yourself to study efficiently and effectively by referring back to and APPLYING the material that I've presented here in Part A. I wish you the best of luck on our journey. Remember:

One must learn by doing the thing;
for though you think you know it,
you have no certainty until you try.

- Sophocles

PART B . . .

B. Mathematic Land . . .

(a primer & reference)

"A more sophisticated tool often simplifies the job. . . once you've mastered the tool!"

Make sure that you've read section A.5. Skim over this section as soon as possible. Then use it as a reference when you find holes in your math. See section A.3.e for two other math references. There are lots out there. Practice/use is, once again, the key ingredient. The fact that this part is pretty long should tell you something. Many, though not all, of the common mathematical stumbling blocks are addressed here. And since mathematical analysis is an integral part of the course, fluency is essential … Hey, was that thunder?

The Journey Began like any other problem, in any other universe …

[MODEL] What's the story? What's the mission? What's the picture?

[SETUP] What have you been given to take along with you? What hidden pearls should you not forget? What are you looking for? What tools and paths are available to you? What methods will you employ?

[SOLVE] You are now entering Mathematic Land. Please adjust your thinking accordingly. If you do not follow the rules, you may be hit with a log, tripped with a trig, vexed with a vector, quashed by a quadratic, or disintegrated by an integral. You have been warned.

[ANALYZE] What did you discover in your explorations?

B.1. SYMBOL SIMON

[Abstract?... Useless? Such is the appearance of subtle wisdom, Grasshopper. Diamonds don't shine until polished. Read on and seek to understand more deeply than you do.] Ever play "Simon Says"? No matter what Simon said, you had to do it. That was the "made-up" rule of the game. Similarly, we (as in human beings) have made up the symbolic language of mathematics which is filled with rules. It represents quantitative relationships succinctly and precisely and allows us to manipulate those relationships while retaining quantitative equality. [Hey! Hey! Get back here and finish reading this section.] As a convenient shorthand, we use symbols to represent concepts. Some symbols are for quantities, some are for "operations," like division and differentiation. We learn the mechanics of mathematics, i.e. how to follow the rules to manipulate the expressions, but frequently lose sight of the meaning behind the symbols. As you develop your ability to use the mathematical tools, not only should you strive to be proficient with the mechanics, but also learn to translate concept to symbol and symbol to concept - much like translating from one language to another.

Behind every symbol is a concept.

We make up symbols as a shorthand way of communicating something. In that way, symbols mean whatever we want them to mean. If you wanted to break with conventional usage, the symbol ÷ could be used for multiplication. Instead we commonly use either of two symbols, x or ·, when we multiply numbers together: $3 \times 2 = 3 \cdot 2 = 5$. (Okay, it's 6. But if I said x and · represent addition, I'd be right. I suppose it is best to stick with convention.) When we multiply "vectors" (things with quantity *and* direction), you'll see that people have defined x and · to mean two *different* types of multiplication! It's not uncommon for one symbol to mean more than one thing. The context of use helps us interpret the concept behind the symbol.

Sometimes the symbols used actually give us a hint at their meaning. The symbol ÷ looks like a fraction: $\frac{dot}{dot}$. The root sign, $\sqrt{\ }$, resembles the symbol used for long division, $2\overline{)6}$ The summation sign, $\sum$, is the Greek capital S (sigma) while the integration sign, $\int$, is the Roman capital S - the former is a discrete (choppy) summation, while the latter is a continuous (smooth) summation. Sometimes the symbolic notation doesn't make much difference; however, **a streamlined symbolic representation can both increase the efficiency of symbolic manipulation and make it easier to see the meaning behind the labels** (the "transparency" of the equations). One dramatic example is Einstein's summation convention. Einstein suggested that we write $\sum_{i=1}^{3} x_i y_i = x_1 y_1 + x_2 y_2 + x_3 y_3$ in shorthand as $x_i y_i$. That is, whenever there is more than one term and a repeated index, in this case "i", we can assume that we are summing over the possible values of that index. In this way we can stop writing the summation sign. Silly? It turns out to make a huge difference in doing calculations using the theory of General Relativity! Another interesting example is in quantum mechanics. Schrödinger and Heisenberg came up with seemingly different ways of doing the math for quantum mechanics. As it turns out, it is only the symbolic system that is different, the meanings are the same. In different situations it is easier to use one symbolic system rather than the other.

Much of symbolic math is an extension of counting sheep. Consider the conceptual basis for some math operations. Multiplication is based on addition; exponentiation is based on multiplication. Subtraction, division, and taking roots are, in a sense, reverse addition, multiplication, and exponentiation, respectively. Fractions were invented by small carnivores who could only eat half a sheep. Take a moment and really get what an enormous conceptual leap it was to introduce the symbol for zero. Do the same for negative numbers. Yet, had people not created the symbols "0" and "-1," the language of math would certainly be much weaker. Differentiation allows us to say quantitatively how much one variable changes as one or more other variables change. Now look around you and notice that any advanced technology could not have come into existence without the language of mathematics ... and curious, innovative, competent scientists and engineers.

[This concludes Math Appreciation 101. Surprisingly, this understanding can make a difference.]

B.2. EVERYONE NEEDS SOME ANALYSIS

In physics, the job is to analyze – conceptually and quantitatively – physical systems. The raw materials are the scientific concepts and data, and the tools are mathematics and measuring equipment (along with your brain, paper and pencil, and your computer). It is interesting to note that there are two broad categories of quantitative representation and analysis:

1. **Graphical** - Provides a way to *visualize* the relationship between two or three variables. We can read numerical values from looking at the points and axes (including max. and min. values), rates of change of one variable as the other changes by looking at the slopes, and the rate of change of the slope by looking at the concavity of the curve. Graphs may be generated from experimental data or from theoretical functions. They may be rough sketches, careful plots by hand, or very accurate computer generated graphs.

2. **Analytical** - Here we operate with equations. We can still relate values of variables (just plug them in), find max. and min. values, find rates of change (take a derivative), and rates of change of rates of change (2nd derivatives). And again, we can make rough calculations, careful calculations, or program a computer to run the calculations (approx. or exact).

In this course, we will lean more heavily on equations. Keep in mind, however, that graphical analysis can be a very powerful tool, especially when there is no known equation to rely on and/or we want to interpret experimental data.

B.3. GRAPHING (Visual representation)

Labeling - A graph could represent just about anything, so it had better be labeled clearly so that the reader can translate the picture to the appropriate concepts and magnitudes.

Isothermal Expansion of an Ideal Gas: $n = 1$ mole, $T = 373$ K

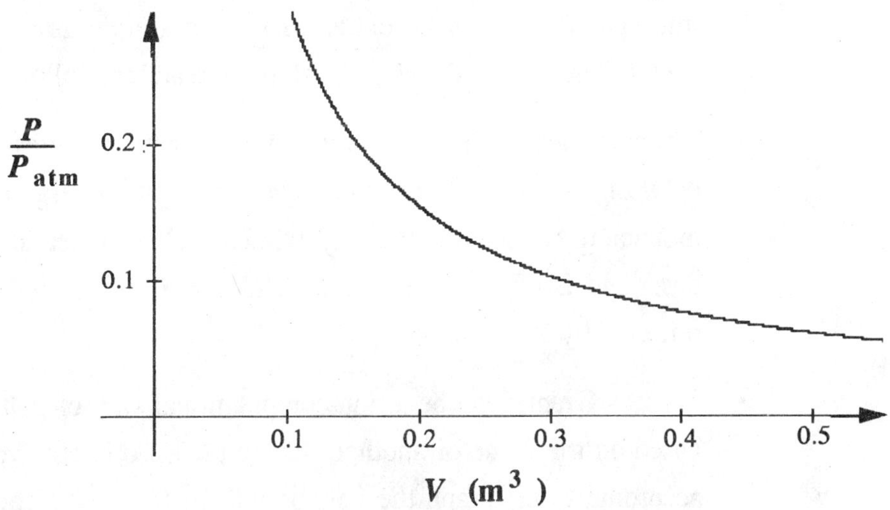

Notice the following elements in the sample graph above. A graph should have:

- A *title* summarizing what the graph is representing.
- *Labeled axes* showing the *variable(s)* and *units* corresponding with each axis.
- *Magnitude increments* marked along the axes.
- *Physical scale* that relates the graphical scale to the physical scale of the quantities involved. (This is only required if it is not clear from the axes.)

Scaling - "Hey! You didn't put units on the vertical axis!" I don't have to! "You said we had to. You think you're special?" No, I *scaled* the vertical axis to a constant–namely the atmospheric pressure, where P_{atm} = 1 atm = 1.01 x 10^5 N/m² = 14.7 psi. You can use whatever units you want for P_{atm}, then P will be in those units. But regardless of the units, 0.1 on the vertical axis means that P is 0.1 P_{atm}. This clearly specifies the quantity of the pressure relative to a good reference amount and makes the graph easier to read. "Oh."

Data Error - If you are plotting a function, you can evaluate some points and draw in a smooth curve that connects the points. If you are plotting data points, you will need to sketch in a "best fit" curve rather than a "connect the dot" curve. Since there is error associated with your data, you can indicate this with error bars, circles, or square around the points. The larger the error symbol, the more error you are reporting. On an informal sketch, these error symbols are drawn in very roughly. For a formal graph, one needs to compute the associated error and scale this to the graph. I will leave a discussion of error analysis to your lab manual or other reference.

Reading a Graph - When you look at a graph, what do you see? There is a lot of information that can be extracted from a simple graph. That's why they can be so useful. You should develop the skill to "read" the following off of a graph.

- *Physical meaning* - Graphs are very abstract representations of the relationships between concepts. For graphs showing spatial relations, such as a 3-D plot of a mountain, the meaning is easily extracted. But connecting a graph to the process being described is often quite a leap. Understand and visualize the variables and process reflected by the graph.

- *Points* - Graphs can be a non-computational way of getting the value of one variable based on the value of another: "if the pressure is 10^4 N/m^2 = 1.5 psi = 0.1 P_{atm}, then according to the graph the volume will be 0.3 m^3." They are also a valuable way of presenting data (EKGs, thermodynamics, economics...you name it).

- *Slope* - Where is the slope positive, negative, or zero (corresponding to the first derivative)? This tells us how one variable changes as the other variable changes in different regions. Additionally, there is likely some concept that is defined by the slope of the curve. For example, the slope of a position vs time curve is the velocity, and the slope of a velocity vs time curve is the acceleration. The slope of a financial graph will display the economic trends.

- *Concavity* - Is the curve like a bowl up (increasing slope, positive second derivative) or a bowl down (decreasing slope, negative second derivative)? This of course tells you how the slope is changing.

  EXAMPLE: on a position vs time graph, the points tell you your position at any given time, the slope tells you your velocity, $\Delta x/\Delta t$ (forward or backward, fast or slow), and the concavity corresponds to the acceleration (the rate of change of the velocity). Sometimes a picture *is* worth a thousand words! (You'll get plenty of practice in Ch.2.)

In closing, let me reiterate: You simply *must* keep the meaning of the variables in mind or you won't be able to make much sense of the graph. This ability to *translate* from graph to concept takes some training, but it is a necessary and useful skill.

B.4. ALGEBRA (Quick & accurate)

B.4.a. Rational Functions

Students often get discombobulated when they are confronted with messy rational functions (some algebraic expression divided by another - remember, "rational" is for "ratio"). You simply need to rewrite and manipulate them using some basic rules.

RULE 1: While combining fractions, you may add or subtract numerators (tops) *only* if you have common denominators; however, you can always multiply numerators with numerators and denominators with denominators. ($2/7+3/7 = 5/7$; $1/5 = 1/(2+3) \neq 1/2+1/3$! ; $2/7 \cdot 3/7 = 6/49$; now just use variables instead of numbers.)

RULE 2: Multiplying both numerator and denominator by the same expression or quantity doesn't change the value of the fraction–it's just multiplying by unity ($A/A = 1$; as an example, $2/3 = 2/3 \cdot 1 = 2/3 \cdot 4/4 = 8/12$, and so on.).

Pretty simple rules. Let's see if you've got them. Work on your *speed* and *accuracy*.

EXERCISE B.1: True Or False? If false, give an expression that would work on the right side of =.

(a) $\dfrac{1}{A+B} = \dfrac{1}{A} + \dfrac{1}{B}$

(b) $\dfrac{A+B}{C} = \dfrac{A}{C} + \dfrac{B}{C}$

(c) $\dfrac{A-B}{C+D} = \dfrac{A}{C+D} - \dfrac{B}{C+D}$

(d) $\dfrac{A-B}{C+D} = \dfrac{A-B}{C} + \dfrac{A-B}{D}$

(e) $\dfrac{A}{1/B} = A \cdot B$

(f) $\dfrac{1/A}{B} = \dfrac{1}{A \cdot B}$

(g) $\dfrac{A}{1/B} \cdot B = A$

(h) $\dfrac{x \cdot y}{z+d^2} = \dfrac{x}{z+d^2} \cdot \dfrac{y}{z+d^2}$

(i) $\dfrac{x \cdot y}{z+d^2} - q = \dfrac{x \cdot y - q(z+d^2)}{z+d^2}$

(j) $\dfrac{F}{\sqrt{\mu\lambda} + \psi^2} \cdot \dfrac{\sqrt{\mu\lambda} - \psi^2}{\sqrt{\mu\lambda} - \psi^2} = \dfrac{F\sqrt{\mu\lambda} - \psi^2}{\mu\lambda - \psi^4}$

(k) $\dfrac{1/B}{A^2 - C^2} = \dfrac{1}{B(A-C)} + \dfrac{1}{B(A+C)}$

(l) $\left(R + r - \dfrac{r^2 - R^2}{R+r}\right) - \left(R - r - \dfrac{r^2 - R^2}{R-r}\right) = 0$

** Watch out for those minus signs! Take it step by step, but do strive to pick up speed.

** Solutions are always in the back of each chapter.

B.4.b. 1 Eq'n, 1 Unk.

It is essential that you are quick and accurate when you solve algebraic equations. Hence, you should neither make too large an algebraic leap, nor pedantically include every minuscule maneuver as a separate step. Keep in mind that the math you write down is both for you and the reader. Communicate clearly and succinctly. Don't be afraid to throw a few explanatory remarks in there–they are highly desirable.

Much of solving one equation for one unknown breaks down to two simple maneuvers. Get to the point where you can perform these two "quick steps" without writing down the elementary steps.

QUICK STEP #1: (+/–) Move a term which is added or subtracted on one side to the other side and change the sign in the process. ($-A$ becomes $+A$ when moved to the other side. Pedantically, you would add or subtract the term from both sides.)

QUICK STEP #2: (X/÷) Move a *common* multiplicative factor to the other side and invert it in the process. Be sure to multiply ALL of the terms on the other side. ($1/A$ becomes A when moved to the other side.)

WARNING! In order to use rule #2, the multiplicative factor must be *common* to all terms on the original side and, once flipped and moved to the other side, becomes common to all the terms on the final side. This is *often* a source of error. You may have to use rule #1 before you can use rule #2.

EXAMPLE 1 - Solve for z quickly by using the two quick steps above.

Starting equation: $\dfrac{F \cdot z}{A} - C^2 D = Q\sqrt{x}$

DON'T do this: $z - C^2 D \neq \dfrac{A}{F} Q\sqrt{x}$, since $\dfrac{F}{A}$ was NOT a common factor to all terms on the left side in the original equation.

Apply rule #1: $\dfrac{F \cdot z}{A} = Q\sqrt{x} + C^2 D$ No need to write anything else. I've moved the term and switched the sign.

Apply rule #2: $z = \dfrac{A}{F}\left(Q\sqrt{x} + C^2 D\right)$ I've flipped $\dfrac{F}{A}$ and made it a factor for *every* term on the left in one quick and easy step.

From the starting equation, it should take you only two lines to isolate z. Now it's time for you to show your stuff.

EXERCISE B.2: Use the "starting equation" given in Example 1. Use the two quick-steps to isolate the variable specified below.

(a) F (b) C (c) x (d) A

FACTORING: If you can set an equation equal to zero and factor it, the solution becomes obvious. (In the following general cases, I resist using x as the variable to break you free from that limited way of thinking about algebra.)

$$t(t - 1/2) = 0 \quad \Rightarrow \quad t = 0 \text{ and } t = 1/2$$
$$(\psi + 3)(\psi - 1) = 0 \quad \Rightarrow \quad \psi = -3,\ 1$$
$$q(q - 8)(q + 3/2) = 0 \quad \Rightarrow \quad q = 0,\ 8,\ -3/2$$
$$(r - 2)(r + 1)(r + \pi) = 0 \quad \Rightarrow \quad r = 2\ -1,\ -\pi$$
$$v^2(v + e) = 0 \quad \Rightarrow \quad v = 0,\ -e$$

WARNING! The solutions above are the algebraic solutions. When the equations reflect physical processes, the algebraic solutions are the *candidates* for the physical solutions. We must go further and check to see if the solutions make physical sense. If one or more of the algebraic solutions does not make physical sense, we throw out the bad and keep the good. Once again you see that you must know what the equations mean.

THINK YOU KNOW THE QUADRATIC FORMULA? **[A surprising application is demonstrated here.]** You probably remember that an equation of the form $ax^2 + bx + c = 0$, can be solved by using the quadratic equation. If it's easily factored, go ahead. But the quadratic formula always works.

$$x = \frac{-b \pm \sqrt{b^2 - 4ac}}{2a}$$

Notice that the "±" gives two solutions, often called "x_+" and "x_-." However, it may turn out that you have only one solution (when the two solutions are equal), or no solutions at all (when you have a negative number under the square root). Graphically, we could plot the function $ax^2 + bx + c$. The function is zero where $y = 0$, i.e. where the parabola crosses the x-axis. Graphically we would see two solutions if the parabola crosses the x-axis twice. One or no solutions are seen if it touches the x-axis at one point, or never crosses it, respectively.

Now, don't be limited or rigid in your thinking about algebra. The "variable" in an equation may be a function as in the following example:

$$3.0[\sin\omega t]^2 - 13.1 \sin\omega t + 9.0 = 0.$$

Here you can define $x = \sin\omega t$ and use the quadratic equation to solve for $\sin\omega t$. If you need to, you can then use the inverse sine to find ωt. Keep in mind that $-1 \le \sin\omega t \le 1$, so any solution outside of this range is not valid. [Use the quadratic equation for the quadratic above. How many valid solutions are there? What is ωt ? ... Using a = 3.0, b = −13.1, and c = 9.0, I get that $\sin\omega t = 3.5$, and 0.85. Only 0.85 is possible, since the sine is always between −1 and +1. Thus, $\sin\omega t = 0.85 \Rightarrow \omega t = \sin^{-1}(0.85) = 58°$.]

DO THE SAME THING TO BOTH SIDES: [Make a mental note of this.] Let "A" and "B" be shorthand for two hideous expressions. Then if A = B, it is also true that A + 34 = B + 34, 0.5 A − 3.3 = 0.5 B − 3.3, sin A = sin B, $\tan^{-1} A = \tan^{-1} B$, $e^A = e^B$, and log A = log B. How about taking the cube root of both sides? Fine! Do whatever you need to do in order to solve for whatever it is that you are trying to find.

B.4.c. Multiple Eq'ns, Multiple Unk.'s

As I'm sure you know, in order to solve for 3 unknowns algebraically, you need 3 *independent* equations. If you only had 2 equations and 3 unknowns, you would be left with one unknown. You could then either find another applicable equation or go measure the unknown quantity in the lab.

RULE: With n *independent* equations you can solve for a maximum of n unknowns

DEFINITION: An equation is INDEPENDENT of the other equations if you *cannot* get it by manipulating the other equations in the set. (In this way it must specify an additional principle, relationship, constraint, or condition for the problem at hand.)

So, if you have 3 unknowns, you'll need to set up ("construct") 3 equations/relations to find all of them. If you have 9 unknowns, curse your instructor and find at least 9 equations/relations. If the equations are independent of one another, you know that, in theory, you could solve for your unknowns. Carrying out the algebra may be a tangled mess, but at least you know that it can be done. [**EXAM TIP:** If you get to the point where you have constructed the equations you need and you know what you're looking for, point it out to the grader. Even if you can't get the algebra to work out during the exam, you may be showered with partial credit!]

EXAMPLE 2 - Are the 3 equations given in each part below "independent" of each other?

(a) Study the following 3 equations:

1. $A + B = C$ 2. $A - B = D$ 3. $2A = C + D$

Can you get one of these equations from either or both of the other two? Of course. Add equation 1 and equation 2 to get equation 3. In fact, you should prove to yourself that you can get any one equation from the other two. These equations are NOT INDEPENDENT. Any two of them do, however, make an independent set. Thus you could solve for a maximum of 2 unknowns.

(b) Now study these 3 equations:

1. $A + \frac{1}{B} = C$ 2. $B\,(A - C) = -1$ 3. $\frac{AB + 1}{BC} = 1$

Can you get one of these equations from either or both of the other two? Indeed. In fact, you can get any one equation from either of the other equations alone. (With eq'n 3, move BC to the other side and then divide by B to get eq'n 1, and so on.) Thus we really only have 1 INDEPENDENT EQ'N. With these three seemingly different equations, we could only solve for one unknown. A different form does not constitute a different equation!

EXERCISE B.3: In each set of equations below, determine how many independent equations there are. How? By seeing if there is a way to get one of the equations from one or two of the others. How many unknowns could you solve for?

(a) $V_1 - I_1 R_1 - I_2 R_2 = 0,\quad V_1 - I_1 R_1 - I_3 R_3 = 0,\quad I_3 R_3 - I_2 R_2 = 0.$

(b) $A + B = C,\quad A - B = D,\quad C + 3D = 7$

(c) $F = m\,a,\quad a = F/m,\quad m = F/a\ .$

Methods For Solving For The Unknowns

So, you've got 2 (or 3, or 4, or 5...) independent equations and 2 (or 3, or 4, or 5...) unknowns to solve for. You've basically got two choices for how you proceed.

1. SUBSTITUTION: Using 1 eq'n, solve for 1 unk. in terms of the other unknown(s). Plug that expression into another equation. Continue until you can solve for an unk. Then work backwards and plug your now known unk. into the simplest equation that will allow you to solve for another unk. (In the best case scenario, one of your equations will only have one unknown and you can solve for it directly. Then substitute that value into your other equation(s).) Whittle away at it!

- Take the equations (1) $A + B = 2$,

  and (2) $A - B = 9 \xrightarrow{\text{solve for A}} A = B + 9$
- Replace A in (1) with $B + 9 \rightarrow B + 9 + B = 2 \rightarrow B = {}^{-7}\!/_{2}$
- Now plug this value for B into either (1) or (2). Using (2) gives $A = {}^{-7}\!/_{2} + 9 = {}^{11}\!/_{2}$.

No matter how ugly the expressions are in your equations, the basic maneuvers are always the same as those demonstrated above.

2. COMBINING EQUATIONS: Consider the two independent equations (templates, really) below.

(1) (expression A) = (expression B)

(2) (expression C) = (expression D)

Surprise! We can do anything – yes, *anything* – we want to these equations as long as we abide by the basic math rules. For example, we could write

$$(1) - (2) \rightarrow (\text{expres. A}) - (\text{expres. C}) = (\text{expres. B}) - (\text{expres. D})$$

$$\frac{29}{R}(2) + (1) \rightarrow \frac{29}{R}(\text{expres. C}) + (\text{expres. A}) = \frac{29}{R}(\text{expres. D}) + (\text{expres. B})$$

$$\frac{(2)}{(1)} \rightarrow \frac{(\text{expres. C})}{(\text{expres. A})} = \frac{(\text{expres. D})}{(\text{expres. B})}$$

[*This move is often overlooked* !]

Do you see? Anything goes. You could even add expression A to expression D and expression B to expression C, since it still maintains a valid equality. You could take the squares or cube roots of the expressions. Whatever it takes to solve for those pesky unknowns. [Is it true that A + B = C + D? If so, only by chance. I hope you see why this equality does NOT follow.] Remember that the point of combining the equations is to get rid of some unknowns so that what you are left with is a simple equation which you can solve. Then, as before, you whittle away as best you can.

Which method you use is up to you. Step back and look at your equations. Which method looks like it will be the easiest in your problem? If you don't know, don't just sit there, jump in and test it out. Get some dirt under your fingernails!

B.5. EXPONENTS, ROOTS, & LOGS (Oh my!)

If you are having trouble with any of these topics, perhaps it is best to review the fundamentals. I hope that the following deepens your understanding.

B.5.a. Know 2, Get 1 Free

[Roots and logs? What is this, botany?] The meaning – the logic – behind exponents, roots, and logs is completely tied to the equation

$$\text{base}^{\text{exp}} = \text{number} \xrightarrow{\text{in short}} b^E = N$$

Given any two of b, E, or N, we can find the other. [bEN there, done that!].

- Exponents: given b and E, we can get N. $b^E = N$
- Roots: given N and E, we can get b. $b^E = N \Leftrightarrow \sqrt[E]{N} = b$
- Logs: given b and N, we can get E. $b^E = N \Leftrightarrow \log_b N = E$

Don't get hung up on the symbolic notation, the fundamental meaning is right there. We'll see that the logic for exponents is closely tied to algebra, the logic for roots is tied to algebra in a somewhat convoluted way, and the logic for log's "transcends" algebra (so it is called a transcendental function), i.e. we can't find the logarithm algebraically. Knowing this will help you understand and remember the basic rules.

So take note that exponents can be found by simple multiplication (or a calculator), while roots and logs can be guessed at in easy cases (or approx. guesses), but usually require a calculator. Before calculators, people had to look these up on tables or graphs!

B.5.b. Exponents ($b^E = N$)

Exponents are a way of counting identical multiplicative factors. Someone made up the notation where you put how many factors there are as a superscript. Sounds good. This turns out to be much more efficient than writing out the factors over and over. Of course if you would rather write them out . . . Here, you choose which notation you like best.

- $10 \times 10 \equiv 10^2$
- $5 \times 5 \times 2 \times 2 \times 2 \equiv 5^2 \times 2^3$
- $4 \times 4 \times 4 \equiv 4^3$
- $9 \times 11 \times 9 \times 11 \equiv 9^2 \times 11^2$

Or how about: $10 \times \pi \times 10 \times a \times a \times \pi \times 10 \times a \times 10 \equiv \pi^2 \times a^3 \times 10^4$? That superscript notation was a great idea! As you review the following rules below, notice that they reflect the basic idea of counting multiplicative factors–the exponents are added and subtracted.

BASIC RULES OF EXPONENTIATION

Let a and b represent any base numbers and x and y be any exponent values.

1. $b^x \cdot b^y = b^{x+y}$
2. $(b^x)^y = b^{x \cdot y}$
3. $b^{-x} = \frac{1}{b^x}$
4. $\frac{1}{b^{-x}} = b^{-(-x)} = b^x$
5. $b^0 = 1$
6. $a^x \cdot b^x = (ab)^x$
7. $\frac{a^x}{b^x} = \left(\frac{a}{b}\right)^x$
8. $b^x + b^x = 2b^x$
9. $a^x + b^x$ ... cannot be reduced if $a \neq b$.
10. $b^x + b^y$ ... cannot be reduced if $x \neq y$.

You really should try putting in 1, 2, and 3 for the exponents in the rules above to help make them transparent to you. In rules 9 and 10, put in easy numbers for the bases (say, $a = 2$, $b = 3$).

If a positive exponent counts how many identical multiplicative factors there are, what could a negative exponent be used to mean? It counts how many identical divisors there are! This reduces (subtracts) the number of multiplicative factors. Check the number of multiplicative factors of 10 in the examples below and notice how well this symbolic notation works to keep track of that number.

- $10^2 \times 10^3 = (10 \times 10)(10 \times 10 \times 10) = 10^5$
- $\dfrac{(10 \times 10 \times 10 \times 10 \times 10)}{(10 \times 10)} = \dfrac{10^5}{10^2} = 10^5 \times 10^{-2} = 10^{5-2} = 10^3$
- $\dfrac{10^5}{10^{-2}} = \dfrac{10^5}{1/10^2} = 10^5 \times 10^2 = 10^7$ ...or... $\dfrac{10^5}{10^{-2}} = 10^5 \times 10^{-(-2)} = 10^7$
- $1 = \dfrac{10^5}{10^5} = 10^5 \times 10^{-5} = 10^{5-5} = 10^0$

FUNCTIONS WITH EXPONENTS: I just want to point out that there are two basic types of functions which involve exponents; there are "power" functions where the variable is raised to some set power (x^2, y^3), and there are exponential functions where the base number is fixed and the variable is in the exponent (2^x, 10^x, or e^x). It's an interesting distinction to keep in mind (variable in the base vs. variable in the exponent). You might play around with some numbers and or graphs to compare and contrast these two types of functions. I'll show you a couple of exponential curves.

BASIC GRAPH CHARACTERISTICS: It is important to be familiar with the behavior of the exponential function as reflected in the following graphs. Both exponential growth and decay are shown.

Graph of: $y = e^x$ and $y = e^{-x}$

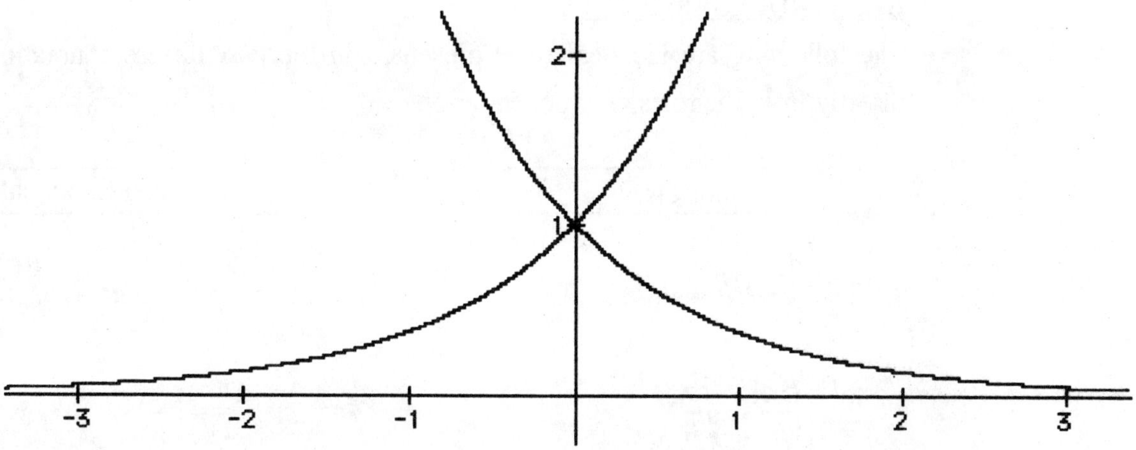

Notice that the y-intercept (at $x = 0$) for both curves is $e^0 = 1$. Further notice that, since e = 2.718... > 0, both curves are above the x-axis ($y > 0$). They don't cross the x-axis, but approach it ($y \to 0$) as the exponent $\to -\infty$; i.e. when $x \to +\infty$ or $-\infty$, depending on the sign of the exponent.

B.5.c. Roots ($b^E = N \Leftrightarrow \sqrt[E]{N} = b$ or $N^{\frac{1}{E}} = b$)

Roots "undo" exponents; the square root of x^2 is $\pm x$, the cube root of 5^3 is 5, and the 20th root of e^{20} is e. So what is the root doing? It's not so much what the root is *doing* as what it has been defined to mean - symbol Simon. Just as the root system is the base of a tree, the root of a number *is* the base number. The cube root of 1000 = 10×10×10 = 10^3 is 10. 10 is the "root" or "base" number. My calculator tells me that the square *root* of 1000 is 31.6227766... i.e. squaring that base number gives me 1000.

This symbol/operation that people defined is a bit harder to grasp because the logic isn't straightforward algebra the way exponentiation is. We can sort of see the algebraic "roots" when we write $\sqrt[4]{\mu \cdot \mu \cdot \mu \cdot \mu}$. We can see that the 4th root is μ. (Actually, it's μ if μ is positive, −μ if μ is negative.) But we can't add, subtract, multiply, or divide numbers to get the roots. We either know the answer or use our calculators.

μ is the Greek lower case mu, pronounced "moo."

As you know, there are two ways that have been devised to symbolize the operation of taking roots. One uses the radical sign, $\sqrt[x]{\ }$. My guess is that this sign is based on the long division sign, $\overline{)\ }$, since it's *sort of* like division in that it reduces the number of factors: μ^4 becomes μ upon taking the 4th root - μ^4 is divided up into 4 equal multiplicative factors "μ" and the 4th root tells us what that common factor is. Taking the cube root of 37.90 gives 3.359 because (3.359)(3.359)(3.359) = 37.90. Since the number of factors is reduced, what is really being divided is the exponent. This leads to the exponential notation where we use a fractional exponent in place of the radical sign. They mean the same thing: $\sqrt[3]{x^3} = \left(x^3\right)^{\frac{1}{3}} = x^{\frac{3}{3}} = x$. The exponential notation tends to be easier to work with.

<u>BASIC RULES OF ROOTS</u>

The following are identical rules presented in the two different notations. They follow directly from the rules of exponentiation.

Radicals	Fractional Exponents
1. $\sqrt[x]{a}\sqrt[x]{b} = \sqrt[x]{ab}$	1. $a^{\frac{1}{x}} \cdot b^{\frac{1}{x}} = (ab)^{\frac{1}{x}}$
2. $\frac{\sqrt[x]{a}}{\sqrt[x]{b}} = \sqrt[x]{\frac{a}{b}}$	2. $\frac{a^{\frac{1}{x}}}{b^{\frac{1}{x}}} = \left(\frac{a}{b}\right)^{\frac{1}{x}}$
3. $\sqrt[x]{b}\sqrt[y]{b}$ ... can't rewrite	3. $b^{\frac{1}{x}} \cdot b^{\frac{1}{y}} = b^{\frac{1}{x}+\frac{1}{y}}$
4. $\sqrt[x]{b} + \sqrt[x]{b} = 2\sqrt[x]{b}$	4. $b^{\frac{1}{x}} + b^{\frac{1}{x}} = 2b^{\frac{1}{x}}$

5. $\sqrt[x]{b} + \sqrt[x]{a}$ ... can't rewrite if $a \neq b$

5. $a^{\frac{1}{x}} + b^{\frac{1}{x}}$ ... can't rewrite if $a \neq b$

6. $\sqrt[x]{b} + \sqrt[y]{b}$ ... can't rewrite if $x \neq y$

6. $b^{\frac{1}{x}} + b^{\frac{1}{y}}$ ... can't rewrite if $x \neq y$

BASIC GRAPH CHARACTERISTICS: It is good to be familiar with the behavior of the root function as reflected in the following graph.

Graph of: $y = \sqrt{x}$

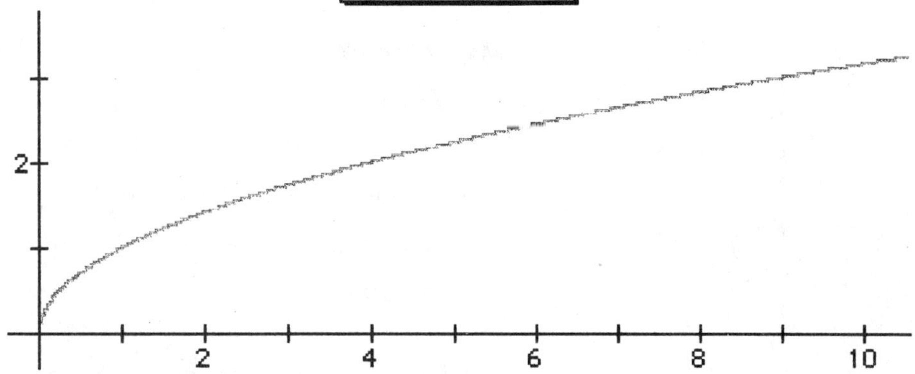

What do you notice about the graph? What values of x are and are not allowed? What about the values of y?

B.5.d. Logarithms ($b^E = N \Leftrightarrow \log_b N = E$)

Ready for a little mental gymnastics? Just keep in mind that the exponent is a way of counting identical multiplicative factors. When you take the log base b of a number N, you are finding the exponent. "What exponent?" Well, if you raise b, the base number, to that exponent, you'll get N. It's simple in that the definition is just based on the meaning of exponents, but I have to admit that the logic is convoluted. It would probably be worth a few minutes to try and write down, on paper and in your own words, an explanation of the definition of the logarithm. Also try plugging in some numbers.

[OK, give me another shot at this convoluted logic!] When you plug "$\log_b N$" into your calculator, you get a number. The number "$\log_b N$" is the exponent which, when you use b as the base, gives back N. Yup, it's definitely time to play with some numbers!

EXERCISE B.4: (Monty Python sings, "Oh I'm a lumberjack and I'm OK. I sleep all night and I work all day. I like to ...")

1. Without using a calculator, deduce the values of the following:

a) $\log_{10} 100 \equiv \log 100 = ?$ (10 to what power gives 100?)
b) $\log_2 32 = ?$ (2 to what power gives 32?)
c) $\log_e e^4 \equiv \ln e^4 = ?$ (e to what power gives e^4?)
d) $\log_\pi 9.8696 = ?$ (Take a guess. $\pi = 3.14...$)
e) $\log(-10) = ?$ (Hmm... What's the meaning of this?)

2. Use a calculator to find the following. Write an expression of the form: $b^E = N$.

a) $\log \pi =$ ____ (Makes sense?)
b) $\ln \pi =$ ____ (Compare with a.)
c) $\log e =$ ____ (Compare with a.)
d) $\ln e =$ ____ (Compare with b.)
e) $\log 200 =$ ____ (Makes sense?)
f) $\ln 200 =$ ____ (Compare with the others.)
g) $\log 1 =$ ____ (Why?)
h) $\ln 1 =$ ____ (You should've known!)
i) $\log(\frac{1}{2}) =$ ____ (Hey!)
j) $\ln(\frac{1}{2}) =$ ____ (Aha.)
k) $\log_2 31 =$ ____ (Try trial and error.)
l) You investigate some!

Notice from your results that $10^{\log\pi} = 10^{0.497} = \pi$, $e^{\ln 200} = e^{5.298} = 200$, and so on. Exponentiation undoes the log *if* we use the same base. If you give this some thought, it makes sense. Similarly, taking the log undoes exponentiation of the same base.

LOGGING RULES: Sometimes you just need to rearrange your logs.

1. $b^{\log_b x} = x$
2. $\log_b x + \log_b y = \log_b(x \cdot y)$
3. $\log_b x - \log_b y = \log_b\left(\frac{x}{y}\right)$
4. $\log_b b^n = n$
5. $\log_b b^n = n$
6. $\log_b\left(\frac{1}{x^n}\right) = \log_b(x^{-n}) = -n \log_b x$

Check your intuition. Do the following make sense?

- $\log 0 = -$, i.e. it is undefined!
- log (negative #) is NOT defined, hence we sometimes put absolute value signs around the argument to force it to be positive.
- $\log(A \pm B)$ cannot be easily reduced!!

RECALL:

$\log_e x \equiv \ln x$; Base e = 2.718... is called the "natural log."

$\log_{10} x \equiv \log x$; Base 10 is called the "common log."

EXERCISE B.5: Use the rules of logs given above.

(a) Change the form of $\ln(b^2\text{-}2b)$

(b) Change the form of $\ln(x/(y+3)^2)$

(c) Solve for t in the equation $N_0e^{kt} = R$. [HINT: take the ln of both sides.]

(d) Challenge!! Prove the following rule: $\log_b N = \dfrac{\log_a N}{\log_a b}$ ... peak at the solution if you get stuck.

BASIC GRAPH CHARACTERISTICS: It is important to be familiar with the behavior of the logarithm function as reflected in the following graph.

Graph of: $y = \ln x$

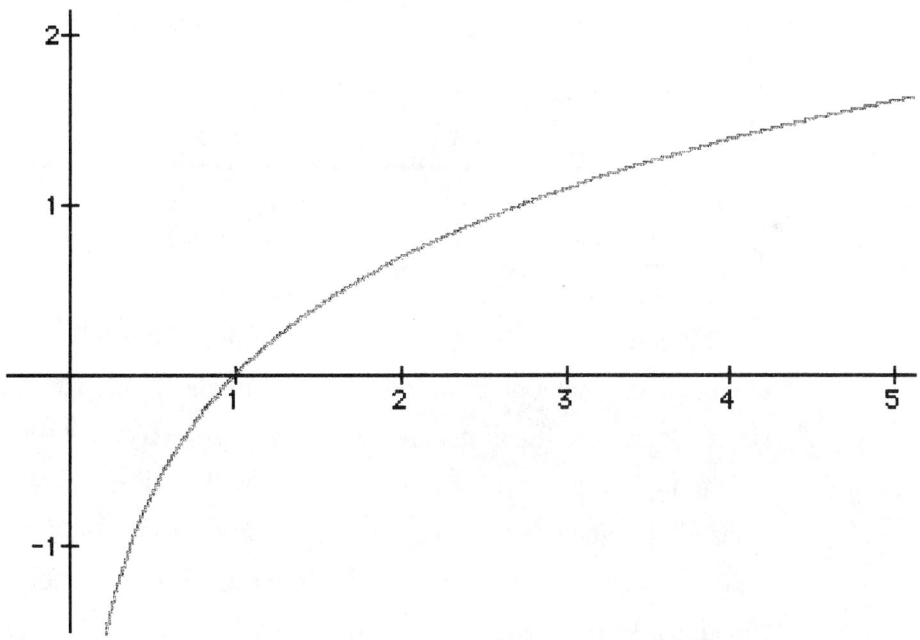

What do you notice about the graph? What values of x are and are not allowed? What about the values of y? Why does this make sense? (Remember, we can express the relationship between y, e, and x in exponential form as $e^y = x$. Now does it make sense that x must be positive? Does it make sense to you that $y = 0$ when $x = 1$? How about the fact that $0<x < 1$ corresponds with a negative y, while $x > 1$ corresponds with a positive y?

B.6. TRIG BASICS (The right & the rad)

WARNING! – Don't skip this section because you think you know it. There is probably something in here that you need to review or speed up on.

B.6.a. The Right Triangle (for angles between 0° and 90°)

If you were to take two similar right triangles and measure the lengths of the sides, what would you discover? The ratios of any two sides of one are identical to the ratios of the similar sides of the other. Get out a ruler and try it out on the similar triangles below. [Come on, it'll be worth it.] How many ratios of two different side lengths can you create for a single right triangle? [Count 'em & list 'em!]

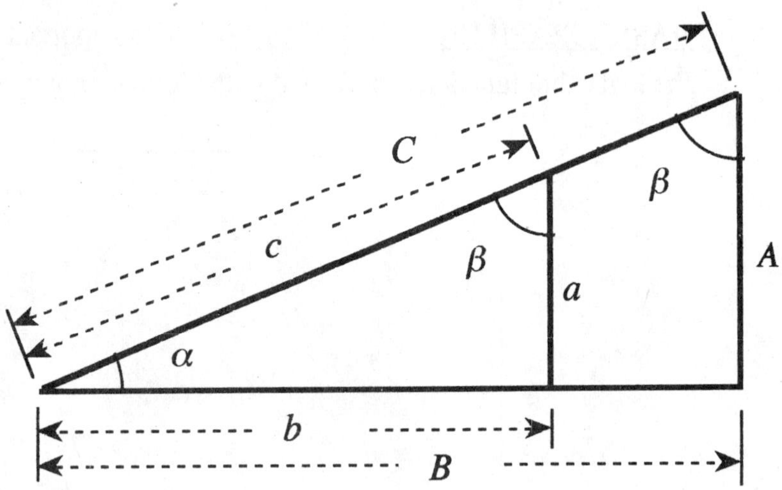

Did you confirm that $A/B = a/b$, $A/C = a/c$, and $B/C = b/c$? Of course this implies that the reciprocal relations, $B/A = b/a$ and so on, also hold. The fundamental definitions of the six trigonometric functions are based on these six ratios (or three and their reciprocals). Any triangle drawn with the same angles will have the same values for the ratios of the sides. (Try some of your own!) You see that the ratios only depend on the angles. Thus the ratios are functions of the angles. You probably remember the mnemonic for the trig functions. What was it? Soak your head, uh, soak your feet, uh, oh yeah, "soak-a-toe-a"!

"SOH–CAH–TOA !"

- SOH: $\sin(\text{angle}) = \dfrac{\text{Opp.}}{\text{Hyp.}}$
- CAH: $\cos(\text{angle}) = \dfrac{\text{Adj.}}{\text{Hyp.}}$
- TOA: $\tan(\text{angle}) = \dfrac{\text{Opp.}}{\text{Adj.}}$

where,

Opp. = length of the side opposite the angle

Adj. = length of the side adjacent to the angle

Hyp. = length of the side opposite the right angle

The reciprocal relations are simply defined as

- cosecant (csc) $\equiv \dfrac{1}{\text{sine}} = \dfrac{\text{Hyp.}}{\text{Opp.}}$
- secant (sec) $\equiv \dfrac{1}{\text{cosine}} = \dfrac{\text{Hyp.}}{\text{Adj.}}$
- cotangent (cot) $\equiv \dfrac{1}{\text{tangent}} = \dfrac{\text{Adj.}}{\text{Opp.}}$

Thus, for the triangles shown at the beginning of the section:

- $\sin(\alpha) = \cos(\beta) = \dfrac{a}{c} = \dfrac{A}{C}$
- $\tan(\alpha) = \cot(\beta) = \dfrac{a}{b} = \dfrac{A}{B}$
- $\cos(\alpha) = \sin(\beta) = \dfrac{b}{c} = \dfrac{B}{C}$

Check these relations by measuring the diagram angles α and β with a protractor and plugging them into your calculator. Do the values match the ratio values that you obtained with a ruler to within the measurement error? (You know, people could have called sine "ophy" for opposite/hypotenuse. Then ophy(30°) = ½. I wonder what the etymologies of the chosen words are … Nevertheless, we know what the definitions are and that provides us with a foundation for understanding these functions.)

A quick note about the **Pythagorean theorem**. Notice that we are told that (*longest side*)^2 = (*short side*)^2 + (*other short side*)^2. Hence it makes sense that (*either short side*)^2 = (*longest side*)^2 – (*other short side*)^2. (It wouldn't make sense to reverse the order of subtraction. Why? Remember, any number squared is positive.) The longest side is always the hypotenuse, thus, $c^2 = a^2 + b^2$, $a^2 = c^2 - b^2$, $b^2 = c^2 - a^2$. [You could check this with your measurements. Have you ever done that? It's worth doing once in your lifetime.]

EXAMPLE 3 – Here are three right triangles that you should be able to sketch quickly. They will give you commonly used values in a flash.

(a) <u>45°–45°–90°, The Isosceles Triangle</u>

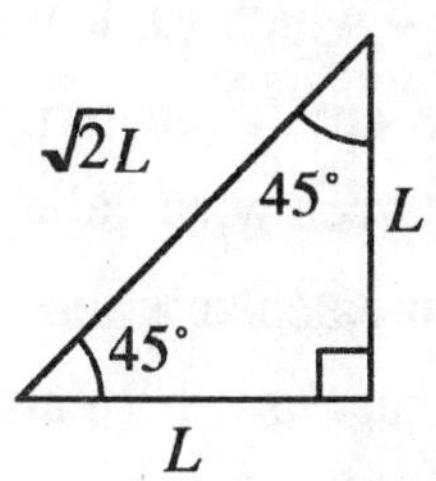

- Hypotenuse = $\sqrt{L^2 + L^2} = \sqrt{2L^2} = \sqrt{2}\,L$
- $\sin 45^\circ = \cos 45^\circ = \dfrac{1}{\sqrt{2}}$
- $\tan 45^\circ = 1$

(b) 30°–60°–90°, Half of an Equilateral Triangle

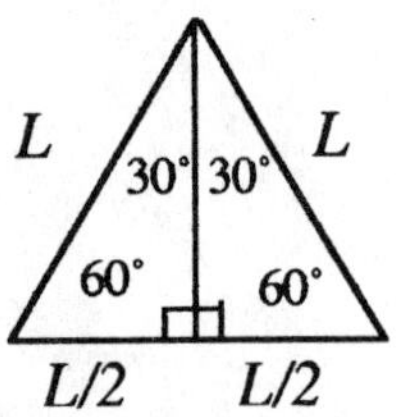

- Bisecting side = $\sqrt{L^2 - \left(\frac{L}{2}\right)^2} = \sqrt{\frac{3L^2}{4}} = \frac{\sqrt{3}}{2}L$
- $\sin 30° = \cos 60° = \frac{1}{2}$
- $\sin 60° = \cos 30° = \frac{\sqrt{3}}{2}$
- What are tan 30° and tan 60° ?

(c) The 3–4–5 Triangle

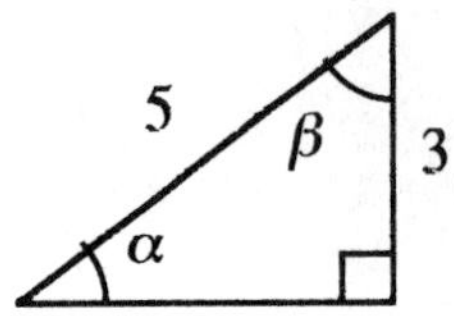

- Check with Pythag.: $5 = \sqrt{3^2 + 4^2}$
- $\alpha = \tan^{-1}(3/4) = \sin^{-1}(3/5) = \cos^{-1}(4/5) =$ _____
- $\beta = \tan^{-1}(4/3) = \sin^{-1}(4/5) = \cos^{-1}(3/5) =$ _____
- Notice: $\sin\alpha = \cos\beta$, $\sin\beta = \cos\alpha$, and
  $\tan\alpha = \cot\beta$, $\tan\beta = \cot\alpha$,

B.6.b. The Unit Circle (a memory aid for any angle)

- Right triangle → trig functions for $0° \le \theta \le 90°$
- Unit circle → trig functions for *any* angle

If you didn't already, I hope you see now that the fundamental, concrete definitions behind the six trig functions can be obtained from the six possible ratios of the side lengths of a right triangle – they're simply fractions. Further notice that the sine and cosine must be less than or equal to 1! It amazes me how such simple definitions can have such wide applicability. Extending these functions, especially the sine and cosine, allows us to describe forces, torques, rotating objects, pendula, oscillating springs, alternating electric current (AC), and the wave phenomena of sound, light, stringed instruments, and earthquakes. Even then, the full power of these simple functions hasn't been seen until one learns about something called Fourier analysis – which I'll leave alone for now.

Happily, the extensions we need to make aren't too terribly difficult. So extend we will. All of the angles in a right triangle are between what values? Right, $0° \le \theta \le 90°$. So what is sin(120°)? Is it undefined? Check your calculator and you'll find that it's the same as sin(60°) and cos(30°). I didn't just know that. I sat here and visualized the unit circle. It's a great memory aid! Make sure you can visualize it quickly and clearly.

Consider a unit circle (radius = 1) drawn on an x–y coordinate system. For starters, lets make right triangles in quadrant I using the radial line as the hypotenuse. We could label any of the angles θ, but we'll use the standard convention:

- \+ Positive θ is measured ccw (counterclockwise) from the $+x$–axis
- – Negative θ is measured cw (clockwise) from the $+x$–axis

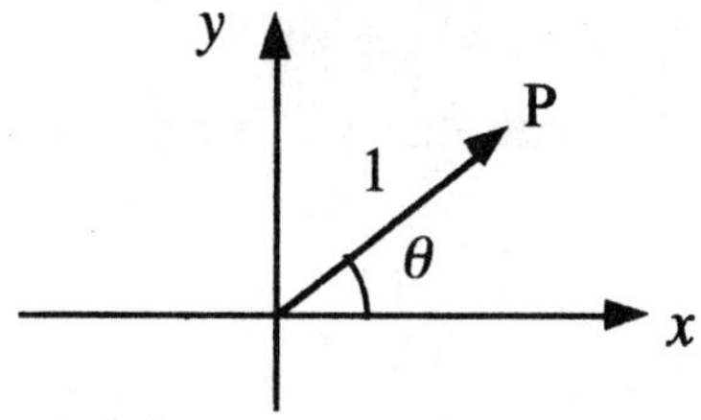

How long is the side opposite from θ, i.e. the vertical side? [HINT: Use a trig function.] How long is the side adjacent to θ, i.e. the horizontal line?

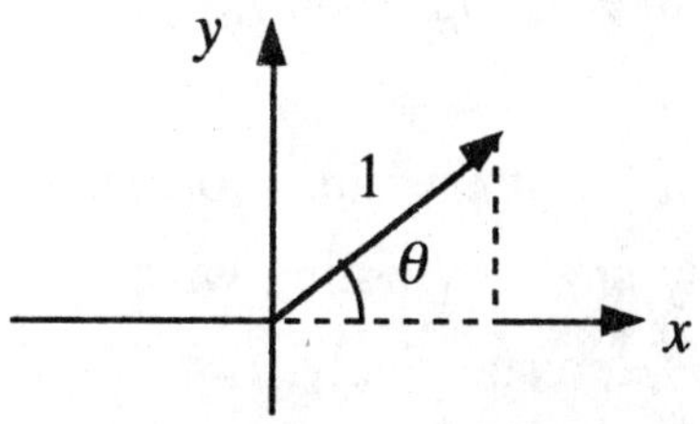

- $\cos\theta = \frac{x}{1} \Rightarrow x = \cos\theta$
- $\sin\theta = \frac{y}{1} \Rightarrow y = \sin\theta$

In terms of the coordinates (x,y) we see that the x part, or "component," or "shadow," is given by the cosine and the y part, or "component," or "shadow," is given by the sine. That's why we choose to use the unit ($r = 1$) circle. If we label the point at the end of the radial line "P," its coordinates are $(x,y) = (\cos\theta, \sin\theta)$.

Now we extend the definitions to angles greater than 90°. For any angle, we *define* the x and y coordinates (or "components") to be $(x,y) = (\cos\theta, \sin\theta)$, with θ measured from the $+x$–axis and ccw being the positive direction. Notice that we've gone beyond the simple rules for right triangles.

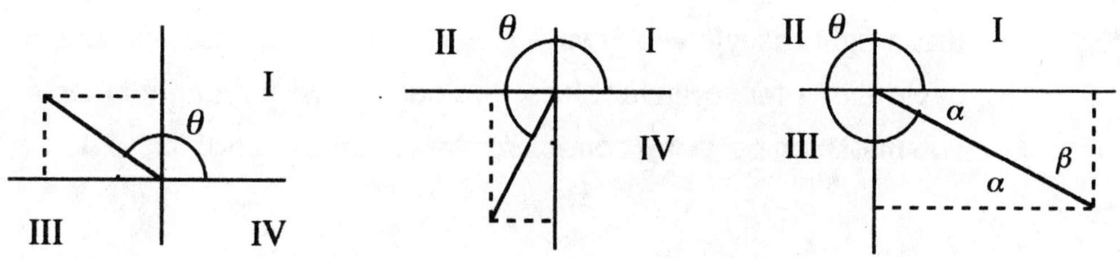

Further notice that we can still form right triangles with the radial line in any quadrant. In fact, we can always form two such triangles: one with the x–axis and one with the y–axis. The angles within the right triangles in quadrants II, III, and IV, however, are *not* θ. Look back on the last diagram where the point is in the fourth quadrant. You understand my point if you can see that, in the fourth quadrant, $x = \cos\theta = +\cos\alpha = +\sin\beta$, and $y = \sin\theta = -\sin\alpha = -\cos\beta$. (Please assume that the radial lines in the 3 pictures are all one unit long.) So, you must draw and see the triangle, determine which angle corresponds to which trig function, and be sure that you've put in the correct signs when you create right triangles this way.

What will we gain by all of this?

1. We extend the trig definitions to include any angle – positive or negative, and as large as you like. (Notice that $450° = 360° + 90°$, so the radial arm is in the same place as $90°$. Thus we know that $\sin 450° = \sin 90°$, and so on.)

2. We can quickly check the sign (+ or −) of the sine, cosine, and tangent (and the 3 reciprocal relations, if we like) for any angle by noting which quadrant it is in. This is given by the sign of x, y, or the ratio y/x, respectively.

3. We can visualize things like $\cos 120° = -\cos 60° = -\sin 30°$ and $\cos\theta = \sin(\theta + 90°)$. These equivalent relations are often helpful in rewriting expressions in forms that are easier to work with.

4. We *could* obtain the values of all of the trig functions by carefully measuring the angles and corresponding x and y coordinates with a protractor and ruler, being careful to note the signs of the coordinates (x,y) and thus the signs for $\sin\theta$ and $\cos\theta$. [I'll leave it to the calculator, but be my guest.]

!

There's no time like the present to get straight on this once and for all. Keep in mind that a right triangle will always give positive values for sine and cosine, whereas the unit circle allows for negative values. **You can always form right triangles in the unit circle, but you must then put in the correct sign which corresponds to the sign of the coordinate.**

EXERCISE B.6: Circle the correct sign. Sketch the unit circle with the radial arm in the appropriate quadrant to obtain your answer.

(a) Quadrant I ($0^\circ \le \theta \le 90^\circ$)

$x = \cos(\theta$ between 0° and $90^\circ)$ is + or –?

$y = \sin(\theta$ between 0° and $90^\circ)$ is + or –?

y/x = slope = $\tan(\theta$ between 0° and $90^\circ)$ is + or –?

(b) Quadrant II ($90^\circ \le \theta \le 180^\circ$)

$x = \cos(\theta$ between 90° and $180^\circ)$ is + or –?

$y = \sin(\theta$ between 90° and $180^\circ)$ is + or –?

y/x = slope = $\tan(\theta$ between 90° and $180^\circ)$ is + or –?

(c) Quadrant III ($180^\circ \le \theta \le 270^\circ$)

$x = \cos(\theta$ between 180° and $270^\circ)$ is + or –?

$y = \sin(\theta$ between 180° and $270^\circ)$ is + or –?

y/x = slope = $\tan(\theta$ between 180° and $270^\circ)$ is + or –?

(d) Quadrant IV ($270^\circ \le \theta \le 360^\circ$)

$x = \cos(\theta$ between 270° and $360^\circ)$ is + or –?

$y = \sin(\theta$ between 270° and $360^\circ)$ is + or –?

y/x = slope = $\tan(\theta$ between 270° and $360^\circ)$ is + or –?

* Check these by plugging sample values of θ into your calculator.

It's hard to memorize all of the information above. Maybe you can come up with another mnemonic that does the trick, but a quick sketch of the unit circle will work every time. The next relations aren't all that easy to remember either. Even the unit circle is a bit tricky. But it does get the job done once you catch on.

EXAMPLE 4: Use the unit circle (r = hypotenuse = 1) to *see* the following.

(a) <u>$\sin 120^\circ = \sin 60^\circ = \cos 30^\circ$.</u>

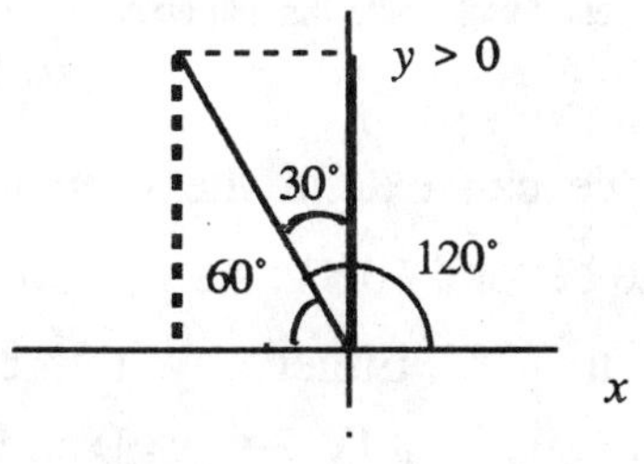

- $\sin 120^\circ = y$–component and is +
- Make a right triangle with the y–axis. You see that the y–length = $\cos 30^\circ > 0$.
- Make a right triangle with the x–axis. You see that the y–length = $\sin 60^\circ > 0$.

Hence, $\sin 120^\circ = \sin 60^\circ = \cos 30^\circ$. Check this result with both the half–equilateral triangle and your calculator.

(b) <u>What is cos 120° equal to if we use angles between 0° and 90°?</u>

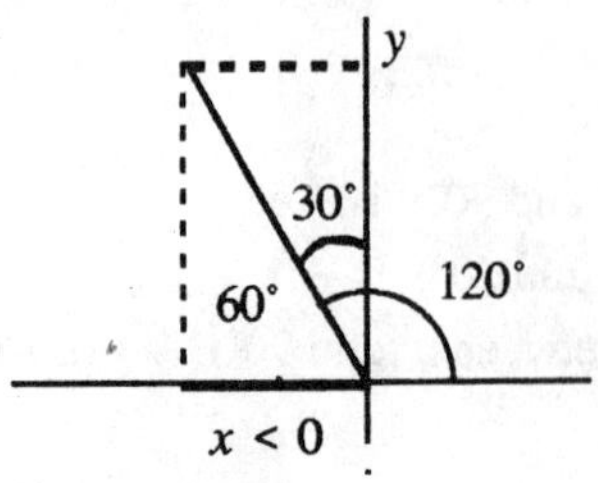

- $\cos 120° = x$–component and is –
- Make a right triangle with the y–axis. You see that the *x–length* = $\sin 30° > 0$.
- Make a right triangle with the x–axis. You see that the *x–length* = $\cos 60° > 0$.

Hence, $\cos 120° = -\sin 30° = -\cos 60°$. Check this result with your calculator.

Using right triangles will always give positive values. However, angles not in the first quadrant can give negative values. The absolute values, i.e. the lengths, may be the same, but the signs may differ. The unit circle helps us see the difference. Thus by creating different triangles with the same point (P) and radial line (r=1), and taking care to adjust the sign (+ or –), we can equate values of trig functions for different angles. This proves quite useful. You try some.

EXERCISE B.7: Use the unit circle to show the following identities. [HINT: Draw the radial line at the first angle given. Then make rt triangles with each axis.]

(a) $\cos 220° = -\cos 40° = -\sin 50°$ and $\sin 220° = -\sin 40° = -\cos 50°$
(b) $\cos 340° = +\cos 20° = +\sin 80°$ and $\sin 340° = -\sin 20° = -\cos 80°$
(c) $\cos\theta = \sin(\theta + 90°)$
(d) $\sin\theta = \cos(\theta - 90°)$

(Okay, parts c and d are tough. Be a detective. Here's a tip: Draw the radial arm at θ in quadrant I and draw another one at $(\theta + 90°)$. Check the lengths. Repeat this for θ in each quadrant. For part d repeat the whole process using $(\theta - 90°)$ as your second radial arm.) [For these identities, I actually prefer to visualize the sine and cosine graphs as demonstrated in B.6.d. But this is great practice. When you finally get this (after you've worked through the frustrating confusion), you'll be quite comfortable with the unit circle.]

Finally, and in the same way, we can extend the domain of the trig functions to negative angles (drawn cw) and those beyond ±360°. As a concrete example, let's rotate the radial arm ccw to $^+330°$ and then cw to $^-30°$. Either way it ends up in the same place and so has the same x and y components: quadrant IV → $x > 0$ and $y < 0$.

$x = \cos 330° = \cos(^-30°) = {}^+0.866$ $\left(= \frac{\sqrt{3}}{2}\right.$, using the half–equilateral triangle)

$y = \sin 330° = \sin(^-30°) = {}^-0.5$ (Use the half–equilateral triangle for the magnitude.)

Placing the radial arm at 380° is the same as placing it at 20°. 270° is in the same position as –90°. And –1800° is in the same position as 360° and 0°. I hope you see that we could rotate around to 1,000,000°, –999.9°, or whatever. As the radial arm rotates, the sine and cosine oscillate in value between $^{+}1$ and $^{-}1$. [See section B.6.d!]

EXERCISE B.8: What is the possible range of values for

(a) $\sin\theta = y$ on the unit circle. Is $\sin\theta$ ever undefined? If so, where?

(b) $\cos\theta = x$ on the unit circle. Is $\cos\theta$ ever undefined? If so, where?

(c) $\tan\theta = y/x$ on the unit circle. Is $\tan\theta$ ever undefined? If so, where?

(d) $\cot\theta = x/y$ on the unit circle. Is $\cot\theta$ ever undefined? If so, where?

(e) $\csc\theta = 1/y$ on the unit circle. Is $\csc\theta$ ever undefined? If so, where?

(f) $\sec\theta = 1/x$ on the unit circle. Is $\sec\theta$ ever undefined? If so, where?

B.6.c. Radians vs Degrees

One name for a unit that comes from dividing a whole into parts is "degree." When you study temperature, you'll see that degrees Celsius (°C) and degrees Fahrenheit (°F) essentially come from dividing up the length of a thermometer in a convenient way. The Kelvin (K) unit of temperature, however, is not based on human choice, but on a physical fact; namely that 0 K is determined by the physical state of virtually no molecular motion. Perhaps because of this distinction, we do not say "degree Kelvin (°K)" but simply "Kelvin (K)." In some cases you can use any temperature scale that you like. In other cases you must use Kelvin because of its deeper physical meaning.

Similarly, we could cut a circle into any size pieces. Cutting up a circle into 360° is a human choice. There is another unit, the "radian," that is based on a physical fact. In many cases you can use either unit of angle. In some cases, only radians will work.

[Drum roll please. You knew it all along, Dorothy. Just click your ruby slippers together and say "There's no circumference without π. There's no circumference without π."] Let's see now ... "radians", "radius", could there be a connection? You bet there is!

Circumference ≡ total perimeter length = (total angle in radians) · (radius)

$$c = (2\pi)\, r \qquad (\textit{not } 360°\, r\,!)$$

Arc length ≡ partial perimeter length = (partial angle in radians) · (radius)

$$s = (\theta)\, r \qquad (\theta \textit{ cannot} \text{ be in degrees here!})$$

"Why can't θ be in degrees, and where does that strange number π come from anyway?" Measurement! The result is set, and the ancient Greeks knew it. No matter what units of length we use, the ratio of circumference to diameter ($2r$) is the fixed, dimensionless number $\pi = 3.14159...$

EXERCISE B.9: Everyone should do this once in their life. I encourage you to participate rather than just read along. It won't take long.

- Get out a piece of string, a pencil, rulers in cm and inches, and a circle template or compass.
- Draw a perfect circle with a template or compass. (Using a can for your circle is actually easier.)
- Use both rulers to measure the diameter. Record these measurements below.

$$2r = d = \_\_\_\_\_\_\_\_\text{cm}, \quad \_\_\_\_\_\_\_\_\text{inches}$$

- Carefully measure the circumference by laying the string along the circle little by little and then measuring the string length with both rulers.

$$c = \_\_\_\_\_\_\_\_\text{cm}, \quad \_\_\_\_\_\_\_\_\text{inches}$$

- Calculate c/d using the measurements in cm and then those in inches.

$$c/d = \_\_\_\_\_\_\_\_\ \text{cm/cm}, \quad \_\_\_\_\_\_\_\_\ \text{inches/inches}$$

- Calculate the % difference with π by using the formula below.

$$\%\ \text{difference} = \frac{\frac{c}{d} - \pi}{\pi} \times 100\%$$

- Multiply one of your length measurements (c or d) by this % difference. Is it possible that your measurements were off by this much? Did you stretch the string on the ruler but not along the circle? Are results within reasonable measurement error?

Using either unit for length, try a circle with a significantly different radius and see what you get. (Work quickly. This exercise need not take long to do.)

The radian unit for angles is based on the fact that π is set by the physical ratio of arc length to diameter length. The total angle for a circle, using physical units, is $c/r = 2\pi$ radians. Now you see why it is utter nonsense to write "s/r = the angle in degrees." The angle *must* be in radians for this expression to be valid. Pretty cool, huh? Since a full circle has an angle of 2π radians, we know that a half circle has π radians, a quarter circle has $\pi/2$ radians, and so on.

HOT POINT!

Sometimes people write in "radians" as a unit, but often people don't. Why wouldn't they? Suppose you measured the circumference in meters and the radius in meters. Since the angle in radians for a full circle is equal to c/r, what would the units be? Yes, meters/meters! This is why **the angle in radians is considered a dimensionless quantity** even though we say π radians, 2.1 radians, and so on. [I often prefer to write in "rad" explicitly when I am writing in the units. I think it adds to the clarity, but some people opt to leave it out for brevity.]

One final quirk about the conventional use of the unit radian follows. If I ask you the angle for one revolution, you would say "2π radians." But if I asked for the circumference of a 2 meter circle, you would say "(2π radians) · (2 meters) = 4π meters." You would not give the circumference in units of radians·meters, even though that is how you found it. Similarly, the arc length ($s = \Delta\theta \cdot r$) is given in "meters", not "radians·meters", since it is just a length. So it shouldn't surprise you to see something like (ω rad/s)·(R m) = v m/s, where the radians just go away without any mention at all. We could say that a radial meter times a radian gives a tangential meter, if you like … I didn't think so.

Now get a feel for the **relation between radians and degrees**. The basic angles are:

$\theta = 2\pi$ radians $\approx$ 6.25 rad $\approx$ 6 rad
$= 360^\circ$

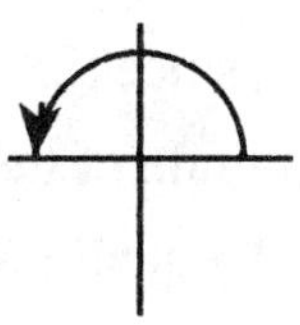

$\theta = \pi$ radians $\approx$ 3.14 rad $\approx$ 3 rad
$= 180^\circ$

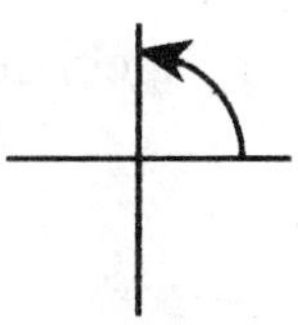

$\theta = \pi/2$ radians $\approx$3.14/2 rad $\approx$ 3/2 rad
$= 90^\circ$

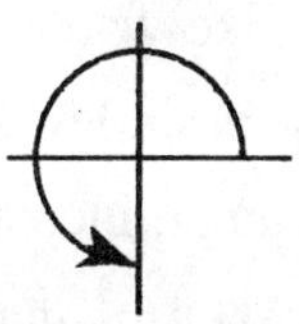

$\theta = 3\pi/2$ radians $\approx$9.4/2 rad $\approx$9/2 rad
$= 270^\circ$

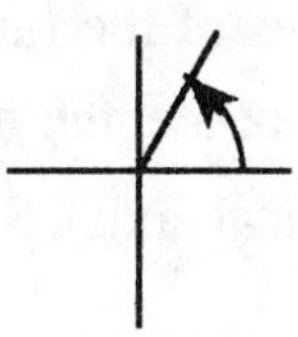

$\theta = 1$ radian $\approx$ 1/3 half circle
$\approx 60^\circ$ (1/3 of a half circle)

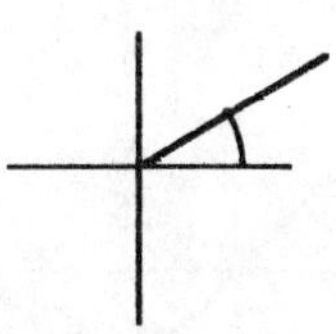

$\theta = 1/2$ radian $\approx$ 1/6 half circle
$\approx 30^\circ$ (1/6 of a half circle)

You can easily visualize 1 radian as roughly 60°, 0.5 rad as roughly 30°, 0.25 rad as roughly 15°, and π rad (roughly 3 rad) as exactly 180°. Converting from one unit to the other is easy. Let's use the physical equality 180° = π rad to convert the following:

- $1 \text{ rad} = 1 \text{ rad} \cdot \frac{180^\circ}{\pi \text{ rad}} = \left(\frac{180}{\pi}\right)^\circ = 57.3^\circ \approx 60^\circ$
- $1^\circ = 1^\circ \cdot \frac{\pi \text{ rad}}{180^\circ} = \left(\frac{\pi}{180}\right) \text{rad} = 0.0175 \text{ rad}$

This confirms that 1 degree is *much* smaller than 1 radian. But that's clear from the pictures. I'll try a few more and then leave it to you to investigate the matter further if you need to.

- $6 \text{ rad} = 6 \text{ rad} \cdot \frac{180^\circ}{\pi \text{ rad}} = \left(\frac{1080}{\pi}\right)^\circ = 344^\circ \approx 360^\circ$
- $360^\circ = 360^\circ \cdot \frac{\pi \text{ rad}}{180^\circ} = 2\pi \text{ rad} = 6.28 \text{ rad} \approx 6 \text{ rad}$
- $88^\circ = 88^\circ \cdot \frac{\pi \text{ rad}}{180^\circ} = \left(\frac{88 \cdot \pi}{180}\right) \text{rad} = 1.5 \text{ rad}$

We see that the $\pi \approx 3$ approximation is certainly limited, since (2X3) rad is only roughly $2\pi \text{ rad} = 360^\circ$. But it helps us get a rough idea of the magnitude of the angle.

Further notice that these equalities are not numerical equalities but rather physical equalities – the physical angle is the same independent of the units used to measure it. (When using conversion factors, the units cancel in an algebraic way. This is true whenever we multiply or divide units. See Ch. 1!)

B.6.d. The Sine & Cosine Graphs

The graphs for the sine and cosine functions look kind of like water waves or waves on a string, but can they really be considered such? At the most basic level these graphs simply reflect the behavior of the mathematical functions; as the angle increases (moving ccw on the unit circle), the values of $\sin\theta$ and $\cos\theta$ oscillate between –1 and +1. Unlike the unit circle which gives us a pictorial way of obtaining the values of the functions by looking at the coordinates ($x = \sin\theta$, $y = \cos\theta$), these graphs relate the value of θ (on the horizontal–axis) to the value of either function (on the vertical–axis). See the difference?

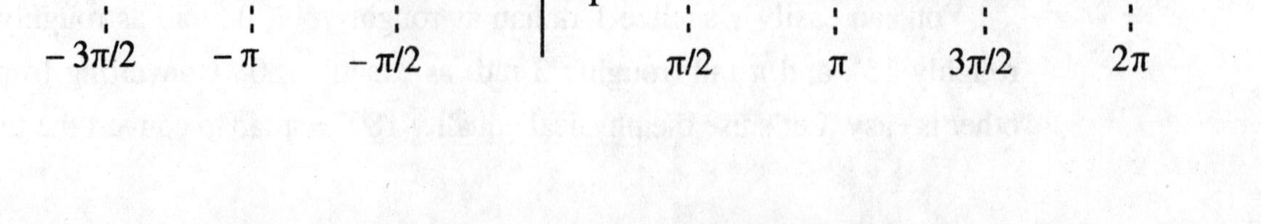

In fact, if you recall that radians are dimensionless quantities, you see that θ doesn't even have to refer to an angle at all; it could be anything with units seconds/seconds, meters/meters, and so on. Hence there will be times when θ is a *geometric* angle and times when it is a *mathematical*, but *not* a geometric, angle. Just plug in the value for θ and punch the function button on your calculator. In this way we continue to abstract the meaning and broaden the usage of the trig functions.

Again, the graphs of $\sin\theta$ or $\cos\theta$ and the value for θ (physical or mathematical angle) are more abstract than a spatial picture. We will be able to relate these graphs to all sorts of cyclic processes: rotating objects; a mass bobbing up and down on a spring; a swinging pendulum; and yes, even to physical waves that look like the graphs themselves.

Along with the unit circle, you should be able to visualize and quickly sketch the graphs of the sine and cosine functions with θ being along the horizontal axis. You will become more familiar with these graphs as you study the various applications in detail later on in the course. For now, give the next two exercises a try and start to get to know these graphs. Make them good friends.

EXERCISE B.10: Compare the slopes of one to the values of the other. Circle the correct answer to each question. Read off the answers from the graph below.

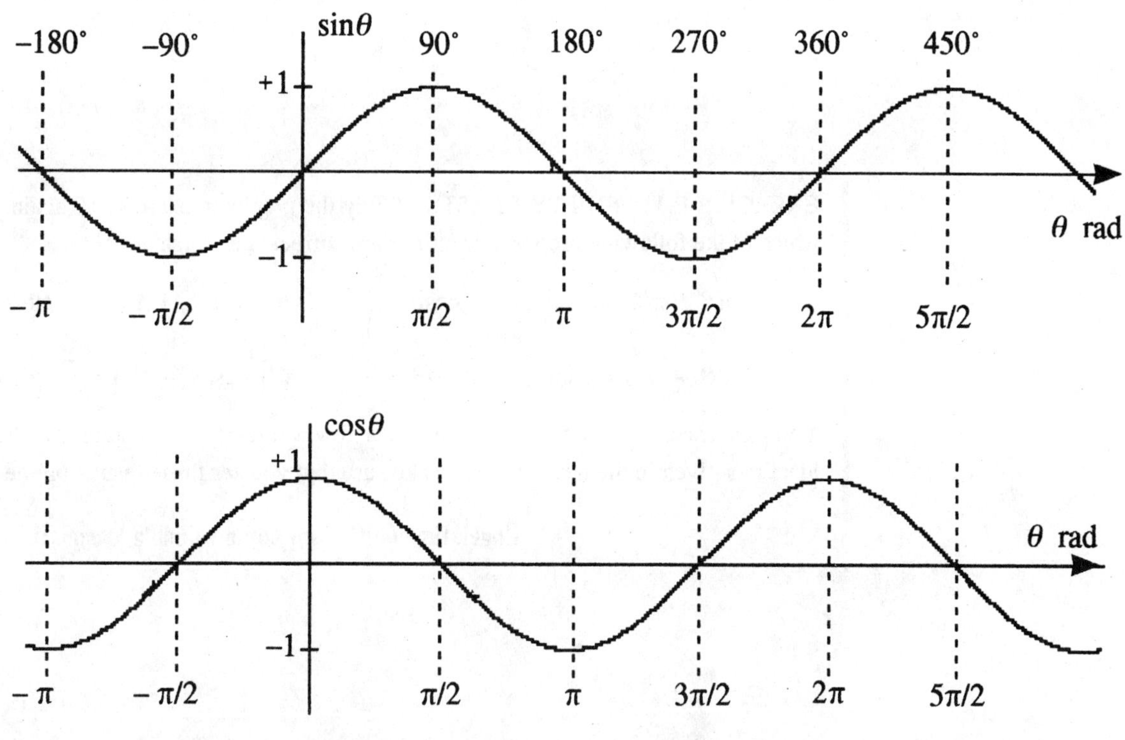

(a) Compare the *slope* of $\sin\theta$ to the *value* of $\cos\theta$.

At $\theta = 0$: Does the slope of $\sin\theta = +1, 0,$ or -1?
Does the value of $\cos\theta = +1, 0,$ or -1?

At $\theta = \pi/2$: Does the slope of $\sin\theta = +1, 0,$ or -1?
Does the value of $\cos\theta = +1, 0,$ or -1?

At $\theta = \pi$: Does the slope of $\sin\theta = +1, 0,$ or -1?
Does the value of $\cos\theta = +1, 0,$ or -1?

At $\theta = 3\pi/2$: Does the slope of $\sin\theta = +1, 0,$ or -1?
Does the value of $\cos\theta = +1, 0,$ or -1?

Does this seem to imply that the derivative of $\sin\theta$ is $+\cos\theta$ or $-\cos\theta$?

(b) Compare the *slope* of $\cos\theta$ to the *value* of $\sin\theta$.

At $\theta = 0$: Does the slope of $\cos\theta = +1, 0,$ or -1?
Does the value of $\sin\theta = +1, 0,$ or -1?

At $\theta = \pi/2$: Does the slope of $\cos\theta = +1, 0,$ or -1?
Does the value of $\sin\theta = +1, 0,$ or -1?

At $\theta = \pi$: Does the slope of $\cos\theta = +1, 0,$ or -1?
Does the value of $\sin\theta = +1, 0,$ or -1?

At $\theta = 3\pi/2$: Does the slope of $\cos\theta = +1, 0,$ or -1?
Does the value of $\sin\theta = +1, 0,$ or -1?

Does this seem to imply that the derivative of $\cos\theta$ is $+\sin\theta$ or $-\sin\theta$?

EXERCISE B.11: SHIFTY BEASTS! Study the graphs in exercise 10 again. From those graphs, decide which of the following proposed identities are correct and which are not.

(a) $\sin(\theta + \pi/2) = +\cos\theta$ or $-\cos\theta$?

(c) $\cos(\theta + \pi/2) = +\sin\theta$ or $-\sin\theta$?

(b) $\sin(\theta - \pi/2) = +\cos\theta$ or $-\cos\theta$?

(d) $\cos(\theta - \pi/2) = +\sin\theta$ or $-\sin\theta$?

You can check yourself by plugging in a few values for θ and/or by checking the angle addition identities given in the next section. Make sure that you *see* these results on the graphs.

Check the solution for some valuable lessons!!

B.6.e Some Trig Identities

Here are some of the trig relations that prove useful to be able to use and to memorize, be able to work out, or at least know where to look up. I've provided a few mnemonics in the parentheses.

MOST BASIC:

- $\sin^2\theta + \cos^2\theta = 1$ (The Pythagorean th'm for the unit circle!)
  - $\Rightarrow \tan^2\theta + 1 = \sec^2\theta$ (Just divide the previous equation by $\cos^2\theta$!)
  - $\Rightarrow 1 + \cot^2\theta = \csc^2\theta$ (Divide the basic relation by $\sin^2\theta$!)
- $c^2 = a^2 + b^2 - 2ab\cos\gamma$ (Law of cosines; valid for any triangle. γ is the angle opposite side c.)

  $\frac{\sin\alpha}{a} = \frac{\sin\beta}{b} = \frac{\sin\gamma}{c}$ (Law of sines; valid for any triangle. Each angle is opposite the side in the same ratio.)

BASIC:

- $\sin(\alpha \pm \beta) = \sin\alpha\cos\beta \pm \sin\beta\cos\alpha$ ("sine" $\Rightarrow$ "same sign," different functions multiplied)
- $\cos(\alpha \pm \beta) = \cos\alpha\cos\beta \mp \sin\alpha\sin\beta$ ("cosine" $\Rightarrow$ "complimentary sign," same functions multiplied)

NOT SO BASIC, BUT USEFUL:

- $\sin 2\theta = 2\sin\theta\cos\theta$ (Check $\sin(\theta + \theta)$ in the angle addition formula.)
- $\cos 2\theta = 2\cos^2\theta - 1$ (Check $\cos(\theta + \theta)$ in the angle addition formula.)
- $\sin\theta = \sqrt{\frac{(1-\cos 2\theta)}{2}}$ (From $\cos 2\theta$ and $\sin^2\theta + \cos^2\theta = 1$.)
- $\cos\theta = \sqrt{\frac{(1+\cos 2\theta)}{2}}$ (From $\cos 2\theta$.)
- $\tan(\alpha \pm \beta) = \frac{\tan\alpha \pm \tan\beta}{1 \mp \tan\alpha\tan\beta}$ $\left(\text{From } \frac{\sin(\alpha\pm\beta)}{\cos(\alpha\pm\beta)}.\right)$

EXERCISE B.12: Practice using these identities. They come in handy when you need to simplify an expression. They may seem unnecessarily ugly, but they really do help.

a) Check that $\sin(\theta + \pi/2) = \cos\theta$ by using the appropriate angle addition identity.
b) Use an angle addition identity to reduce $\cos(\theta - \pi/2)$.
c) Derive the $\tan(\alpha \pm \beta)$ identity using the hint given next to it.
d) Derive the law of sines. [Draw a non–rt triangle and create some rt triangles!]

B.7. COORDINATES (Where's that fly?)

Funny how "carte" is French for map!

I remember being told that René Descartes, the famous French philosopher, came up with his Cartesian (from "Descartes"), AKA rectangular, coordinate system to describe space while pondering the how to describe the position of a fly buzzing around the room. He established, the story goes, one corner to be the origin, and imagined measuring the distances from two walls and the floor, thereby creating a coordinate grid to specify the fly's position. [The story creates such a nice image that it ought to be true!].

Being able to describe the location of an object relative to some chosen fixed reference point could be considered the first step toward quantitatively describing and understanding the physical universe. As you can imagine, there are lots of ways to quantitatively describe the location of an object, and hence a lot of possibilities for coordinate systems. The three most widely used systems are the rectangular (AKA Cartesian), cylindrical, and spherical coordinate systems. The polar coord. system is essentially the cylindrical system in 2–D.

2–D:

Rectangular

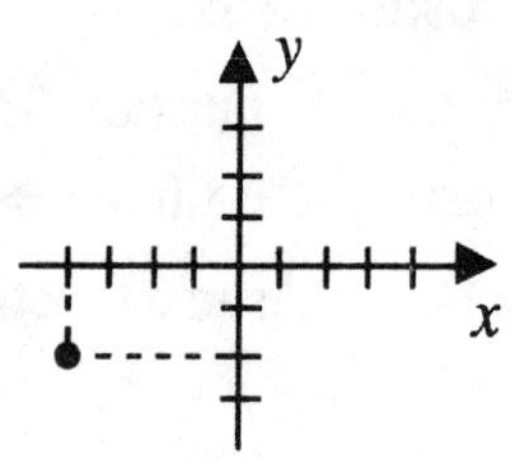

$(x, y) = (-4, 2)$

Polar

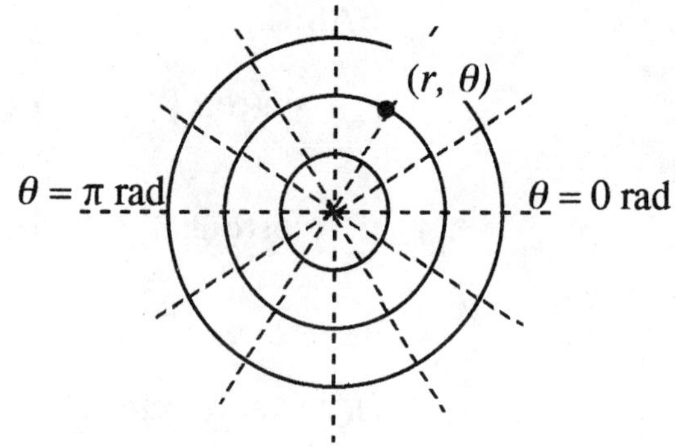

3–D:

Rectangular

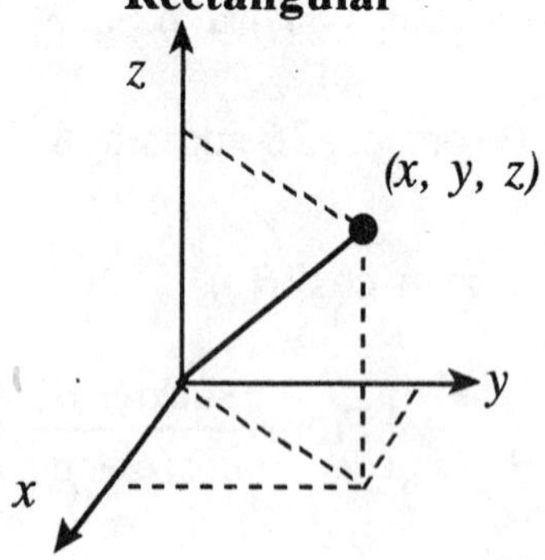

Cylindrical

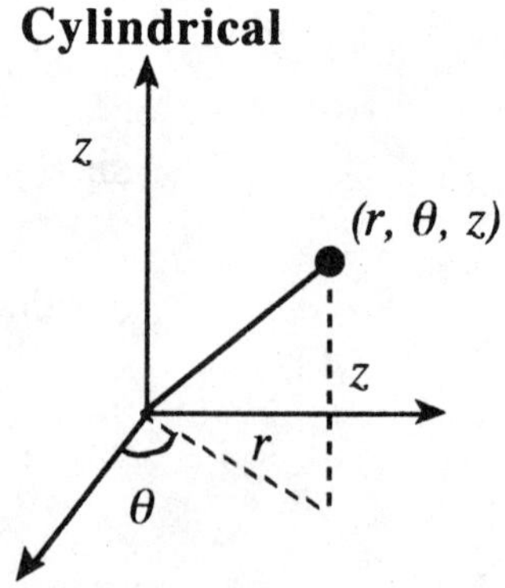

Spherical

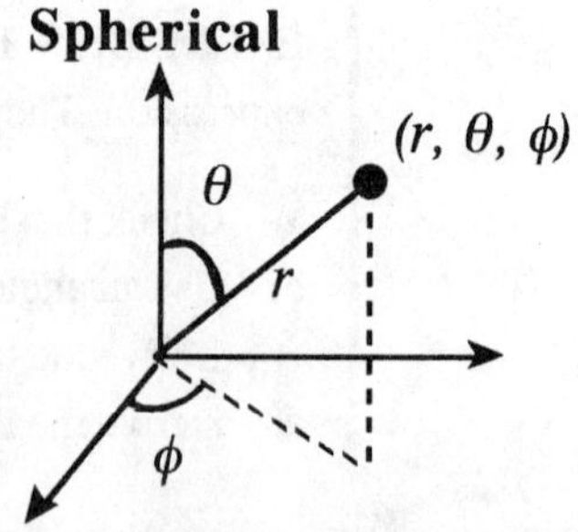

Getting from one of these three coordinate systems to the other is "just" a matter of using the Pythagorean theorem and the basic trig relations. Sure, they're ugly at first. But give them a chance. They're actually rather cute and very useful in the right circumstances. If you can draw the pictures, you can figure out the relationships needed. With practice, and only with repeated practice, you'll be able to make these quick sketches in a snap.

Rect. → Polar/Cyl.

$$r = \sqrt{x^2 + y^2}$$

$$\theta = \tan^{-1}\left(\frac{y}{x}\right)$$

$$z = z$$

Polar/Cyl. → Rect.

$$x = r\cos\theta$$

$$y = r\sin\theta$$

$$z = z$$

Rect. → Spherical

$$r = \sqrt{x^2 + y^2 + z^2}$$

$$\theta = \tan^{-1}\left(\frac{\sqrt{x^2 + y^2}}{z}\right)$$

$$\phi = \tan^{-1}\left(\frac{y}{x}\right)$$

Spherical → Rect.

$$x = (r\sin\theta)\cos\phi$$

$$y = (r\sin\theta)\sin\phi$$

$$z = r\cos\theta$$

These all stem from the Pythagorean theorem and the basic trig definitions for right triangles. The key is seeing the triangles. For instance, try to see that

$$r\sin\theta = \sqrt{x^2 + y^2} = \text{the diagonal component in the } x - y \text{ plane}.$$

Look at the dark lines:

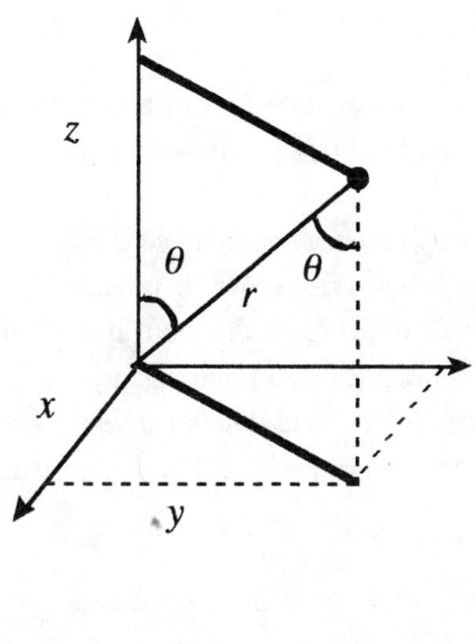

B.8. VECTORS (Which way did they go?)

You MUST understand vectors or you will struggle throughout the course.

Vectors get a bad rap. Why? Because they're hard as titanium to get used to since: they seem incredibly abstract at first; they have unusual rules for addition and subtraction; they have two different rules for multiplication; you can't really divide by them, but you can divide them by numbers; and you can analyze them graphically, geometrically, or by using coordinate systems and trigonometry to break them down into manageable parts – just for starters. But I have to tell you that it is absolutely NOT beyond you *if* you have a solid grasp of the trig basics and algebra. So wipe the sweat from your brow and dig in. Vector analysis will come to make sense with practice. You may even come to appreciate the power and elegance of vectors. [Hey, some students that have struggled with and finally gotten them have said just that!]

Bounce back and forth between this section and Ch. 1.

Your text covers vectors rather completely, and we'll visit them again in Ch. 1, so I'll review a couple of things and hone in on the typical stumbling blocks. I'll also try to give you another perspective on how to think about vectors. As you study the text and the other parts of this guide, you will have lots of opportunity to become ... VECTOR PERSON!!! With newfound direction in your life!

Just about everything that we want to discuss in physics has a quantity, or "magnitude," associated with it. Even "blue" can be related to certain wavelengths and corresponding frequencies. But not all concepts are tied to a direction. What's the direction of the money in your pocket? Yeah, yeah, I know the quantity is decreasing. And yes, we could put it on a number line and assign it a direction – kind of a neat idea. But it doesn't have a spatial direction. Neither does temperature. "I'm really sick. I've got a temperature of 102°F... and it's pointing north!" But some concepts are defined to include the spatial direction along with the magnitude. We categorize these concepts as "vectors." We could also say that we *represent* the magnitude and direction of these concepts by vectors.

EXERCISE B.13: Which of these are vectors? (It's OK if you don't know the last three.)

a) Time
b) Distance (between your nose and the page number on this page)
c) Position (of the page number on this page relative to the tip of your nose)
d) Speed (on your speedometer)
e) Velocity (when traveling in your car)
f) Force (on you from your chair seat)
g) Number of people in the room at the moment
h) Momentum (of a bull running at you)
i) Energy (of that bull running at you)
j) The earth's magnetic field (at any particular place, like the tip of your nose).

When the directions are always along a straight line, you can get away with using + to designate one direction (prescribed or of your choice) and – to mean "the other way." In this case, a negative number only carries the meaning of which *direction* we are referring to. This is very different than saying that you have – 439 dollars, which means it's time to visit the financial aid office again!

In general, any vector quantity can be drawn graphically as an arrow or expressed analytically. Let's take a brief look at each type of representation and see how we perform both graphical and analytical vector analysis. This is essential when we are dealing with vectorial concepts (ones where spatial direction matters).

B.8.a. Graphical Representation

At least two of the points stated here are commonly not understood by students.

Certainly, drawing vectors as arrows (graphical representation) has limited use; but it is often essential in visualizing the vectors before jumping into the equations. Remember that vector quantities are anything with magnitude (okay, everything has magnitude) and direction (that's the difference!). Hence, an arrow " → " can represent any number of different things.

- <u>LENGTH OF ARROW</u> – represents the magnitude of the quantity (velocity, force, momentum, electric field, etc.). The length can be drawn to an unspecified rough scale on a sketch (a force that's twice as great as another force is drawn roughly twice as long), or you can choose a convenient scale (for example 1 cm = 10 meters/sec.) and get out your ruler and graph paper. (On very rough sketches, you don't even need to worry about the length.)

- <u>DIRECTION OF ARROW</u> – represents ... the direction of the quantity. Again you can illustrate the direction roughly on a sketch that is meant to help orient your thinking, or actually measure the angle between the arrow (vector) and a convenient reference line.

- <u>"HEAD" OF ARROW</u> – is the tip (pointed end) of the arrow. When you draw the vector as *coming out* of the page, draw a *dot* inside a circle (⊙), where the dot is the tip, or "head," of the arrow. (To see this, stand your pencil on the paper with the tip up. The direction is "perpendicular to and out of the page.")

- <u>TAIL OF ARROW</u> – is just the back end. When you draw the vector as *going into* the page, draw a *cross* inside a circle (⊗), where the cross can be thought of as the tail *feathers* of an arrow (as used with a bow).

- <u>LOCATION OF ARROW</u> – [Ah, here's the rub!] You have some choices. Perhaps more than you would like. Let me give you two examples:

Velocity vectors: When we speak of velocity, we're speaking of the velocity of what? Some object. That object is moving at some speed and in some direction; hence we *usually* draw the *tail* of the vector at the position of the object. (It doesn't matter where on the object you place the tail.) Note that you could just as well draw the vector above or below the object. The direction, and perhaps to some degree the magnitude, is then adequately specified. [Velocity arrows remind me of the motion lines on cartoons.]

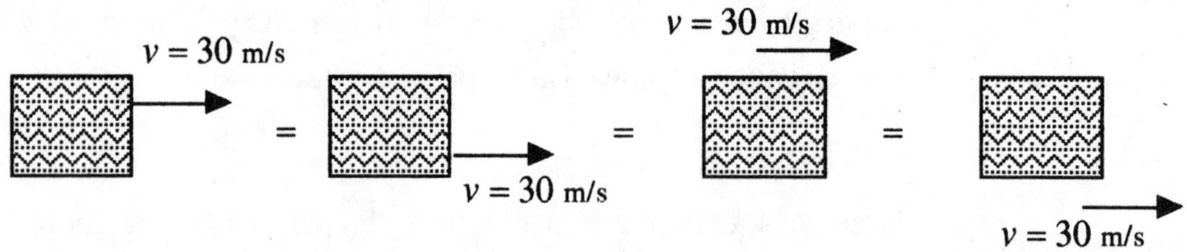

Forces – You may want to draw the arrow tip at the point of application of the force on the object "receiving" the force. In this way the arrow appears to be *pushing* the "receiver" of the force. Then again, you could draw the tail at the point of application of the force; this looks like a *pulling* force. Whether or not the actual force is a push or a pull, you are free to draw in your vector either way, since the physical effect on the "receiver" is the same. Consider a refrigerator wearing ice skates on an ice rink. Push somewhere on one of its sides. To represent this on paper, you could draw a vector with the tip exactly at the place you pushed or draw it as a pull with the tail directly opposite the point of application. **SPECIAL CASE**: When, and *only* when, you are not worried about tipping the object over or rotating the object, you can draw the vector with the tip or tail anywhere on the object receiving the force (and even above or below the object). [You'll study the associated rotations when you get to "torques."]

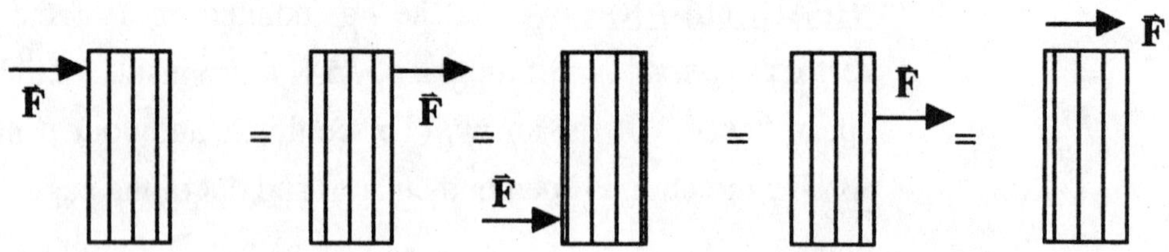

We'll deal with torque, angular momentum, electric field, and magnetic field vectors when the time comes. I hope you get the idea. Studying various situations and problem solving will bring out more subtleties.

B.8.b. Analytical Representation

Vector notation is an ingenious bookkeeping device and a powerful, simplifying tool when learned well.

We denote vector variables with arrows over the top $\left(\vec{\mathbf{v}}, \vec{\mathbf{F}}, \vec{\mathbf{A}}, \vec{\tau}, \text{etc.}\right)$. This screams "Hey, wait! You have to worry about the direction too! You *must* use vector methods rather than good old numerical arithmetic!" We'll get to vector analysis in a moment. For now, don't be fuzzy wuzzy about putting arrows above your variables or not. If you are including, or must include, the direction, denote the variable as a vector. If you are only dealing with the magnitudes (quantities), leave the arrow off (v, F, A, τ, etc.) [Did William Tell's son have direction?] or use absolute value signs $\left(|\vec{\mathbf{v}}|, |\vec{\mathbf{F}}|, |\vec{\mathbf{A}}|, |\vec{\tau}|, \text{etc.}\right)$. The latter notation clearly communicates that it is the magnitude $\left(|\vec{\mathbf{A}}|\right)$ of a vector $\left(\vec{\mathbf{A}}\right)$, while the former has the advantage of being easier to write quickly.

"But Master, can we not separate the direction of the vector from the magnitude of the vector? Or will this reductionist thinking divide the unity of all, thereby disturbing our gastrointestinal tract?" You again? Well, at first this stuff is hard to swallow (unless you're in a circus). But with time you will see that it isn't too difficult to separate the magnitude from the direction. We begin by developing a way to indicate direction without magnitude.

Unit vectors are what we use to designate a direction without reference to the magnitude. A unit vector simply points in some particular direction and has a magnitude (length) of unity. To distinguish these vectors as unit, or "directional," vectors, we often use a caret top rather than an arrow above the variable; $\hat{\mathbf{x}}$ is called "x hat" and points in whatever direction you have called $+x$. Rather than $(\hat{\mathbf{x}}, \hat{\mathbf{y}}, \hat{\mathbf{z}})$, many people prefer to use $(\hat{\mathbf{i}}, \hat{\mathbf{j}}, \hat{\mathbf{k}})$ to designate the 3 unit vectors pointing in the rectangular (a.k.a. Cartesian) directions. $\hat{\mathbf{r}}$ points radially outward from some chosen origin, and $\hat{\mathbf{n}}$ is the "normal" vector which is perpendicular to any surface that we happen to be interested in.

"Normal" is another word for perpendicular.

Look at the 2–D rectangular and polar coordinate systems (grids) below. Notice that I've placed the tails of the unit vectors at any point of interest, that the unit vectors point in the direction of increase for the corresponding variable, and that they are all the same length – unity.

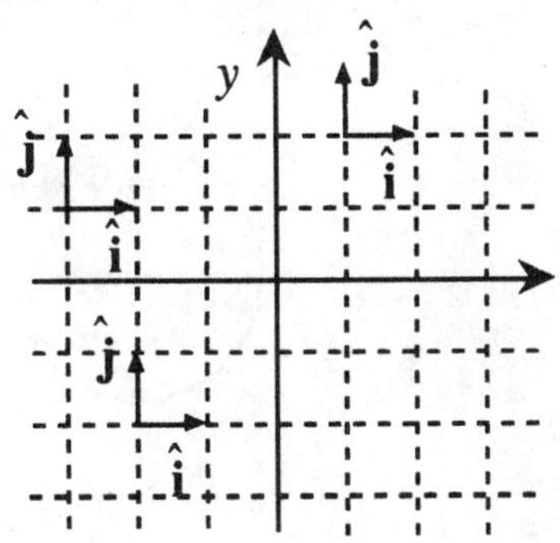

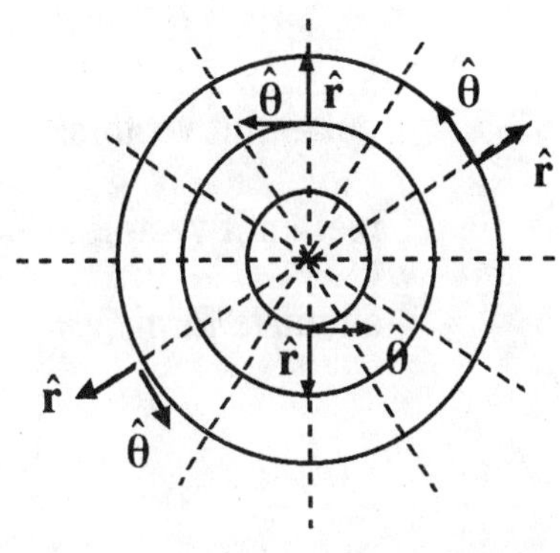

$\hat{\mathbf{i}}$, $\hat{\mathbf{j}}$, and $\hat{\mathbf{k}}$ are simply treated as constant, multiplicative factors when differentiating and integrating. This is not true of the polar or spherical unit vectors.

What's different about the rectangular and polar unit vectors as you look at different points? Do you see that $\hat{\mathbf{i}}$ and $\hat{\mathbf{j}}$ everywhere have the same direction, while $\hat{\mathbf{r}}$ and $\hat{\theta}$ change orientation as you move from point to point? This is a very important observation. The rectangular unit vectors are constant vectors because their magnitudes and directions don't change; whereas the polar unit vectors are *not* constants. This means that differentiation and integration are not as easy if we use $\hat{\mathbf{r}}$ and $\hat{\theta}$. We can work around this, but for now your first choice should be to use $\hat{\mathbf{i}}$, $\hat{\mathbf{j}}$, and $\hat{\mathbf{k}}$, since these can simply be treated as constants. We can easily augment the 2–D rectangular and polar unit vectors to the 3–D rectangular and cylindrical unit vector system, respectively, by adding $\hat{\mathbf{k}}$ perpendicular to the 2–D plane. The spherical unit vectors are shown below.

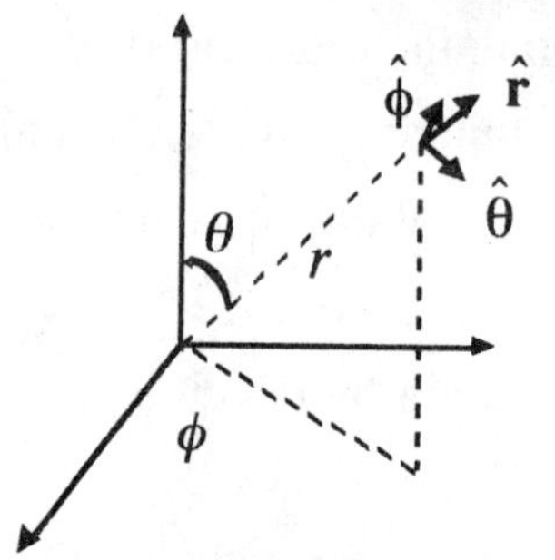

Can you see that $\hat{\mathbf{r}}$, $\hat{\theta}$, $\hat{\phi}$, and are all ⊥ and in the direction of increasing r, θ, and ϕ, respectively?

Suppose an object's velocity has a magnitude (called "speed") of 4 m/s and is traveling in the $+y$ direction. You could denote this simply by writing $\vec{\mathbf{v}} = (4\ \text{m/s})\ \hat{\mathbf{j}}$. A force of 10 pounds in the negative x direction can be symbolized as $\vec{\mathbf{F}} = (10\ \text{lb})\ (-\hat{\mathbf{i}}) = -\ (10\ \text{lb})\ \hat{\mathbf{i}}$. But what if the vector that you're describing isn't going in the x, y or z directions, but rather is going at some angle, i.e. partly in each direction? You could either choose another coordinate system – polar if it's traveling in a circle, cylindrical if it's traveling in a helix, and so on – or, if you want to use the rectangular unit vectors, say just how much the object is traveling in each direction. Thus an object traveling in a circle at a constant speed v could be described in either way shown below. (This is called *uniform* circular motion. Notice that there is no arrow over the "v"!)

polar unit vectors: $\vec{\mathbf{v}} = v\ \hat{\theta}$

rect. unit vectors: $\vec{\mathbf{v}} = v_x\ \hat{\mathbf{i}} + v_y\ \hat{\mathbf{j}}$

alternate form: $\vec{\mathbf{v}} = \vec{\mathbf{v}}_x + \vec{\mathbf{v}}_y$

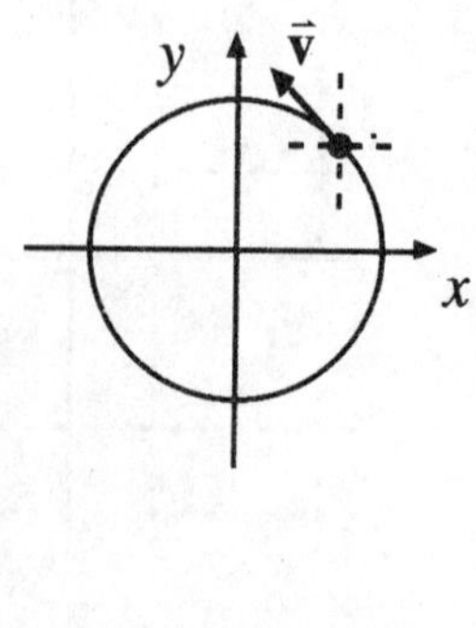

All of the speed (v) is directed around the circle ($\hat{\theta}$, tangent to the circle, in the direction of increasing θ), but we can say that the object is traveling "partly in the $-x$ direction and partly in the $+y$ direction." The "components" of any vector (velocity, acceleration, force, torque, angular momentum, electric field, magnetic field, etc., etc.) tell us how much the vector is directed in each direction. Now the trick is . . .

B.8.c. Getting Components

Just as a good musician practices scales slowly and precisely at first, be methodical and you will soon be a whiz.

[Psst! This is my secret miracle–method. If you train yourself to use it, you will save yourself an *enormous* amount of time throughout the course! Hot tip at the end of this section!] I have seen students struggle with getting the components of vectors up to the final exam of the last semester of the physics sequence. There is a straightforward way to get them, but you must be methodical. Make sure that you are clear on how to do this as soon as possible and practice, practice, practice.

* Method For Finding A Vector's Components *

- **Sketch the arrow** on the associated object or wherever is appropriate. *[The weight (force of gravity) of an object on an inclined surface is shown here. Either diagram is fine.]*

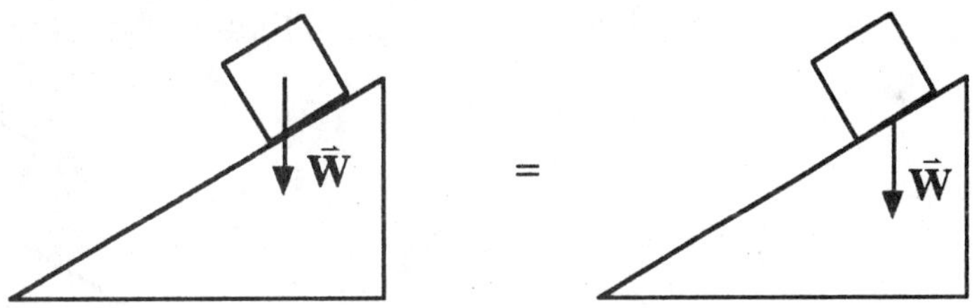

- **Choose the coordinate system** and unit vectors you want to use (usually rectangular at this stage of the game).

  "Use the rectangular coordinates (x,y) and rectangular unit vectors $(\hat{\mathbf{i}}, \hat{\mathbf{j}})$."

- **Choose the orientation** of your coordinate system. You may want to: tilt the coordinates so that the x–axis is at an angle (up or down along a surface); have the $+y$ direction be downward; have $+z$ be vertically up, $+y$ be horizontal and to the north, and $+x$ be horizontal and to the east; or choose any other directions that suits the spatial patterns of the physical process that you are analyzing. *[I've selected the direction that the block might move to be the $+x$ direction (down along the surface). Then the $+y$ direction is perpendicular to the surface, NOT NOT NOT vertically up or down!!]*

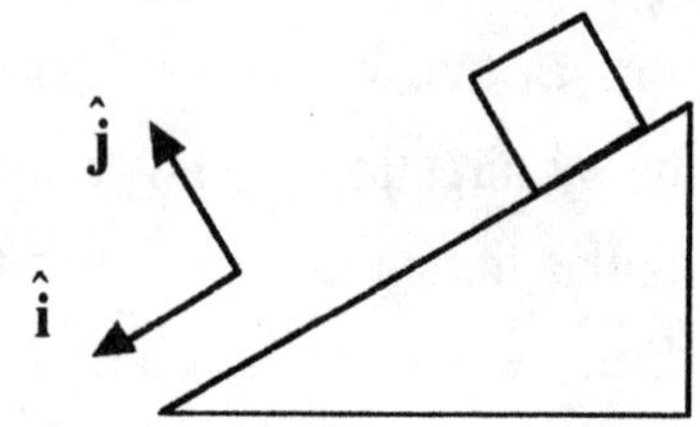

- **Imagine a grid**, as on graph paper, that parallels the coordinate axes. (Your vector will be the diagonal of a coordinate grid box.)

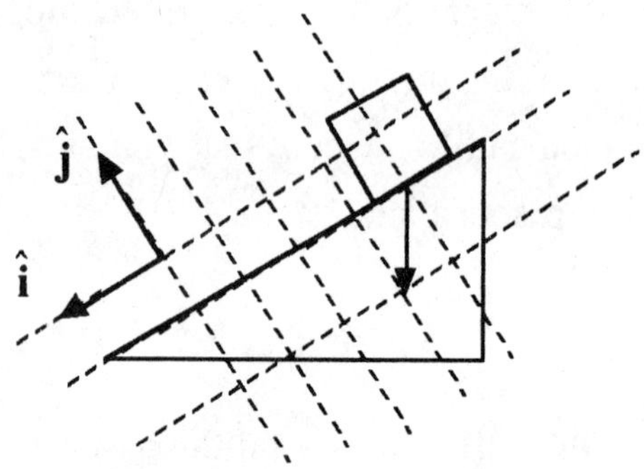

- **Construct a coordinate rectangle** around the vector. Draw lines parallel to your grid, i.e. parallel to the axes, and through both the **tip** and **tail** of the vector.

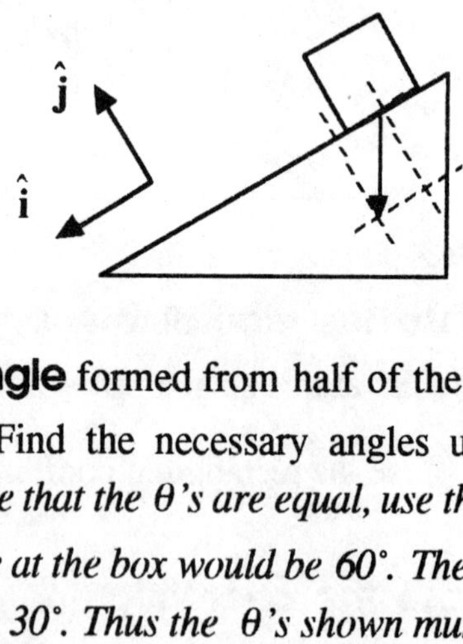

- **Choose either right triangle** formed from half of the rectangle. Use whichever is easiest to find the angles. Find the necessary angles using the basic rules of the geometry of angles. *[To prove that the θ's are equal, use the darkened right angle. If θ were 30°, the unlabeled angle at the box would be 60°. The remaining angle that forms the right angle would then be 30°. Thus the θ's shown must always be equal.]*

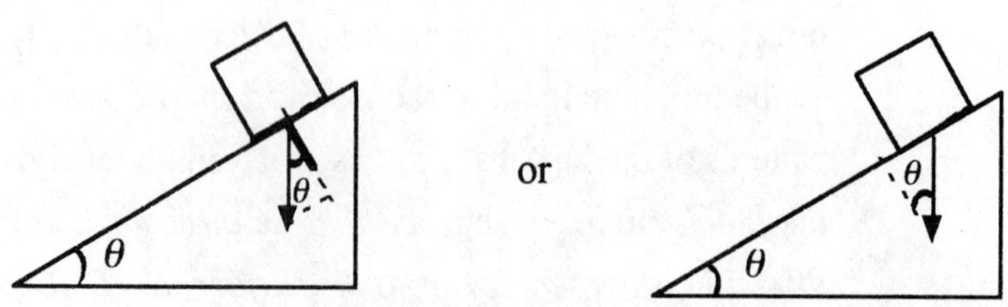

Notice that the components are along the grid and FROM the tail TO the tip. This is essential!!

- **Isolate that right triangle** in your mind or on paper. You are now back to basic trig and the Pythagorean theorem. The hypotenuse is the total vector itself, while the sides show how much the vector is going in each of your coordinate directions; that is, they are the *components* of the vector in the prescribed directions.

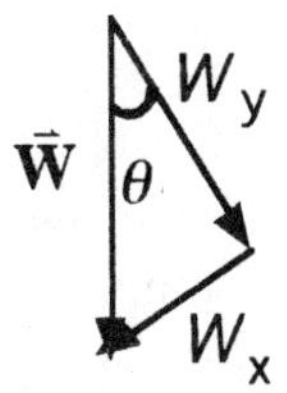

Use either triangle.

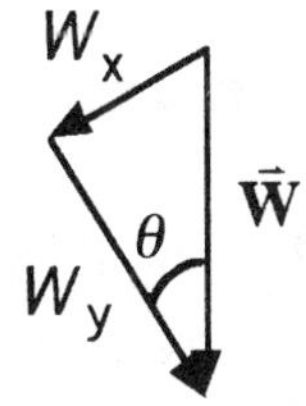

- **Solve for the unknowns** that you need. Using the magnitude of the vector as the "length" of the hypotenuse, we can obtain the component magnitudes with the basic trig functions. In fact, given at least two bits of information about the triangle, you can easily find out everything else using the following relations. *[Notice that the tangent is NOT always the y-component divided by the x-component. You simply need to make a sketch.]*

$$W_x = W\sin\theta \qquad W = \sqrt{(W_x)^2 + (W_y)^2}$$

$$W_y = W\cos\theta \qquad \theta = \tan^{-1}\left(\frac{W_x}{W_y}\right)$$

!

HOT TIP: Actually, you needn't draw the coordinate *rectangle* around the vector; you only need to draw one of the two possible coordinate triangles around the vector. I recommend starting at the tail of the vector, "walk" your pencil in a direction parallel to either coordinate axis until you can make a perpendicular "turn" and continue on toward the tip and parallel to the other coordinate axis. This is like going for a walk along a rectangular grid of streets. The crow flies straight from start to finish (along the vector) and you walk down the component streets–**FROM the tail TO the tip along the coordinate grid**.

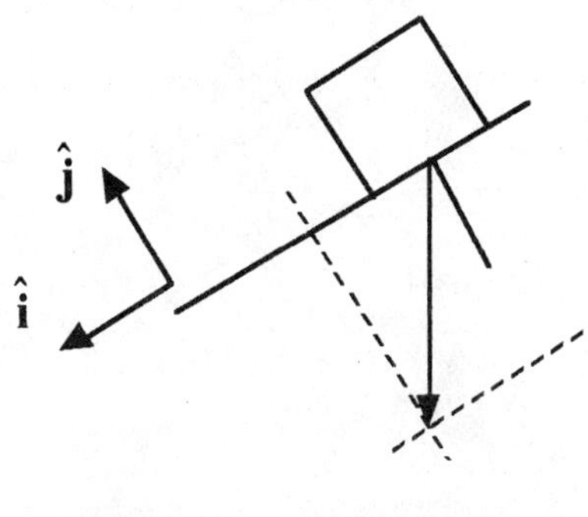

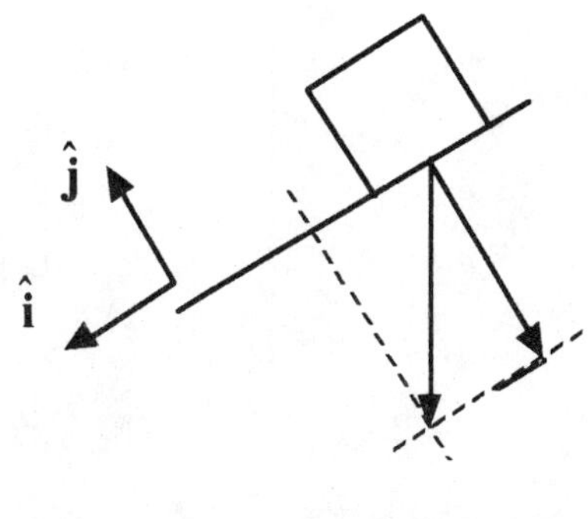

Summarize the steps on an index card and practice. You *will* need to work out a number of applications in order to become proficient at getting the components of vectors. Just be methodical and talk yourself through it with probing questions: "Okay, what *are* my coordinate directions? How do I draw lines parallel to the coordinate directions?" If you get stuck, formulate some questions, enter them in your journal, and go bug your instructor. Pronto!

EXERCISE B.14: Draw in and calculate the components of the vectors shown. Use the coordinate system specified. When you can do this, you'll be in GREAT shape! [Note: I've used some units and concepts that you're not familiar with. Just use them as given and get the components.]

(a)

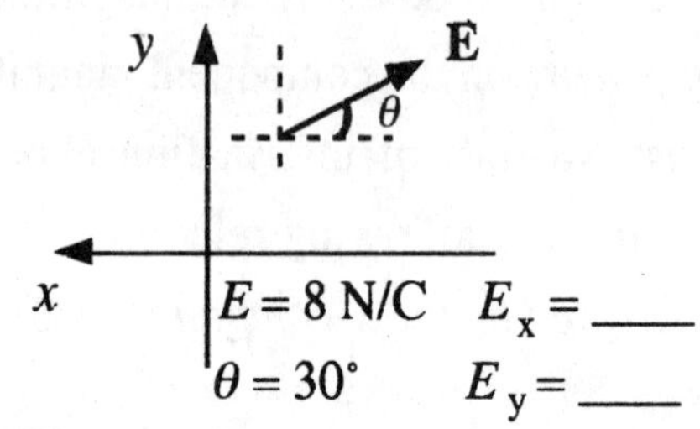

(b)

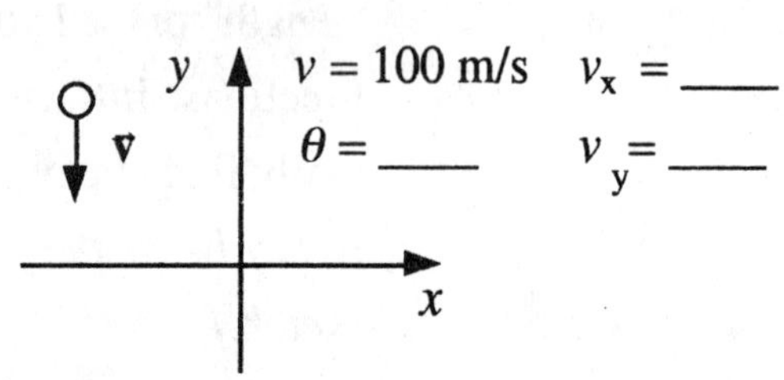

(c)

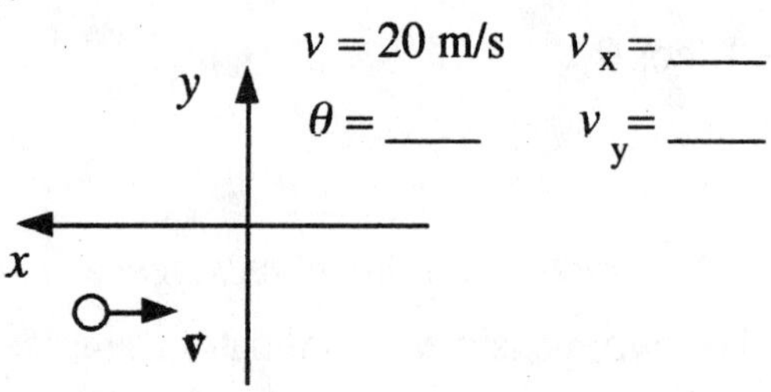

(d)

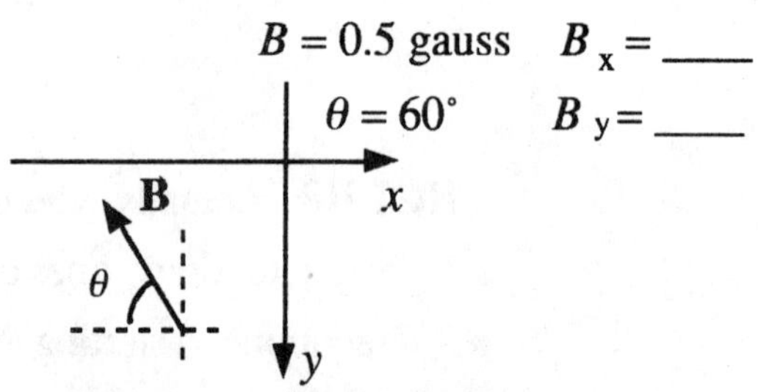

(e) The test: Draw in the components of $\vec{\mathbf{F}}$. If the smaller angle of the component triangle is 20° and the force is 222 N, how would you write this vector using unit vectors? [Suggestion: make a sketch that's easier to see!]

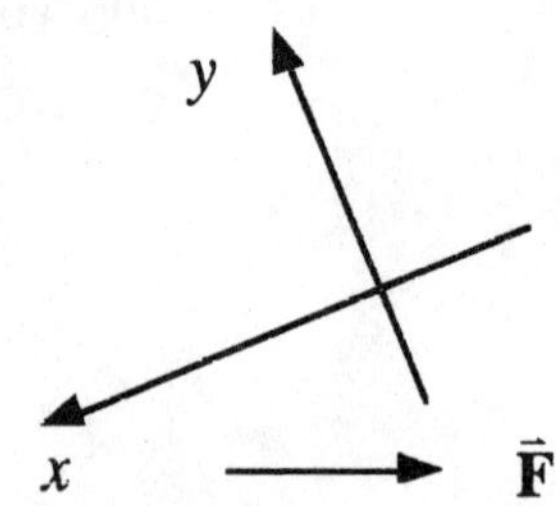

B.8.d. Vector Addition & Subtraction

You must account for the directions when adding, subtracting, and, as you will soon see, multiplying vectors.

If you can get the components of any vector, you're in great shape. If not, go ahead and do a preread of this section, but make sure that you get good at getting components!

Vector arithmetic is very different from numerical arithmetic. The difference is elucidated by a simple right triangle. Consider the "3 4 5" right triangle. The shorter sides do not *numerically* add up to the hypotenuse ($3 + 4 \neq 5$). Yet the sides do "add up" to the hypotenuse if you add the sides in a rt. triangular, i.e. vectorial, sense. Of course the tool for right triangles is the Pythagorean theorem. However, if you have a non right triangle, you can employ graphical means, geometric relations such as the law of cosines, or break the vectors down into components and add parallel parts numerically. So there are three approaches to adding quantities where direction is a concern:

1. Make a careful, scaled diagram and use a protractor and ruler to measure the lengths and angles.
2. Make a sketch (or a scaled diagram if you wish to go to the trouble) and use geometric relations such as the law of cosines to obtain the result.
3. Make a sketch, break the vectors down into components, add parallel components, and piece them back together to find the result.

But let's take it from the top. As you might expect, adding or subtracting vectors in parallel or anti parallel directions is quiet easy. You simply add or subtract their magnitudes.

EXAMPLE 5: Consider two forces of equal strength and same direction pulling you forward (one to an arm). What is the net, or "resultant," pull on you? Assume that the forces have a magnitude of 100 newtons (= 22.5 pounds or very roughly 25 pounds).

CONCEPTS – Thinking physically, as you always should at the start of a physics problem, you already know the answer. Working this problem out both graphically and analytically will bridge the gap to more difficult problems. We can model this situation nicely with two vectors pointing in the same direction. I'll say they are pulling to the right and call this the $+x$ direction, i.e. the direction of $\hat{\mathbf{i}}$.

GRAPHICAL ANALYSIS

Draw the vectors on the object

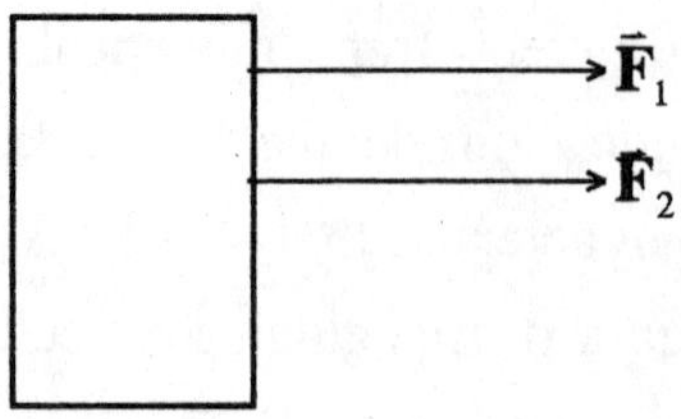

Draw the vectors "head to tail." That is, place the tail of *either* one at the head of the other one. Always keep the vectors parallel to the original direction.

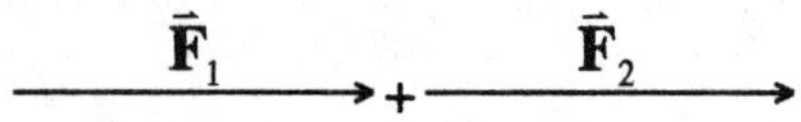

The sum, or *resultant*, is the vector that starts at the tail of the 1st vector and ends at the head of the last vector.

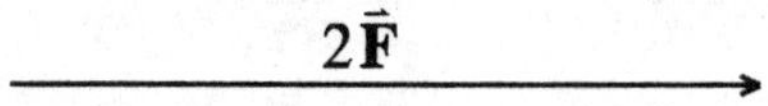

The total length is twice the original length, so the net force is 200 newtons in the same direction as the originals. But you knew that.

ANALYTICAL ANALYSIS

Represent each vector using unit vectors to specify the direction.

$$\vec{F}_1 = F_1\hat{i} = (100\text{ N})\hat{i}$$

$$\vec{F}_2 = F_2\hat{i} = (100\text{ N})\hat{i}$$

Add the two vectors. Since they are both in the same direction (parallel to $\hat{i}$) we simply add the magnitudes and treat the $\hat{i}$ as a common factor.

$$\vec{F}_1 + \vec{F}_2 = F_1\hat{i} + F_2\hat{i}$$

$$= (F_1 + F_2)\hat{i}$$

$$= (200\text{ N})\hat{i}$$

This, of course, matches the result obtained graphically.

So this case is simple – if one force were 10 times as strong as the other, the sum would simply be 11 times the strength of the weaker force.

Now what about subtraction? In a directional sense, multiplying by –1 simply means "the other way." This gives meaning to negative vectors and the subtraction of vectors.

EXAMPLE 6: Now suppose the two FORCES in example 5 are in opposite directions. There are a couple of lessons to be extracted if we work this out formally.

CONCEPTS – You know the answer is zero; the two forces balance (equilibrate) each other. The method is what's important here. The point is to get used to the different representations.

GRAPHICAL ANALYSIS

Draw the vectors on the object

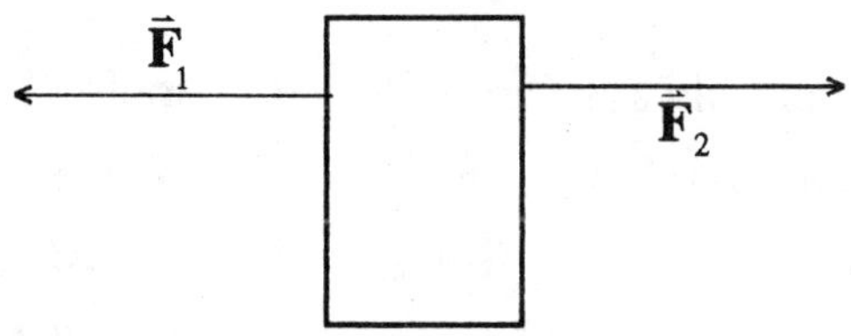

Draw the vectors "head to tail." This is tricky since they overlap. I'll draw one just above the other.

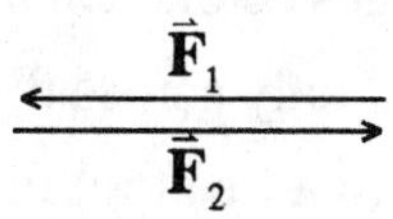

The *resultant* is the vector that starts at the tail of the 1st vector and ends at the head of the last vector. But there is no distance between these points so the resultant is 0. Without a magnitude, there is no direction.

ANALYTICAL ANALYSIS

Represent each vector using unit vectors to specify the direction.

$$\vec{\mathbf{F}}_1 = |\vec{\mathbf{F}}_1|(-\hat{\mathbf{i}}) = -(100\text{ N})\hat{\mathbf{i}}$$

$$\vec{\mathbf{F}}_2 = F_2\,\hat{\mathbf{i}} = (100\text{ N})\hat{\mathbf{i}}$$

Add the two vectors. Since they both lie along the same line (parallel to $\hat{\mathbf{i}}$) we simply add the magnitudes and treat $\hat{\mathbf{i}}$ as the common factor that it is.

$$\vec{\mathbf{F}}_1 + \vec{\mathbf{F}}_2 = (-100\text{ N})\hat{\mathbf{i}} + (100\text{ N})\hat{\mathbf{i}}$$

$$= 0$$

Again, the two methods produce identical results.

Adding vectors in the same direction demonstrates that multiplying a vector by a positive number changes the magnitude (length): $\vec{\mathbf{F}} + \vec{\mathbf{F}} = 2\vec{\mathbf{F}}$. Subtracting vectors illustrates that multiplying by a negative number reverses the direction and changes the magnitude. (Multiplication by –1 simply reverses the direction of the vector.) A velocity of (–10 m/s, north) is physically the same as a velocity of (10 m/s, south). The negative tells us that the vector is pointing in the direction opposite that which we chose to be the positive direction.

EXERCISE B.15: You should actually try this sometime, but for now imagine that you are walking down the aisle of a bus, train, or plane. Let's say it's a train moving east at 10 meters/second (= 22 miles/hour, or roughly 20 miles/hour).

a) If you are walking slowly in the *same* direction that the train is moving and at a speed of 1 m/s (≈ 2 miles/hour), how fast and in what direction are you traveling over the train tracks?

b) If you are walking slowly in the *opposite* direction that the train is moving and at a speed of 1 m/s (≈ 2 miles/hour), how fast and in what direction are you traveling over the train tracks?

c) If the bus or train is slowing to a stop, could you walk in such a way that your speed over the ground is 0? Explain. [TRY THIS!! Pick an object (sign, person, garbage can) that you are passing and walk so that you don't pass it.]

It's too bad that vectors aren't always parallel or anti parallel, but we can look at the parallel *parts* of vectors and add or subtract those. This is where "Getting Components" becomes an important and powerful tool that makes the seemingly undoable, doable. [If you can come up with another way, let me know! ... Okay, besides graphically! ... Okay, besides using geometric relations such as the law of cosines ... but that's all ... isn't it?] First, look at what happens if the vectors are perpendicular.

EXAMPLE 7: Adding perpendicular vectors. On a boat moving with velocity 10 m/s east, you walk 2 m/s, north; that is, directly across the boat relative to the boat deck. A person on shore (or looking down from a tree top) measures your velocity (speed and direction) to be...what?

<u>TOP VIEW (looking down)</u>

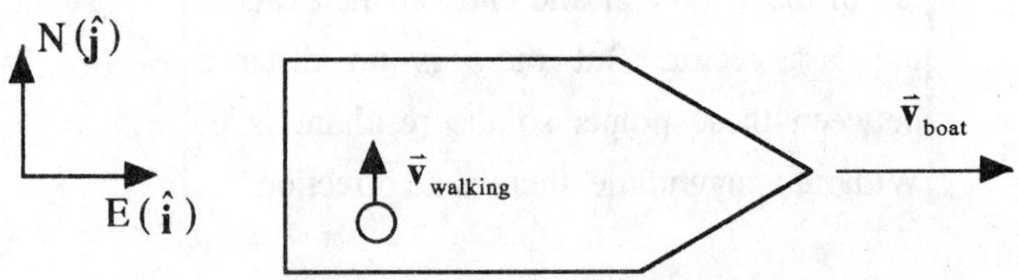

CONCEPTS – Can you imagine the path through space that you are traveling? What would that person in the tree top above see? They would see you move both downstream and towards the other side of the boat. Can we say that the speed is v_{total} = 10 m/s + 2 m/s = 12 m/s? No, in this case you are walking neither in the same nor the opposite direction as the boat. Here we are really forced to use triangular (vectorial), not numerical, addition.

Once again I'll do the graphical and analytic analyses side by side. You'll see that the graphical sketches help us visualize what the equations are saying. That's why we usually diagram physics problems.

GRAPHICAL ANALYSIS

Draw the vectors on the object (you):

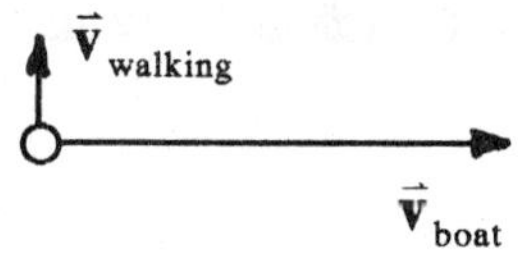

Draw the vectors "head to tail." That is, place the tail of *either* one at the head of the other one. Always keep the vectors parallel to the original direction. Notice that the order does NOT matter, since the result will always be the same.

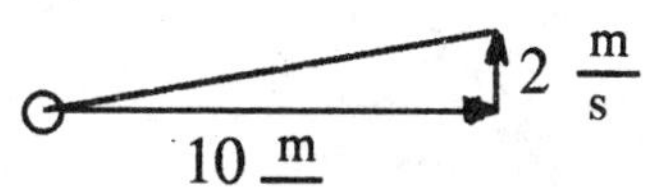

or

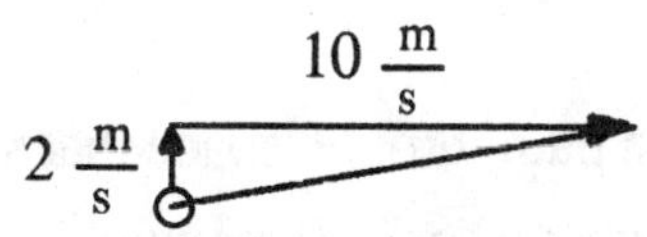

The sum is the vector that starts at the tail of the 1st vector and ends at the head of the last vector. The total length is clearly not the numerical sum of the two speeds. But we do know how to deal with right triangles! Perhaps it's best to peek in on the other column and see what equations we can borrow, after all, they've been peeking in on our sketches!

ANALYTICAL ANALYSIS

Represent each vector using unit vectors to specify the direction:

$$\vec{v}_{walking} = (2\ \text{m/s})\ \hat{j}$$

$$\vec{v}_{boat} = (10\ \text{m/s})\ \hat{i}$$

Add the two vectors. Since they are perpendicular, we don't have a common unit vector. Hence we simply write:

$$\vec{v}_{over\ ground} = \vec{v}_{walking} + \vec{v}_{boat}$$

$$= (2\ \text{m/s})\ \hat{j} + (10\ \text{m/s})\ \hat{i}$$

Notice from the sketches that the two components of the total velocity can be thought of as forming the two sides of a right triangle. The magnitude of this vector can be found by using what we know about right triangles.

Pythagorean Th'm → magnitude:

$$v_{total} = \sqrt{\left(2\ \tfrac{m}{s}\right)^2 + \left(10\ \tfrac{m}{s}\right)^2} = \sqrt{104}\ \tfrac{m}{s}$$

Any inverse trig function → direction:

We need a sketch! Defining θ as the angle opposite the 2 m/s side in the component triangle shown in the graphical column, we have:

$$\theta = \tan^{-1}\left(\frac{v_{walking}}{v_{boat}}\right) = \tan^{-1}\left(\frac{2}{10}\right)$$

From the first triangle shown, it is clear that θ is "North of East."

Notice that we use v_{total} to refer to the magnitude alone, whereas $\vec{v}_{total}$ includes the direction. Be careful to make this distinction.

Finally, what if the two (or more) vectors are not parallel, anti parallel, or perpendicular? One thing you can do is to make them parallel (or anti parallel), and perpendicular. That is, you **break one or more of the vectors down into parts (components) that *are* parallel and perpendicular.** What are your remaining choices?

"30° S of E" means that you start out pointing east and turn 30° toward the south. You've moved south "of," or "from," east.

EXAMPLE 8: Let's try adding vectors which are neither perpendicular nor parallel. On the same boat moving with velocity (10 m/s, east), you walk (2 m/s, 30° south of east ... denoted "S of E") relative to the boat's deck. Find your $\vec{v}_{total}$ (magnitude, a.k.a. speed, and direction) relative to the ground.

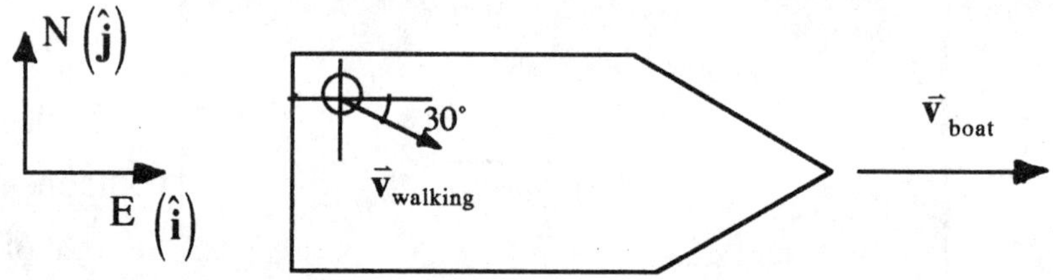

CONCEPTS: Again, imagine the path through space that you are traveling. What would that person in the tree top above see? They would see you move toward the front of the boat and toward the south side as the boat moves forward. Your path through space results from the vector addition of your walking velocity on the boat and the boat's velocity. Notice that you move forward (eastward with the boat) faster than if you were either standing still, walking perpendicular to the boat's motion, or walking toward the back of the boat. This is because a *component* of your walking velocity is parallel to the boat's velocity – it adds to the boat's velocity. The rate at which you walk *across* the boat (south) does not effect the rate at which you move east. Hmmm, what does this mean about the angle of your path? On the boat, you're facing 30° S of E and headed straight for the ginger ale, but what angle does the observer in the tree top see? Are you traveling over the Earth at 30° S of E?

Now that we have a sense of what's going on, let's solve the problem. You may prefer to read one column at a time to see the separate analyses, or look back and forth to compare the analyses. It's probably good to read through this a few times and do both.

GRAPHICAL ANALYSIS	ANALYTICAL ANALYSIS

Sketch the vectors on the object of interest (you).

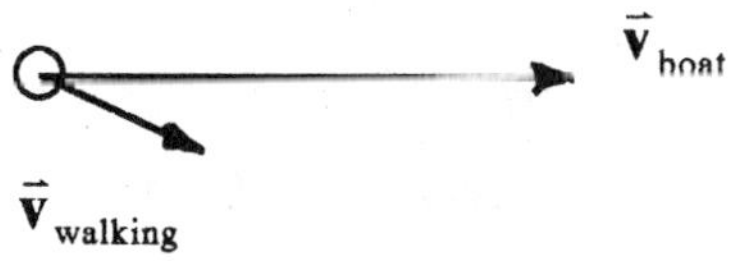

Draw the vectors "head to tail." That is, place the tail of *either* one at the head of the other one. You MUST keep the vectors parallel to the original direction, but you CAN move the vectors to different places for the sake of adding them graphically! Notice that the order does NOT matter, since the geometry, and hence the result, will always be the same.

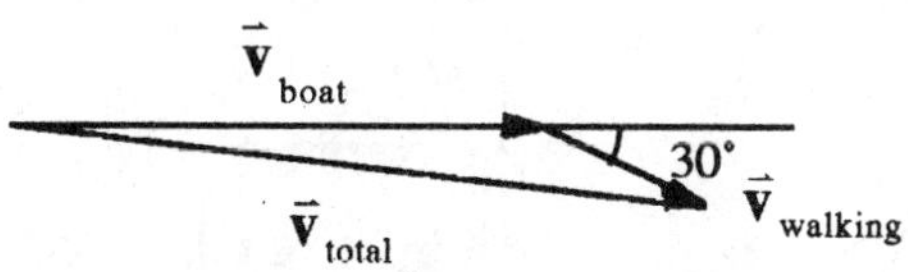

The **sum** is the vector that starts at the tail of the 1st vector and ends at the head of the last vector. The total length is clearly not the numerical sum of the two speeds. In a purely graphical analysis, we would use a ruler and protractor to measure the length and angles, and convert the length to a speed using whatever length to speed scale (for example, 2 cm : 10 m/s) we chose to use in our diagrams. This is a little difficult here, so let's use geometric relations to finish off the problem.

Since we don't have a right triangle, it makes sense to choose to use the law of cosines to get the magnitude and angle.

Law of Cosines → magnitude (v_{total}):

Represent each vector using **unit vectors** to specify the direction. Let's choose the positive *x* direction , noted by $\hat{\mathbf{i}}$, to be east, and the positive *y* direction, noted by $\hat{\mathbf{j}}$, to be north. (Actually, I'd rather use south to be the $\hat{\mathbf{j}}$ direction, but this will illustrate a point about signs.) The boat's velocity is easy.

$$\vec{v}_{boat} = (10\ ^{m}/_{s})\ \hat{\mathbf{i}}$$

$$\vec{v}_{walking} = v_{w,x}\ \hat{\mathbf{i}} + v_{w,y}\ \hat{\mathbf{j}}$$

Get the components of your walking velocity that are parallel and perpendicular to the boat's velocity, denoted $v_{w,x}$ and $v_{w,y}$, respectively, since we know how to add parallel and perpendicular vectors. Visualize an *x* and *y* grid on top of the vector and make a right triangle. Use trig to get the magnitude of the sides.

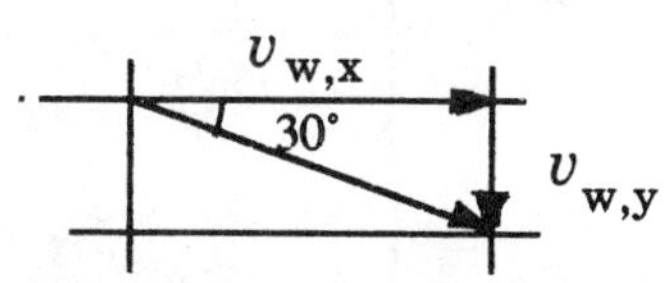

$$v_{w,x} = (2\ ^{m}/_{s})\cos 30° = 1.7\ ^{m}/_{s}$$
$$v_{w,y} = (2\ ^{m}/_{s})\sin 30° = 1.0\ ^{m}/_{s}$$

Okay, I stuck a sketch in there. I needed to see what I was doing! For clarity, you've got to do what you've got to do.

Now we can **add the corresponding components** of the two vectors:

$$v_{total,x} = v_{b,x} + v_{w,x} = (10\ ^{m}/_{s}) + (1.7\ ^{m}/_{s}) = 11.7\ ^{m}/_{s}$$

$$v_{total,y} = v_{b,y} + v_{w,y} = (0\ ^{m}/_{s}) + (1.0\ ^{m}/_{s}) = 1.0\ ^{m}/_{s}$$

Law of Cosines → direction (θ):

The angle that we seek is the angle between the total velocity vector and east (the boat velocity vector). That angle is opposite $\vec{v}_{walking}$, which has a magnitude of 2 m/s. Thus,

$$2^2 = 10^2 + 12^2 - 2(10)(12)\cos\theta \Rightarrow$$

$$\theta = \cos^{-1}\left(\frac{10^2 + 12^2 - 2^2}{2(10)(12)}\right) = 0°!!!$$

That is clearly wrong! What happened? Round-off error! Actually, we get 11.77…m/s for v_{total}. You can play with it a bit more if you like.

We can always get right triangles if we break the vectors down into parallel and perpendicular components, as is shown in the component analysis on the right.

A sketch of the total velocity components looks like:

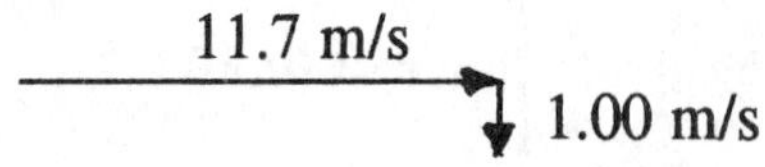

Notice from the sketch that the two components of the total velocity can be thought of as forming the two sides of a right triangle. The magnitude of this vector can be found by using what we know about right triangles.

Pythagorean Th'm → magnitude (v_{total}):

$$v_{total} = \sqrt{(11.7\ \tfrac{m}{s})^2 + (1.00\ \tfrac{m}{s})^2}$$
$$= 11.7\ \tfrac{m}{s}$$

Ahhh! Even 3 sig. fig's won't save us from a round-off predicament. Surely the hypotenuse is greater than either side, but the difference only shows up with one more sig. fig. (See Ch. 1 for what to do.)

Any inverse trig function → direction (θ):

I can avoid the hypotenuse value by using the components:

$$\theta = \tan^{-1}\left(\frac{1.00}{11.7}\right) = 4.89° \text{ S of E}$$

Wow! I thought you were walking at 30°! How could that be? Even though you are facing 30° S of E, the boat carries you eastward as you walk. This makes your path relative to the ground have a different angle than your path relative to the boat!

In the last example, I showed you two ways to solve the same problem. I didn't do a purely graphical analysis because I elected not to draw everything perfectly to scale and then measure with ruler and protractor. Instead I implemented the geometric law of cosines to calculate the result. Similarly, in the second column I did not analyze the problem purely with equations. I could have, but I would have needed to visualize those sketches that I drew to help guide my use of the equations. When using either the method of components or the laws of sines and cosines, clear pictures are usually essential. Your sketches need not be exactly to scale unless you are using a ruler and protractor, but some semblance of scale usually helps you get a feel for what to expect.

* General Methods For Vector Addition *

KEY POINTS

- No matter what the vectors represent, the techniques for adding them are identical. Be methodical and you will get the hang of it.
- You can only add/subtract quantities with the same units. This includes vectors. You cannot add a force vector to a velocity vector, nor can you add a position vector to a velocity vector.
- There are 3 basic methods: graphical (measuring scaled diagrams), geometric (using the laws of cosines and sines), and component analysis (choosing axes and breaking the vectors down to add or subtract parallel parts).
- When using geometric laws or component addition, sketches will help us visualize the lengths, angles, and components.

COMPONENT ADDITION

- Break down all vectors into components (see B.8.c)
- Add/subtract corresponding components. This gives the components of the resultant vector: $\vec{\mathbf{R}} = R_x\,\hat{\mathbf{i}} + R_y\,\hat{\mathbf{j}}$, where $R_x = A_x + B_x + C_x + \ldots$ and $R_y = A_y + B_y + C_y + \ldots$
- Get R (magnitude of $\vec{\mathbf{R}}$) using the Pythagorean th'm: $R = \sqrt{R_x^2 + R_y^2}$
- Get θ (the direction of $\vec{\mathbf{R}}$) by using any inverse trig function and any two of R_x, R_y, and R. A rough sketch of R_x, R_y, and R is very helpful here.

We add the components of vectors because only parallel (and anti parallel) components may be added numerically. Unit vectors help to distinguish the perpendicular components. We can also create tables that separate the components. (Next page.)

Sample table to organize your vector component addition:

Vectors	x-components	y-components
$\vec{A}$	A_x	A_y
$\vec{B}$	B_x	B_y
$\vec{C}$	C_x	C_y
$\vec{R} = \vec{A} + \vec{B} + \vec{C}$	add the components in this column (some may be –)	add the components in this column (some may be –)

Remember, these vectors must all represent the same concept and have the same units!!

ADDITION BY GEOMETRIC LAWS

- The law of cosines and the law of sines provide nice tools for obtaining the magnitude and angle of the sum of two vectors. (Review their use in the trig section.)
- As usual, a rough sketch using the graphical techniques will assist in getting the correct angles and corresponding sides properly placed in the equations.

GRAPHICAL ADDITION

Draw the vectors. Using an accurate scale will allow us to measure the result from the diagrams. A rough sketch will give us a sense of what to expect from our calculations using either geometric relations or the component method.

Head To Tail Method

- Place the tail of one of the vectors to be added at the origin of a coordinate system.
- Place the tail of any other vector to be added at the head of the first one. Keep the added vector the same length as and pointing in a direction parallel to it's original.
- Place the tail of any other vector to be added at the head of the last one. Repeat until all of the vectors have been added.
- The resultant is drawn straight from the beginning to the end; that is, it's tail is at the tail of the first vector and it's head is at the head of the last vector added.

Moving vectors around like this may seem a bit odd at first. After all, a parallel direction is not strictly the same direction. It is, however the same orientation as seen on a compass. Keep in mind that this is a graphical *representation*, not a physical picture, of what is going on.

When subtracting vectors, first reverse the direction of the one being subtracted and then add the vectors as before: $\vec{\mathbf{A}} - \vec{\mathbf{B}} = \vec{\mathbf{A}} + (-\vec{\mathbf{B}})$.

If you draw in the components of the vectors and then add parallel components, you will end up with one x and one y component. If you then add these head to tail to form a right triangle, you will obtain the same result as you would if you added the vectors head to tail without breaking them into components. (As you see, there are many ways to obtain the same result!)

Parallelogram Method

- Draw two vectors, only two, **tail to tail**.
- Form a parallelogram using these vectors as two of the sides and two parallel vectors as the other sides.
- The resultant is the diagonal from the initial tails to the final tips

This is really equivalent to the head to tail method as seen below. Adding vectors $\vec{\mathbf{A}}$ and $\vec{\mathbf{B}}$

results in the diagonal vector below.

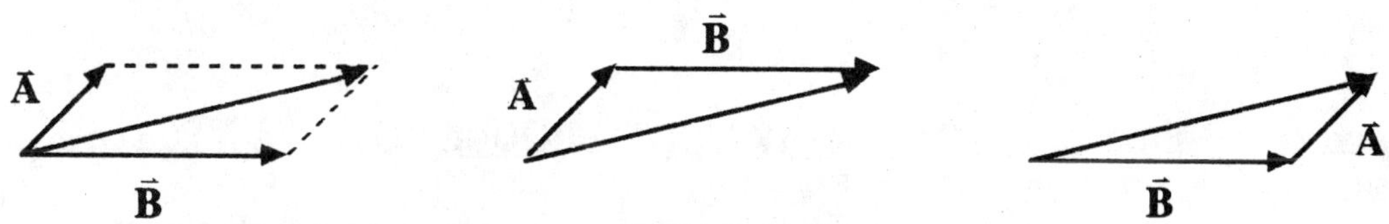

The resulting magnitude and direction is the same any way you construct it!

Phew! All that for addition and subtraction. You'll soon see that these few techniques will go a long way, and with practice, it'll be like driving a car – you basically do the same thing each time, keeping alert to changing road conditions. To complete vector arithmetic, we now turn our attention to how we can multiply two vectors together.

Try adding a few vectors using the head to tail and parallelogram methods in your study journal. Just sketch two, three, or four arrows and put them together by carefully following each step. Compare and contrast the "head-to-tail" and the "parallelogram" methods. Like most things, it's slow going at first, but you soon become proficient. It's worth the small investment of time now.

B.8.e. Vector Multiplication

Preview now. Come back and refer to this when you need it.

"We know that multiplying a vector by a number just changes the magnitude (length, when represented as an arrow) of the vector (unless it's a negative number, which reverses the direction), but how do we multiply a vector by another vector?" In one sense, however we want. That is, however is useful and consistent with the other rules in our mathematical language. Remember, mathematics is a *constructed* language which serves the purpose of quantitative analysis. People made–up (some would argue for "discovered") vectors to deal with directional quantities. People made–up the rules for vector addition to correspond with the physical addition of directional quantities such as displacement, force, and velocity. And people made the rules for vector multiplication to be an additional, useful tool for analyzing vector quantities. So what is a useful way to "multiply" vectors?

It turns out that there are 2 different, useful ways to multiply vectors together. They are based on the idea that sometimes the *parallel* parts of two vectors are the only things that matter and sometimes only the *perpendicular* parts matter. (Parallel parts and perpendicular parts – a recurring theme when dealing with vectors.) Now how are we going to symbolize these two different types of multiplication? Well, we already have two symbols for multiplication that we use interchangeably. What are they? Yes, the dot (·) and the cross (x). When multiplying vectors, these two symbols have been assigned to mean *very* different things.

<u>THE SCALAR (AKA DOT) PRODUCT OF TWO VECTORS</u>

- The *parallel* parts of the two vectors are what matters.
- Use the *dot* multiplication symbol.
- The result is a *scalar* (number), not a vector.
- There are two approaches: using magnitudes and the angle between them, or using the components. Sometimes we'll use one, sometimes both.

As noted above, there are basically two ways to multiply the parallel components together. The result is the same either way, but there are times when one way is easier than the other. The first way is to multiply the magnitude of either one of the vectors times the other vector's *component* that is *parallel* to the first one. Look at the pictures below and reread the last sentence until it makes sense.

The order of multiplication does NOT matter with the dot product.

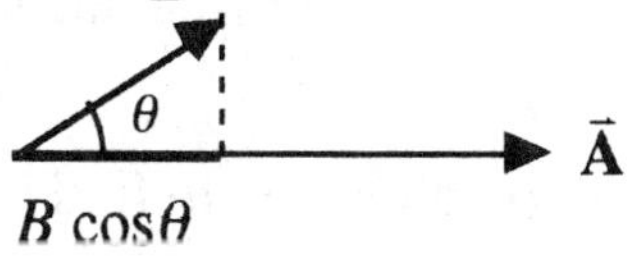

$$\vec{\mathbf{A}} \cdot \vec{\mathbf{B}} = \left|\vec{\mathbf{A}}\right|\left(\left|\vec{\mathbf{B}}\right|\cos\theta_{\text{between } \vec{\mathbf{A}} \text{ and } \vec{\mathbf{B}}}\right)$$

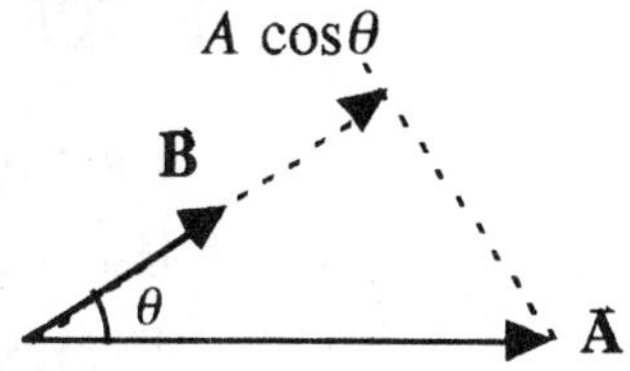

$$\vec{\mathbf{A}} \cdot \vec{\mathbf{B}} = \left|\vec{\mathbf{B}}\right|\left(\left|\vec{\mathbf{A}}\right|\cos\theta_{\text{between } \vec{\mathbf{A}} \text{ and } \vec{\mathbf{B}}}\right)$$

First, notice that the angle, θ, is specifically the smallest angle between $\vec{\mathbf{A}}$ and $\vec{\mathbf{B}}$ when they are **placed tail to tail**. See it? Further notice that $\left|\vec{\mathbf{B}}\right|\cos\theta_{\text{between } \vec{\mathbf{A}} \text{ and } \vec{\mathbf{B}}}$ is the component of $\vec{\mathbf{B}}$ that is parallel to $\vec{\mathbf{A}}$, while $\left|\vec{\mathbf{A}}\right|\cos\theta_{\text{between } \vec{\mathbf{A}} \text{ and } \vec{\mathbf{B}}}$ is the component of $\vec{\mathbf{A}}$ that is parallel to $\vec{\mathbf{B}}$. Either way you look at it, the result is the same: $\vec{\mathbf{A}} \cdot \vec{\mathbf{B}} = \vec{\mathbf{B}} \cdot \vec{\mathbf{A}}$.

This works well if you know the vector magnitudes and the angle between them. Of course if you know the components of the vectors you can always get the magnitudes by using the **Pythagorean theorem**:

$$\left|\vec{\mathbf{A}}\right| \equiv A = \sqrt{A_x^{\,2} + A_y^{\,2} + A_z^{\,2}}$$

$$\left|\vec{\mathbf{B}}\right| \equiv B = \sqrt{B_x^{\,2} + B_y^{\,2} + B_z^{\,2}}$$

But if the task is to multiply parallel parts, and we already know the components of the vectors, why not just go ahead and **multiply parallel parts**, i.e. multiply parallel components? This provides the second method for obtaining the scalar (dot) product.

$$\vec{\mathbf{A}} \cdot \vec{\mathbf{B}} = A_x B_x + A_y B_y + A_z B_z$$

EXERCISE B.16: Use either of the two methods for obtaining the scalar product to compute the following. [NOTE: the physical meaning of $\vec{\mathbf{F}} \cdot \vec{\mathbf{d}}$ is work, W.]

(a) Given $F = 40$ newtons, $d = 2$ meters, and the angle between $\vec{\mathbf{F}}$ and $\vec{\mathbf{d}}$ is $\theta = 60°$, find $\vec{\mathbf{F}} \cdot \vec{\mathbf{d}}$.

(b) Given $\vec{\mathbf{F}} = (2\text{ N})\,\hat{\mathbf{i}} + (10\text{ N})\,\hat{\mathbf{j}} - (3\text{ N})\,\hat{\mathbf{k}}$ & $\vec{\mathbf{d}} = (-5\text{ m})\,\hat{\mathbf{i}} + (1\text{ m})\,\hat{\mathbf{j}} - (4\text{ m})\,\hat{\mathbf{k}}$, find the work.

(c) Given that $\vec{\mathbf{F}} = (2\text{ N})\,\hat{\mathbf{i}} + (10\text{ N})\,\hat{\mathbf{j}} - (3\text{ N})\,\hat{\mathbf{k}}$ & $\vec{\mathbf{d}} = (-5\text{ m})\,\hat{\mathbf{i}} + (1\text{ m})\,\hat{\mathbf{j}} - (4\text{ m})\,\hat{\mathbf{k}}$, find the angle between the vectors. [Hint: set the two equations for the dot product equal and solve for θ. Don't forget to use the Pythagorean th'm to find F and d!]

THE VECTOR (AKA CROSS) PRODUCT OF TWO VECTORS

- The *perpendicular* parts of the two vectors are what matters.
- Use the *cross* multiplication symbol.
- The result is a *vector*, so we need to know how to find both the magnitude and the direction.
- Again, there are two approaches: using magnitudes and the angle between them, now coupled with the right hand rule to tell us the direction, or using the components. Sometimes we'll use one, sometimes both.

The first method is similar to that for the scalar (dot) product except that we now have an additional rule for finding the direction. To find the magnitude, we must multiply the *perpendicular* component of one times the magnitude of the other. To find the direction we use a convenient helper called the "right hand rule."

MAGNITUDE OF $\vec{\mathbf{A}} \times \vec{\mathbf{B}}$

Multiply the magnitude of either one of the vectors times the other vector's *component* that is *perpendicular* to the first one. Look at the pictures below and reread the last sentence until it makes sense.

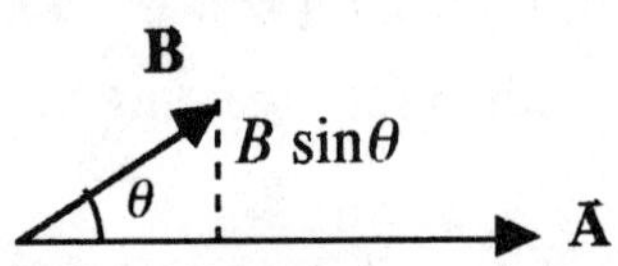

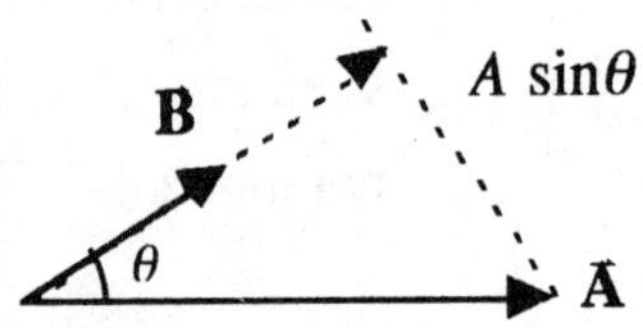

$$\left|\vec{\mathbf{A}} \times \vec{\mathbf{B}}\right| = \left|\vec{\mathbf{A}}\right|\left(\left|\vec{\mathbf{B}}\right|\sin\theta_{\text{between } \vec{\mathbf{A}} \text{ and } \vec{\mathbf{B}}}\right)$$

$$\left|\vec{\mathbf{B}} \times \vec{\mathbf{A}}\right| = \left|\vec{\mathbf{B}}\right|\left(\left|\vec{\mathbf{A}}\right|\sin\theta_{\text{between } \vec{\mathbf{A}} \text{ and } \vec{\mathbf{B}}}\right)$$

Notice that $\left|\vec{\mathbf{B}}\right|\sin\theta_{\text{between } \vec{\mathbf{A}} \text{ and } \vec{\mathbf{B}}}$ is the component of $\vec{\mathbf{B}}$ that is perpendicular to $\vec{\mathbf{A}}$, while $\left|\vec{\mathbf{A}}\right|\sin\theta_{\text{between } \vec{\mathbf{A}} \text{ and } \vec{\mathbf{B}}}$ is the component of $\vec{\mathbf{A}}$ that is perpendicular to $\vec{\mathbf{B}}$. Either way you look at it, the result is the same: $\left|\vec{\mathbf{A}} \times \vec{\mathbf{B}}\right| = \left|\vec{\mathbf{B}} \times \vec{\mathbf{A}}\right|$. Notice that I've used absolute values. This is because it is only the *magnitudes* that are equal. When you look at the direction, you'll see that $(\vec{\mathbf{A}} \times \vec{\mathbf{B}}) = -(\vec{\mathbf{B}} \times \vec{\mathbf{A}})$, i.e., the directions are exactly opposite.

DIRECTION OF $\vec{A} \times \vec{B}$ – THE RT. HAND RULE

This is a necessary skill which is a stumbling block for far too many students.

I know of 5 variations of the right hand rules that will give you the direction of a vector (cross) product! Of course they all give the same result, so you just need to **pick the one you think is easiest.** You will then need to practice the *methodical* application of the rule until it takes you very little time to determine the resulting *direction* of the vector (cross) product.

Before we begin, how many directions will we need to be concerned with? We have the directions of each of the two vectors in the product and the direction of the resulting vector. So there are 3 directions to pay attention to. Further notice that any two *intersecting* lines, intersecting pencils, intersecting arrows, or intersecting vectors will specify a plane – we can always make them lie along a flat piece of paper. If you think of the two vectors in the product as forming a plane, the vector (cross) product of the two will always be a vector perpendicular to that plane. Why? Because it is useful to define it that way. Burn that idea into your memory banks! **The vector (cross) product is always perpendicular to the plane formed by the two vectors being multiplied. Always. Always always.**

1. The Curling Right Hand – often the preferred method
2. The Pushing Right Hand – a close variation of the curling hand
3. The Right Handed Screw – used a screw driver much?
4. The 3 Finger Axes – very close to 1 above
5. The Flat Right Hand – quite different, same result (I like this one)

WARNING: You must be *careful* when you point your fingers, thumb, and/or hand to determine the direction using the right hand rule. Any sloppiness may well lead to frustration.

1. **The Curling Rt Hand** – This is the method chosen by your text and so may be the logical choice for you. Now, the vector (cross) product always results in a vector that is perpendicular to what?

   - Draw the vectors tail to tail. It doesn't matter where the vectors are on the page, just place them tail to tail and keep them pointing in a direction that is exactly parallel to their original directions.

     – DRAW THE TWO VECTORS TAIL TO TAIL

- Visualize the plane made by the two vectors as a sheet of clear plastic wrap stretched across the vectors (or as a sheet of paper).

– VISUALIZE THE PLANE MADE BY THE VECTORS

- Place your thumb in a direction perpendicular to that plane. You have two choices. Make a guess and check it in the next step. (Imagine punching a hole through the plane with your thumb.)

– THUMB ⊥ TO PLANE OF VECTORS

- Straighten your fingers so that they are parallel to the 1st vector in the cross product. (Keep your thumb in the right direction!)

– FINGERS PARALLEL TO 1st VECTOR

- If your *palm* faces the second vector, your thumb is in the right direction. (Notice that you could curl your fingers into the 2nd vector, hence the name of the rule.) If the back of your hand faces the 2nd vector, you simply need to reverse the direction of your thumb.

– PALM FACING 2nd VECTOR?

- If you've been careful, your thumb, sticking straight up from your fist, is now pointing in the direction of the vector (cross) product.

– RESULT PTS IN DIRECTION OF THUMB

Unlike the dot product, the cross product depends on the order of multiplication.

Notice that using the other vector as the 1st vector gives you the exact opposite direction. This nicely reflects that we *do* need to distinguish between $\vec{\mathbf{A}} \times \vec{\mathbf{B}}$ and $\vec{\mathbf{B}} \times \vec{\mathbf{A}}$ – while they have the same magnitude, they have opposite directions. $\vec{\mathbf{A}} \times \vec{\mathbf{B}} = -\vec{\mathbf{B}} \times \vec{\mathbf{A}}$.

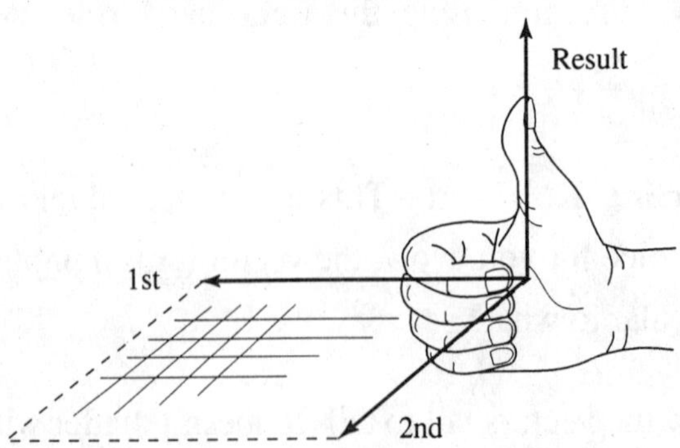

<u>STOP!</u> Skip the next 4 right hand rule variations if you are fine with the one just explained. You may want to read the alternatives if your instructor likes to use one, or if you just want to see them and possibly find one you like better.

2. **The Pushing Rt Hand** – Again, follow the steps *carefully* when you point your fingers, palm, and thumb in the prescribed directions.

- Draw the vectors tail to tail. It doesn't matter where the vectors are on the page, just place them tail to tail and keep them pointing in a direction that is exactly parallel to their original directions.

  – DRAW THE TWO VECTORS TAIL TO TAIL

- Open up your *right* hand and flatten your palm and fingers

  – FLATTEN YOUR RT HAND

- Place your hand so that your fingers are parallel to the 1st vector.

  – FINGERS PARALLEL TO 1st VECTOR

- Keeping your fingers parallel to the 1st vector, rotate your hand around the vector so that your palm faces the 2nd vector; that is, so that you are in a position to "push" the 1st vector onto the 2nd vector.

  – ROTATE PALM TO FACE 2nd VECTOR

- Push the 1st vector onto the 2nd vector. As you "push" the 1st vector, you can curl your hand so that it mimics that shown in the text.

  – PUSH 1st VECTOR ONTO 2nd VECTOR

- Your thumb, sticking straight up from your fist, is now pointing in the direction of the vector (cross) product . . . if and only if you have been very careful!

  – RESULT PTS IN DIRECTION OF THUMB

Once again, notice that pushing in the reverse direction, from $\vec{\mathbf{B}}$ to $\vec{\mathbf{A}}$, points your thumb in exactly the opposite direction. That is the definition we want: $\vec{\mathbf{A}} \times \vec{\mathbf{B}} = -\vec{\mathbf{B}} \times \vec{\mathbf{A}}$.

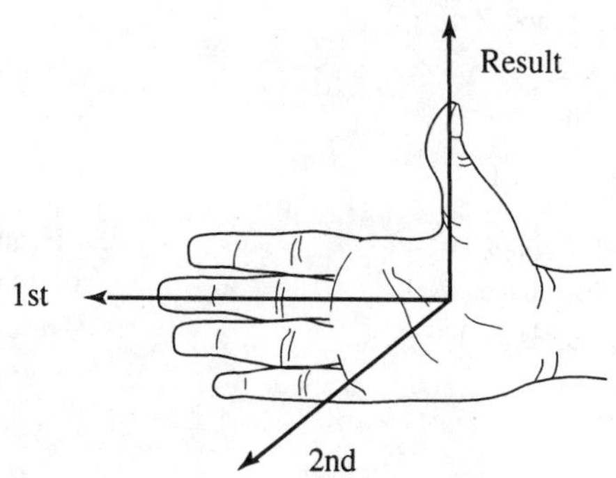

3. **The Rt Hand Screw** – If you've used a screwdriver quite a bit, this may be the one for you! Most screws follow the simple mnemonic: "right to tighten, left to loosen." Thus if you are looking at the head of the screw and you twist your screwdriver *cw*, so that the top of the screw moves *right*, the screw will move *INTO* the wood or metal. If you twist your screwdriver *ccw*, so that the top of the screw moves *left*, the screw will move *OUT OF* the wood or metal. If this is intuitive to you, try the following:

- Draw the vectors tail to tail. It doesn't matter where the vectors are on the page, just place them tail to tail and keep them pointing in a direction that is parallel to their original direction.

  – DRAW THE TWO VECTORS TAIL TO TAIL

- Imagine that a screw is at the place where the tails connect, and is perpendicular to the plane created by the two vectors.

  – IMAGINE A SCREW AT THE TAILS

- Now, with an imaginary screwdriver, rotate the screw (through the smaller angle) from the 1st vector in the product to the 2nd vector

  – TURN THE SCREW

- The screw will move either in or out of the plane created by the two vectors: cw in, ccw out. That is the direction of the vector (cross) product. Notice that if I reverse the direction, corresponding to a change in the order of multiplication, I get the opposite direction. This is as it should be.

  – RESULT PTS IN DIRECTION SCREW MOVES

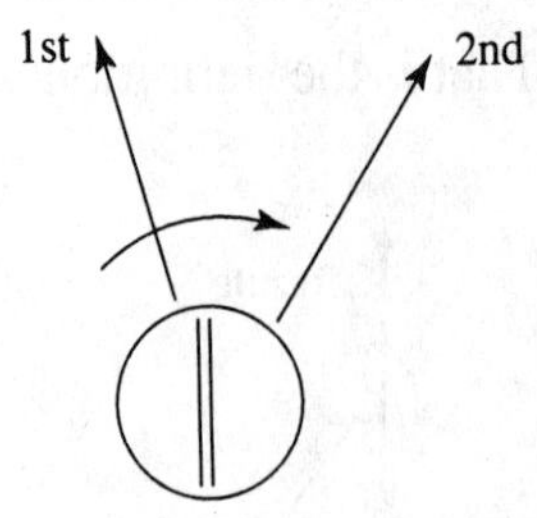

Head of screw goes **into** page; resultant is (x)

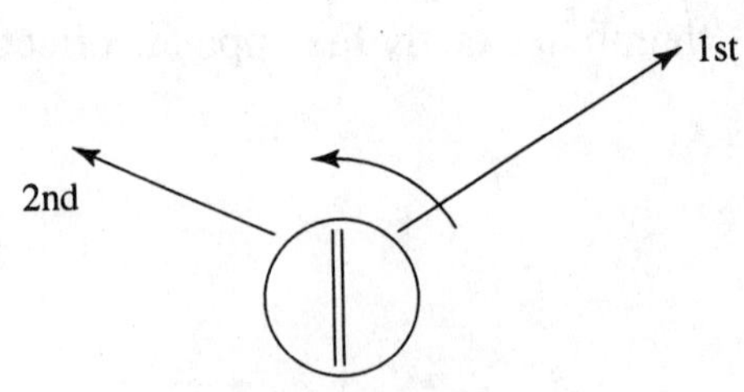

Head of screw comes **out** of page; resultant is (•)

4. **The 3 Finger Axes** – This is much like methods 1 and 2, except that there are no moving hand parts. Here we specify the 3 directions with the index finger (1st vector), middle finger (2nd vector), and thumb (result) of the right hand.

- Draw the vectors tail to tail. It doesn't matter where the vectors are on the page, just place them tail to tail and keep them pointing in a direction that is parallel to their original directions.

  – DRAW THE TWO VECTORS TAIL TO TAIL

- Make a fist with your right hand. Lift your thumb up. Point your index finger as if to make a "play" gun. Straighten your middle finger so that it makes a right angle with your index finger. Keep the two smaller fingers curled. Your hand should look something like the three x–y–z coordinate axes. This 3 finger pointer is what you can use to relate the 3 directions of interest.

  – FORM A 3 FINGER POINTER

- Point your index finger parallel to the 1st vector.

  – INDEX FINGER PARALLEL TO 1st VECTOR

- Point your middle finger parallel to the 2nd vector. (Now it doesn't have to be perpendicular to the first.)

  – MIDDLE FINGER PARALLEL TO 2nd VECTOR

- The plane created by your index and middle fingers should now be parallel to the plane created by the two vectors being multiplied. Your thumb should be perpendicular to this plane and thus in the direction of the vector (cross) product.

  – RESULT PTS IN DIRECTION OF THUMB

(You could also have your thumb point in the direction of the 1st vector, your index finger point in the direction of the 2nd vector, and your middle finger point in the direction of the result. Check for yourself to see that this is equivalent the to the flat rt hand technique below.)

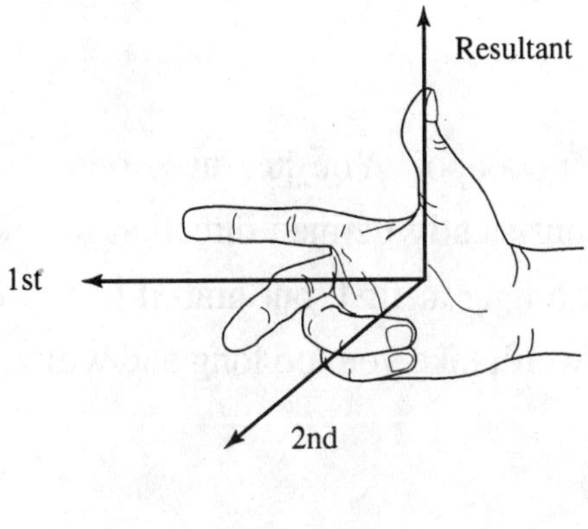

5. **The Flat Rt Hand** – This is a nice option. This will give the same result, but in this case the direction is NOT given by the thumb. To find the direction of the vector (cross) product of any two vectors:

- Draw the vectors tail to tail. It doesn't matter where the vectors are on the page, just place them tail to tail and keep them pointing in a direction that is parallel to their original direction.

  – **DRAW THE TWO VECTORS TAIL TO TAIL**

- Place your thumb along the 1st vector in the product

  – **THUMB PARALLEL TO 1st VECTOR**

- Keep your thumb pointing in the correct direction and make your hand perfectly flat

  – **FLATTEN PALM & FINGERS**

- Still keeping your thumb in a fixed direction, rotate your hand (like a door) around your thumb (acting like the hinge) until your fingers are parallel to the 2nd vector

  – **FINGERS PARALLEL TO 2nd VECTOR**

- The vector (cross) product points in the direction out of and perpendicular to your palm

  – **RESULT PTS OUT OF PALM**

Again check that reversing the order of the cross product reverses the direction of the resulting vector.

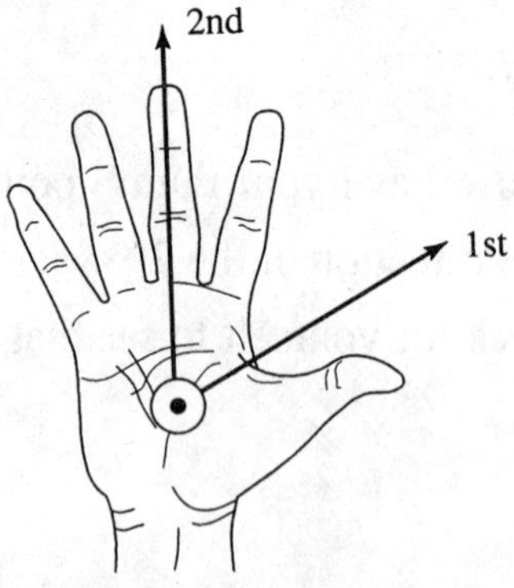

Sooooooooooooooooo! You just need one way to skin this cat. What's important is that we have a consistently defined direction for the vector (cross) product. Yes, I know this cat is a bit of a tiger at first, but grab it by the tail and don't let go until you've got it under control. It won't take you too long and will very quickly save time and aspirin.

COMPONENT METHOD

"Is there another way to get the vector (cross) product if we know the components of the vectors, as we did with the scalar (dot) product?" Well, what do you think? Of course there is! With the scalar (dot) product we multiplied parallel components, $A_x B_x$ etc., whereas here we multiply perpendicular components: $A_x B_y$, $A_x B_z$, $A_y B_x$, $A_y B_z$, $A_z B_x$, $A_z B_y$. There are six different combinations for 3–D, but only two combinations for 2–D (all of the z terms are then zero).

In this second way of obtaining the vector (cross) product, you don't have to use the right hand rule to determine the direction, the unit vectors account for the direction. (But you may still want to use it to help you see the direction.) I'll grant you that this multiplication rule looks unwieldy at first, but with practice you'll be rattling it off in your sleep. Besides, there are a couple of nice mnemonics for generating the expression.

2–D: $$\vec{\mathbf{A}} \times \vec{\mathbf{B}} = (A_x B_y - A_y B_x)\,\hat{\mathbf{k}}$$

3–D: $$\vec{\mathbf{A}} \times \vec{\mathbf{B}} = (A_y B_z - A_z B_y)\,\hat{\mathbf{i}} + (A_z B_x - A_x B_z)\,\hat{\mathbf{j}} + (A_x B_y - A_y B_x)\,\hat{\mathbf{k}}$$

Once again you see that the vector (cross) product gives us a vector, hence its name. Here we see that the magnitude of the z component is determined by products of the x and y components of the vectors being multiplied, and so on. (Multiplying perpendicular parts, remember?) We can get the magnitude of the resulting vector by applying the Pythagorean theorem, and the direction using the inverse trig functions. [More on this in Part 1.]

Something tells me that you would like to see those mnemonics I spoke of. There are 2: permutations and the determinant. First just note that we are indeed multiplying only perpendicular terms. Second notice that it would be easy to get a minus sign in the wrong place. Be careful. Inaccuracy can be expensive in science and engineering.

Permutations mnemonic (my favorite): Permutations are much easier than they sound. They are simply the reordering of any objects. For example, the only possible permutation of xy is yx. However, xyz can be ordered in $3! = 3 \cdot 2 \cdot 1 = 6$ ways: xyz, yzx, zxy, and yxz, xzy, zyx. Notice that I've separated these into two groups. The first group consists of "cyclic" AKA ""right–handed" permutations of the starting order xyz. These can be achieved by taking the front letter and moving it to the back so that the letters "cycle" around. The second group, "left–handed permutations," is achieved by first switching any two letters in xyz and then cycling around.

It's actually pretty easy to spot the difference between the two groups. We started with the alphabetic ordering of xyz. If you "read" to the right, and then start over when you get to the end, you keep reading xyz xyz... Notice that this is also true of the orderings yzx and zxy. For the other orderings (yxz, xzy, and zyx) we would need to read backwards to read "x-y-z." That is, rather than reading to the right, we would read to the left. This gives you an idea why they are sometimes called right–handed and left–handed permutations or orderings. Here is an easy way to visualize the difference:

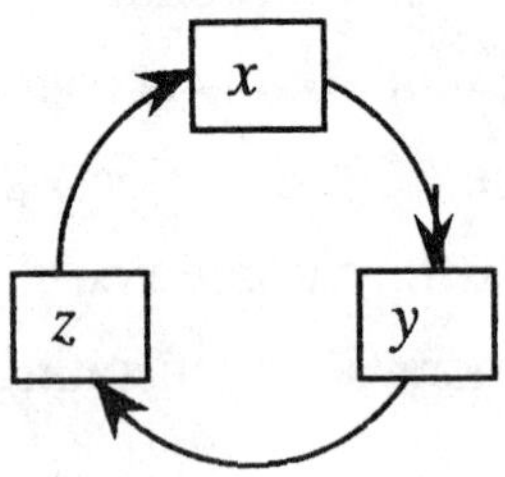

(+) Right–handed order: start at any one and cycle **cw** to the other two. (xyz, yzx, zxy).

(–) Left–handed order: start at any one and cycle **ccw** to the other two. (xzy, zyx, yxz).

Since, for the vector (cross) product, a change in order results in a change in sign, it may not be too surprising to see that the terms in the component expression for the vector (cross) product have + and – signs depending on the order of xyz. Study the expression until you see that the cyclic (rt–handed) orderings of xyz have + signs while the non–cyclic (left–handed) orderings get the – signs. Here we are treating $\hat{\mathbf{i}}$ as x, $\hat{\mathbf{j}}$ as y, and $\hat{\mathbf{k}}$ as z. Take a close look. You can actually write down that ugly component expression in a flash without the need for rote memorization! Start by writing the first line and then fill in the blanks.

$$\vec{\mathbf{A}} \times \vec{\mathbf{B}} = \hat{\mathbf{i}}\left(A_{\_}B_{\_} - A_{\_}B_{\_}\right) + \hat{\mathbf{j}}\left(A_{\_}B_{\_} - A_{\_}B_{\_}\right) + \hat{\mathbf{k}}\left(A_{\_}B_{\_} - A_{\_}B_{\_}\right)$$

$$\vec{\mathbf{A}} \times \vec{\mathbf{B}} = \hat{\mathbf{i}}\left(A_yB_z - A_zB_y\right) + \hat{\mathbf{j}}\left(A_zB_x - A_xB_z\right) + \hat{\mathbf{k}}\left(A_xB_y - A_yB_x\right)$$

$$\rightarrow \ \mathrm{x\,(yz \ - \ zy) \ + \ y\,(zx \ - \ xz) \ + \ z\,(xy \ - \ yx)}$$

Just write down $\hat{\mathbf{i}}$ and treat it as an x. Then you've got yz and zy left. Which gets the minus sign? Whichever gives the left handed permutation! → x(yz – zy). Next write $\hat{\mathbf{j}}$ and treat it as y. I just think "What comes after y?" and start there to give y(zx – xz). Finally, write down $\hat{\mathbf{k}}$ for z. What comes after z? For us, it's x! So you have z(xy – yz). If you get the pattern, you can write it down and never have to sweat it again! Try it now by filling in the subscripts of the first line above without looking at the second line. Then, take out a fresh sheet of paper and do it yourself! Ah, the ugly now becomes beautiful.

Determinants can also be used to get the order and sign right for each of the terms. (In fact, determinants are a nice booking device created to keep track of all of the permutations of a set.) You've probably seen determinants before, but if not, we'll go through it here. There are some very interesting properties associated with determinants, but all we need them for now is to remind us of how to write out the vector (cross) product in terms of the components of the two vectors being multiplied. Using the determinant notation we can write:

$$\vec{\mathbf{A}} \times \vec{\mathbf{B}} = \begin{vmatrix} \hat{\mathbf{i}} & \hat{\mathbf{j}} & \hat{\mathbf{k}} \\ A_x & A_y & A_z \\ B_x & B_y & B_z \end{vmatrix}$$

Evaluate by writing $\hat{\mathbf{i}}$ and crossing off (mentally) the terms in the same column and row.

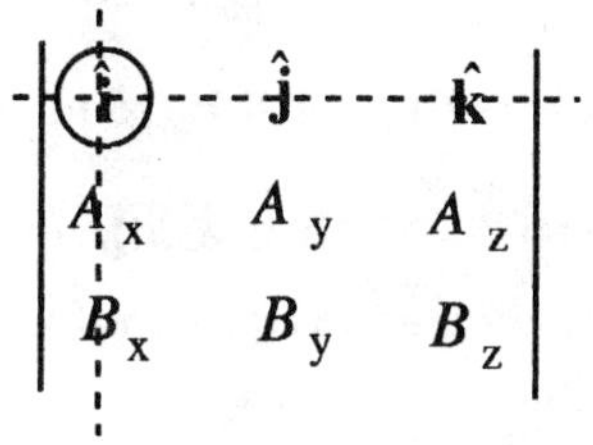

$= \hat{\mathbf{i}}$

Then multiply the remaining diagonals. Move from A_y down and to the right → A_yB_z. Move from A_z down and to the left → $-A_zB_y$. Notice that I put in a minus sign when I move down and to the *left*.

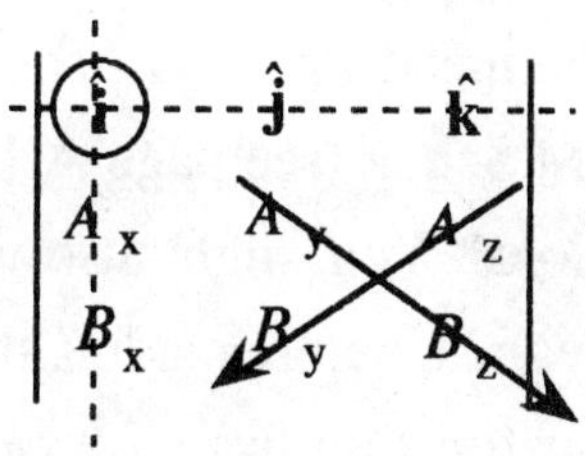

$= \hat{\mathbf{i}}\,(A_yB_z - A_zB_y) + \ldots$

Next comes the $\hat{\mathbf{j}}$ term. You can get the signs right if you subtract this term and keep the pattern of making the product "down and to the right," A_xB_z, positive, and the "backwards" product, A_zB_x, negative. (Or add the term and reverse the signs on the products, whatever!)

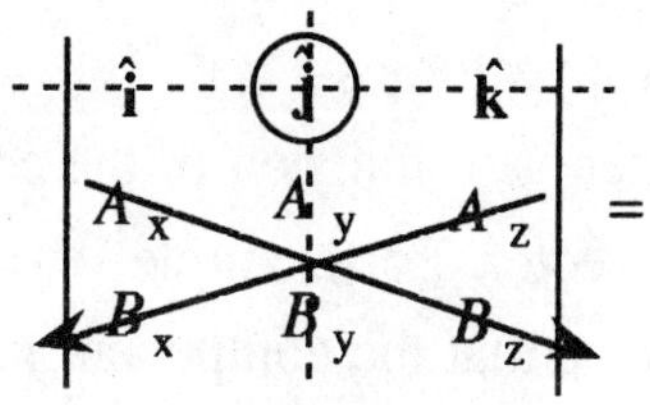

$= \hat{\mathbf{i}}\,(A_yB_z - A_zB_y) - \hat{\mathbf{j}}\,(A_xB_z - A_zB_x) + \ldots$

Finally, the $\hat{\mathbf{k}}$ term

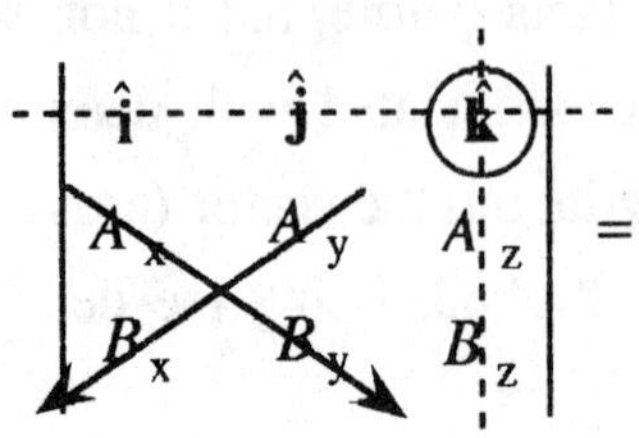

You finish it off and see that it is the same result as that obtained by the permutations.

EXERCISE B.17: Use either of the two methods for obtaining the vector product to compute the following. [NOTE: the physical meaning of $\vec{\mathbf{d}} \times \vec{\mathbf{F}}$ is torque, $\vec{\tau}$.]

(a) Given $F = 40$ newtons, $d = 2$ meters, and the angle between $\vec{\mathbf{F}}$ and $\vec{\mathbf{d}}$ is $\theta = 60°$, find $\vec{\mathbf{F}} \times \vec{\mathbf{d}}$.

(b) Given $\vec{\mathbf{F}} = (2\text{ N})\,\hat{\mathbf{i}} + (10\text{ N})\,\hat{\mathbf{j}} - (3\text{ N})\,\hat{\mathbf{k}}$ & $\vec{\mathbf{d}} = (-5\text{ m})\,\hat{\mathbf{i}} + (1\text{ m})\,\hat{\mathbf{j}} - (4\text{ m})\,\hat{\mathbf{k}}$, find the torque.

(c) Given that $\vec{\mathbf{F}} = (2\text{ N})\,\hat{\mathbf{i}} + (10\text{ N})\,\hat{\mathbf{j}} - (3\text{ N})\,\hat{\mathbf{k}}$ & $\vec{\mathbf{d}} = (-5\text{ m})\,\hat{\mathbf{i}} + (1\text{ m})\,\hat{\mathbf{j}} - (4\text{ m})\,\hat{\mathbf{k}}$ find the angle between the vectors. [Hint: set the two equations for the dot product equal and solve for θ. Don't forget the Pythagorean th'm!]

B.8.f. Vector Functions

There is no way that I can cover this entire subject here, but I would like to point out a few things to get you started. I highly recommend that you track down the little book called *Div, Grad, Curl, And All That*, by H.M.Schey (see page A.14 of this guide) by the time you get to electricity. It really is exceptional. You might also want to read the sections B.9 and B.10 below, on derivatives and integrals, before reading this section.

We first need to distinguish a *scalar* function from a *vector* function. Scalar functions are just the ordinary functions that you are used to. They don't tell us anything about direction. Hence $f(x) = x^2$ is a scalar function of one variable; $g(x,y) = x^2 + y^2$ is a scalar function of two variables; $h(x,y,z) = \sin(z) - e^y + 2x - 9$ is a scalar function of three variables. In these examples, the value of the functions are associated with a spatial position given by the coordinates. If you plug in values for x, x and y, and x, y, and z, respectively, you will get a number, NOT a vector.

A vector function might also be written as a function of the coordinates (x, y, z), but when we plug in numbers we will get a vector. For example, $\vec{\mathbf{F}}(x, y, z) = (x^2 + y^2)\hat{\mathbf{i}} + \sin(z)\hat{\mathbf{j}} + 35\,\hat{\mathbf{k}}$ is a vector function. Notice that the component functions can depend on any of the coordinates, or they may be constant like the z component in this example. The point is that each component is expressed as a function; when we plug in values for the variables, we obtain a vector. Just like with scalar functions, some vector functions are easy to visualize and some are nearly impossible. Nevertheless, we can apply the appropriate math rules to manipulate the equations and see what we come up with.

Preview For The Bold

I'm going to explain something here that you may not understand. So what's the point, right? It's a preview; it's bound to be confusing. In fact, you may not even use this idea until you do your upper division work. So I'll keep it short and sweet. If you ever need to use this, you'll have seen it. With time, it'll begin to make complete sense and you'll wonder what the problem was. For now, just read on, don't get hot and bothered if you don't understand something fully, and see what ideas you can pick up that further your understanding of the language of mathematics..

In writing an expression for a vector function, you can use one system for the coordinates and another for the unit vectors. [I'm sure that went over big.] For example, it's not uncommon at all in physics to use the spherical spatial variables (r, θ, ϕ) with the rectangular unit (direction) vectors $\left(\hat{\mathbf{i}}, \hat{\mathbf{j}}, \hat{\mathbf{k}}\right)$. (Example: $\vec{\mathbf{v}} = r\sin\theta\ \hat{\mathbf{i}} + r\cos\theta\ \hat{\mathbf{j}} + \sin\phi\ \hat{\mathbf{k}}$.) Let's say that we are trying to describe the velocity of the air flowing around a car. At each point, the air may have very different velocity magnitudes and/or directions. We can locate a point in space that we are interested in by using any coordinate system that we like: P = (x, y, z) or $(r, ,z)$ or (r, θ, ϕ). Since the velocity is a vector, we'll probably want to break it down into components. We could write $\vec{\mathbf{v}} = v_x\ \hat{\mathbf{i}} + v_y\ \hat{\mathbf{j}} + v_z\ \hat{\mathbf{k}}$. As I said, the magnitudes of the components depend on the point in space that we are interested in. And as I said, we can use rectangular, cylindrical, or spherical coordinates to describe that point. Hence the component functions, v_x, v_y, and v_z, may be written as functions in *any* coordinate system. Naturally we'll choose the one that is easiest to work with – the one which best reflects the "symmetry" or geometrical patterns (rectangular, cylindrical, or spherical) of the system that we are describing.

Recall that the beauty of using rectangular unit vectors is that they don't change direction from point to point like radial and angular unit vectors do. Thus they are constants and need not be worried about when taking derivatives or integrating. A nice feature!

On that point I will leave you with more questions than answers. You will get a full treatment when you study vector calculus and if you take more advanced engineering and/or physics courses.

B.9. COMPLEX NUMBERS (SQRT -1)

NOTE: You actually don't need this for this course (unless your instructor wants to expose you to it). However, if you do go on to upper division engineering or physical science courses, you will certainly use these very handy tools. I've provided you with a short preview of these ideas because they often seem to come from left field and leave students wondering what hit them. It's really not bad stuff. In fact, it ultimately simplifies calculations – why do it otherwise? (The next section, B.13, *will* be used in this course.)

B.9.a. The Idea

[Refer to B.1. Symbol Simon.] I want to solve the equation: $x^2 + 1 = 0$. "But you can't! There is no solution." Hey, if math is made up, then I can make up a symbol that lets me do it. "Well, there already is a symbol, it just doesn't make any sense. In fact, you could solve this three different ways: factor it as $(x + \sqrt{-1})\ (x - \sqrt{-1}) = 0$, use the quadratic equation, or just subtract 1 from each side to get $x^2 = -1$. But the solution, $\sqrt{-1}$, doesn't make any sense. It would be a silly waste of time." $\sqrt{-1}$ doesn't make sense for counting objects (integers: 1,2,3,...), parts of object (fractions: $\frac{1}{2}$, $\frac{3}{4}$, $\frac{11}{9}$), or a deficit of objects (negative numbers), but we've transcended sheep herder mathematics long ago. Consider the irrational numbers (π, e, $\sqrt{2}$, etc.). These are numbers are not only "irrational" in the sense that they are neither whole (1, 62, -11) nor can be expressed as a ratio (hence, "rational") of whole numbers, but also in the sense that they don't make much sense ... in the shepherd's mind.

So $\sqrt{-1}$ isn't a "real" number in the sense that it can be used to count real objects, but we can introduce the symbol as an extension of our mathematical language. "But is it useful?" It let me solve my equation didn't it? Okay, I admit that's a pretty meager accolade. As it turns out, $\sqrt{-1}$ is a quite surprisingly, perhaps as yet unimaginably to you, powerful concept to add to the language of mathematics A full appreciation can only be had with a study of differential equations, Fourier transforms, and the like. But let me say that I'm not sure how physical science and engineering would have progressed to the degree they have without it!

B.9.b. Imaginary Numbers

As I've tried to get across, these numbers are called "imaginary" because $\sqrt{-1}$ is an extension of the language of math that doesn't make sense in regards to counting or measuring objects. After all, what number times itself gives -1? If you think of this in terms of counting sheep, it's absurd. And yet this number system turns out to be a powerful tool for analyzing physical systems. Go figure.

An imaginary number is quite simply any "real" number times the "artificially" constructed "number" i. (In electrical engineering, they sometimes use j and reserve i for electric current. The meaning is still the same.) Since there's nothing more to it, take a gander at these little examples:

$2i + 3i = 2\sqrt{-1} + 3\sqrt{-1} = 5\sqrt{-1} = 5i$

$\pi i + ei = (\pi + e)i$

$\sqrt{-4} = \sqrt{(4)(-1)} = \sqrt{4}\sqrt{-1} = 2\sqrt{-1} = 2i$ (meaningless in the "real" number system)

$i^2 = i\,i = \sqrt{-1}\sqrt{-1} = -1$, a real number! i undid itself. Sassy!

$i^3 = i\,i\,i = i^2\,i = -i$

$i^4 = i\,i\,i\,i = i^2\,i^2 = 1$

$i^5 = i^2\,i^3$ or $i\,i^4 = i$, and we're back.

$2i\,\pi i = -2\pi$

All I'm doing is pretending $\sqrt{-1}$ makes sense. All the other rules of mathematics still apply. One question remains: how do we add a real number (like 9, $5^{1/2}$, or π) to an imaginary number? The answer is we just do ... and we don't.

B.9.c. Complex Numbers

A complex number is a bit more general than imaginary numbers and answers the last question in a somewhat less cryptic fashion. Here are some complex numbers

$2 + i = (2 + i)$, no more reduction is possible.

6i, imaginary numbers are included in the complex number system.

12, real numbers are included in the complex number system

That about handles the basic definition. A complex number is any combination of real and imaginary numbers. Happily, the basic rules of algebra still apply, except that now we can say that we have a solution for equations like $x^2 + 1 = 0$.

$(2 + 3i)(1+4i) = 2 + 3i + 8i + -12 = -10 + 11i$

$(a + bi)(a - bi) = a^2 + b^2$!! Prove it. These two factors are called "complex conjugates" of one another. The product always gives a real number.

$(3i + 10) - (1 - 8i) = 9 + 11i$

Much ado about nothing? Well, you're right to think that there isn't much to the basic definitions. But the power as a computational tool is certainly veiled by those simple definitions. The following section will start to give you an idea of what uses brilliant minds have concocted. When you incorporate these numbers into calculus, you really find some astounding results. But don't worry, we won't do that in this course. If you're interested, take a complex analysis course.

B.9.d. The Complex Plane & Phasors

"We're going to be entering the complex plane Spock, so put your phasors on stun." [I've gone too far, haven't I?]

Just notice that a complex number has a real part (the part without the i) and an imaginary part (multiplied by i). When we add or subtract complex numbers, we add or subtract the real parts and the imaginary parts separately. This is much like adding vectors in component form: $(3\,\hat{\mathbf{i}} - 2\,\hat{\mathbf{j}}) + (\hat{\mathbf{i}} + 7\,\hat{\mathbf{j}}) = 4\,\hat{\mathbf{i}} + 5\,\hat{\mathbf{j}}$. Since we treat the real and imaginary part as separate parts (components), we can *represent* a complex number graphically by letting the real and imaginary parts correspond with different axes. [It's a made-up representation.]

The "Complex Plane"

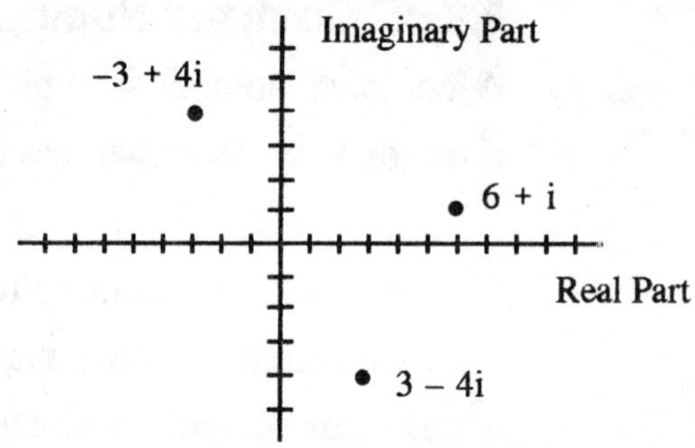

If we draw in a radial arm that extends out to the point, we see that it can be thought of as having real and imaginary components. This radial arm is sometimes called a *phasor*.

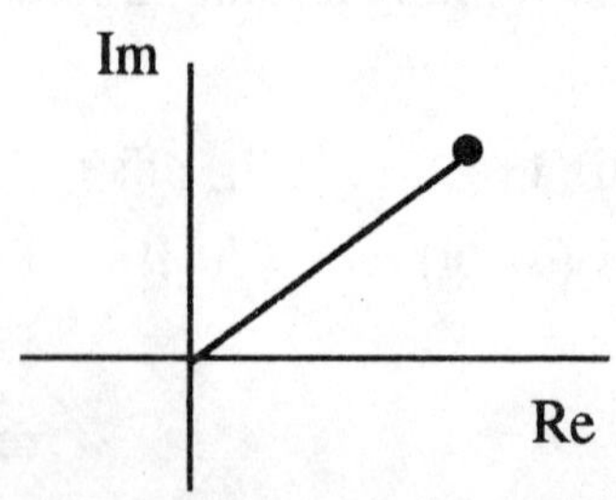

Imagine the phasor rotating around in a circle as time progresses. Imagine the shadow (real component) on the horizontal axis. Do you see that the real component varies from positive to zero to negative to zero to positive as the phasor moves from the first through the fourth quadrants? You could say that the real value oscillates between positive and negative values. (So does the imaginary part.) Thus we could, if we found it useful, use the real part of a rotating phasor to represent any quantity that varies cyclically between positive and negative values. You can bet it's useful or I wouldn't be wasting ink. I will leave the specific applications for your later studies. I simply wanted to introduce this idea now as a preview.

If the phasor has a magnitude (length) of r, it will trace out a circle of radius r as it rotates around. If $r = 1$ we have a unit circle. Either way, I hope it is easy for you to see that the components can also be written in terms of the angle that the phasor makes with the positive real (x) axis. This is just like what we did before with the unit circle, only now we aren't using unit vectors to distinguish the components but are using that simple little addition of "i." We can then write a complex number, represented by a point in the "complex plane" with sine and cosine functions as shown below.

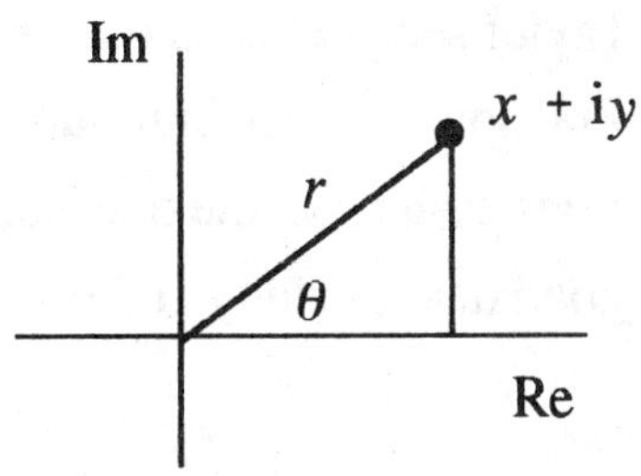

$$x + \mathrm{i}y = r\cos\theta + \mathrm{i}\, r\sin\theta$$

$$\text{where } r = \sqrt{x^2 + y^2}$$

This is just a useful representation that allows us to visualize some processes more easily. It also allows us to visualize computations using the complex numbers system. I realize that the point of going to all of this trouble is probably not clear to you now. But if you ever take a course where you use this stuff extensively, you will come to appreciate it. (That's enough of an introduction for now.)

B.10. SERIES EXPANSIONS (Cut!)

There has been much concern on the cutting edge of theoretical baseball that the World Series is too short. It has been suggested that the series be expanded to 8 games. This would allow for a tie and perhaps stimulate greater sportsmanship by de-emphasizing the "winning is everything" mind set. In physics and math, however, we suggest expanding the series to infinity and cutting it off wherever it is convenient and appropriate. Somebody stop me!

What I'm talking about here is making life easier in the long run (not to be confused with the home run). Suppose you were analyzing a situation and you came up with an equation that was very difficult, even impossible, to solve. What would you do? (All right, all right, no computers allowed just yet.) How about if you could write a simpler equation that gave results that were good enough for your purposes? ("Good enough" means that if you are measuring the dimensions of a room to install some carpeting, you needn't measure to within 10^{-4} meters.) This approach is often taken in physics and applied math, for instance, in writing computer programs for numerical analysis.

If you haven't studied Taylor series expansions or Fourier series expansions yet, that's okay, you eventually will. For now I'll just introduce you to a few identities that will, on occasion, make our calculations easier. On the one hand they are really quite incredible. On the other hand, and after a good deal of study and use, they make perfect sense.

$$\text{For all } x, \quad \sin x = x - \frac{x^3}{3!} + \frac{x^5}{5!} - \frac{x^7}{7!} + \ldots$$

$$\text{For all } x, \quad \cos x = 1 - \frac{x^2}{2!} + \frac{x^4}{4!} - \frac{x^6}{6!} + \ldots$$

$$\text{For all } x, \quad e^x = 1 + x + \frac{x^2}{2!} + \frac{x^3}{3!} + \frac{x^4}{4!} + \ldots$$

$$\text{For } |x| < 1, \quad (1+x)^n = 1 + nx + \frac{n(n-1)}{2!}x^2 + \ldots$$

$$\text{For } |x| < 1, \quad \frac{1}{\sqrt{1-x}} = 1 + \frac{1}{2}x + \frac{1\cdot 3}{2\cdot 4}x^3 + \frac{1\cdot 3\cdot 5}{2\cdot 4\cdot 6}x^5 \ldots$$

Pretty weird, huh? You'll study where these came from in your third semester of calculus. For now, just look at what these expressions can do for us. As is often the case, clever advanced math can make complex problems much easier and previously unsolvable problems solvable.

EXERCISE B.18: Are these really identities? How good are they if we "truncate" them, i.e. cut them short by dropping the "higher order" terms? Check it out!

(a) We will often use the approximation that $\sin\theta \approx \theta$ for small angles. See how well $\sin\theta \approx \theta$ works for $\theta = 5°$.

(b) I tricked you! Did you catch it? θ better be in radians or the relation won't work. Now convert 5° to radians and see how well the approximation $\sin\theta \approx \theta$ works. Also try $\theta = \pi/2$. For what range of angles do you think the approximation $\sin\theta \approx \theta$ is reasonable?

(c) Now check out that e^x expansion. Try $x = 0, 0.5, 1, -2$, and 10. Do you see that a larger exponent requires more terms for the right hand side expansion to approach the value of the function? Thus $e^x \approx 1 + x$ might be an acceptable approximation if x is small enough, where "small enough" is determined by the accuracy that you need to maintain.

(d) Still hanging in there? Okay, one more. As it turns out, $(1 - x)^{-1/2}$ is a very important function in Einstein's theory of special relativity. So let's take a look. First try $x = 10$. ¿Que pasa? We'd better stick to the restricted domain. Just for kicks, let $x = v^2/c^2$, where c = speed of light = 3×10^8 m/s. This gives you the relativity relation that you'll see later on. Since nothing travels faster than the speed of light, try $v = 0.5$ c How many terms satisfy your need for accuracy? [Did you plug in the value for c or did you just cancel the c's?]

[I leave the rest to you. Play around with some numbers.]

EXERCISE B.19: Use only the terms shown in the series expansions above.

(a) Show that the derivative of $\cos x$ is $-\sin x$ and the derivative of $\sin x$ is $\cos x$. You could also work backwards and show that the integral of $\sin x$ is $\cos x$, etc.

(b) Show that the derivative of e^x is e^x. Show that the derivative of e^{9x} is $9\,e^{9x}$.

Bonus Preview Finale (not required)

Want to see something clever that's used all of the time in more advanced work? You know how easy it is to work with e^x, right? After all, $d(e^x)/dx = e^x$, and so on. Well, if you plug in $i\theta$ for x in the series expansion for e^x, look what you get!

$$e^{i\theta} = \cos\theta + i\sin\theta$$

$$= x + i\,y \text{ , on the unit circle } (r = 1) \text{ in the complex plane.}$$

I just grouped the terms with even powers of x and got the cosine expansion (since $i^2 = -1$, $i^4 = i^2\,i^2 = +1$, etc.) and the terms with odd powers of x and got i times the sine expansion (since $i^3 = -i$, $i^5 = +i$, etc.). Thus we end up with an alternate way of describing the position of a point in the complex plane that we just studied. This turns out to be eggggstreeeemely helpful for more advanced analysis. It's called Euler's equation. [If you're looking for a tattoo, that might be the one.]

Hey, good work maestro!

Exercise Solutions

EX. B.1:

(a) **False**. You just can't split this up. If you had your heart set on splitting this fraction up into two fractions, you could multiply top and bottom by (A - B) to obtain:

$$\frac{1}{A+B}=\frac{1}{A+B}\cdot\frac{A-B}{A-B}=\frac{A-B}{A^2-B^2}=\frac{A}{A^2-B^2}-\frac{B}{A^2-B^2}$$

(b) **True**.

(c) **True**.

(d) **False**. You can't treat the denominator like the numerator. That is, you can't just split it up. I hope you understand this more deeply than merely remembering the rules. Just make up a simple example. Is $^1/(2+3) = {}^1/2 + {}^1/3$? Of course not. Try a few more if you still don't get it. The answer to (c) is still the correct one.

(e) **True**! This needs to be obvious! Just multiply the top and bottom by B (do it in your head).

(f) **True**! This must also be basic. Multiply top and bottom by A.

(g) **False**. The left side is equivalent to $A \cdot B^2$.

(h) NO WAY! It's **False**. Look at the right side. When you multiply those two fractions together, you multiply the numerators to get $x \cdot y$ (which is fine), and you multiply the denominators together to get a denominator of $(z + d^2)^2$. Better to leave the left side as it is.

(i) **True**. In order to add q to the first term, you need to get a common denominator. To do this, we multiply q by $(z + d^2)/(z + d^2)$. Then you can put it all over the same denominator.

(j) Almost, but no cigar. **False**. To multiply fractions together, multiply the tops together and multiply the bottoms together. The expression on the bottom is correct (needs to be obvious). The top is wrong because the F would need to multiply both term.

(k) **False**. Take a look:

$$\frac{1/B}{A^2-C^2}=\frac{1}{B\left(A^2-C^2\right)}=\frac{1}{B(A+C)(A-C)}$$

(l) Can you trust your algebra? When the expressions are long and ugly, just take them one step at a time. Proceed carefully, but do develop speed. You could see right away that the first R's cancel and the r's add up to $2r$, and then deal with the rational expressions. That's fine. I'll do it as follows:

$$\left(R+r-\frac{r^2-R^2}{R+r}\right)-\left(R-r-\frac{r^2-R^2}{R-r}\right)=\left(R+r-\frac{(r-R)(r+R)}{R+r}\right)-\left(R-r-\frac{(r-R)(r+R)}{R-r}\right)$$

$$=(R+r-(r-R))-(R-r+(r+R))$$

$$=(R+r-r+R)-(R-r+r+R)$$

$$=(2R)-(2R)=0\ !$$

EX. B.2:

(a) $$\frac{F\cdot z}{A}-c^2D=Q\sqrt{x}\xrightarrow{\text{quick \#1}}\frac{F\cdot z}{A}=Q\sqrt{x}+c^2D\xrightarrow{\text{quick \#2}}F=\frac{A}{z}\left(Q\sqrt{x}+c^2D\right)$$

(b)

$$\frac{F\cdot z}{A}-c^2D=Q\sqrt{x}\xrightarrow{\text{quick \#1}}\frac{F\cdot z}{A}-Q\sqrt{x}=c^2D\xrightarrow{\text{quick \#2}}c^2=\frac{1}{D}\left(\frac{F\cdot z}{A}-Q\sqrt{x}\right)$$

$$\xrightarrow{\text{take root}}C=\pm\sqrt{\frac{1}{D}\left(\frac{F\cdot z}{A}-Q\sqrt{x}\right)}$$

(c) $$\frac{F\cdot z}{A}-c^2D=Q\sqrt{x}\xrightarrow{\text{quick \#2}}\frac{1}{Q}\left(\frac{F\cdot z}{A}-c^2D\right)=\sqrt{x}\xrightarrow{\text{square}}x=\frac{1}{Q^2}\left(\frac{F\cdot z}{A}-c^2D\right)^2$$

(d) $$\frac{F\cdot z}{A}-c^2D=Q\sqrt{x}\xrightarrow{\text{quick \#1}}\frac{F\cdot z}{A}=Q\sqrt{x}+c^2D\xrightarrow{\text{quick \#2}}\frac{F\cdot z}{Q\sqrt{x}+c^2D}=A$$

EX. B.3: What did you discover? I can't show you every possible way of doing this, so if you did it another, that's fine ... if you were careful.

(a) I notice that the first equation has an I_2R_2 but no I_3R_3. The second equation is close to the first, but with an I_3R_3 and no I_2R_2. The third equation has both I_2R_2 and I_3R_3. Play around a bit and discover that (eq'n 2) + (eq'n 3) = (eq'n 1). Alternatively, (eq'n 1) - (eq'n 3) = (eq'n 2), (eq'n 1) - (eq'n 2) = (eq'n 3).

Only 2 independent equations ⇒ can solve for only 2 unknowns.

(b) The last equation doesn't have an A or a B. So let's try to combine the first two to get rid of those variables. Hmm... (eq'n 1) + (eq'n 2) ⇒ $2A = C + D$. Further playing (or is it obvious?) shows that there is no way to get any of these from the others.

3 independent equations ⇒ can solve for 3 unknowns.

(c) Don't jump to conclusions. Are you sure that these are all variations of the same equation? Maybe one was written incorrectly. Well, OK, no algebra errors here.

Only 1 independent equation ⇒ can solve for only 1 unknown.

EX. B.4:

1. Guessing the logarithm without a calculator.

a) $\log 100 = \mathbf{2}$, since $10^2 = 100$.

b) $\log_2 32 = \mathbf{5}$, since $2^5 = 32$.

c) $\ln e^4 = \mathbf{4}$, of course.

d) I'd guess $\log_\pi 9.8696 = \mathbf{2}$, since $3^2 = 9$, π^2 must be a little more than 9. Of course, that's right.

e) Absurd! 10 to what power is -10? No power will make it negative. In fact, no positive number can be raised to a power to obtain a negative number ... Unless you want to try and advance a new concept in mathematics.

2. Using a calculator and doing a rough mental check as in d above.

a) $\log \pi = 0.50 \Rightarrow 10^{0.50} = \pi$. Since $10^1 = 10$, we'd expect a fractional exponent.

b) $\ln \pi = 1.1 \Rightarrow e^{1.1} = \pi$. This makes sense too, since now the base is less than π. Here, $e^1 = e = 2.7$.

c) $\log e = 0.43 \Rightarrow 10^{0.43} = e$. Again, the base (10) is less than the number e.

d) $\ln e = 1 \Rightarrow e^1 = 1$. Now obvious?

e) $\log 200 = 2.3 \Rightarrow 10^{2.3} = 200$. Since $10^2 = 100$, and $10^3 = 1000$, we'd expect a result just over 2.

f) $\ln 200 = 5.3 \Rightarrow e^{5.3} = 200$. Very roughly, $e \approx 3$, and $3^2 = 9$, so $3^4 = 3^2 \cdot 3^2 = 9 \cdot 9 = 81$,

so $3^5 = 3 \cdot 81 \approx 240$, so we'd expect it to be around 5!

g) $\log 1 = 0 \Rightarrow 10^0 = 1$. You knew that $10^0 = 1$, so did you realize that $\log 1 = 0$?

h) $\ln 1 = 0 \Rightarrow e^0 = 1$. You knew that $b^0 = 1$, so did you realize that $\log_b 1 = 0$ for any base?

i) $\log(1/2) = -0.30 \Rightarrow 10^{-0.30} = 1/2$. Notice that $10^1 = 10$, $10^0 = 1$, $10^{-1} = 0.1$, and $10^{-2} = 0.01$

j) $\ln(1/2) = -0.69 \Rightarrow e^{-0.69} = 1/2$. In general, $\log_b$(positive number, less than 1) = neg. #!!

l) I'll just remind you here that $\log_b$(negative number) is UNDEFINED. Makes sense, no?

EX. B.5:

a) Hey, wait! First you say "Sorry, no can do!! Just can't split additive terms." True, but are there any common multiplicative factors? Yes indeed:

$$\ln (b^2 - 2b) = \ln[b(b-2)] = \ln b + \ln (b - 2)$$

b) Just how far are you willing to go with this?

$$\ln \lfloor x/(y + 3)^2 \rfloor = \ln x - \ln\lfloor (y + 3)^2 \rfloor = \ln x - 2\ln\lfloor (y + 3) \rfloor$$

Notice that you could have used $\ln[(y + 3)^2] = \ln[(y + 3)] + \ln[(y + 3)]$ ... that's where the rule for exponents in the argument ($\log_b x^n = n \log_b x$) comes from!

c) To pull that t out, we must undo the exponentiation. So, we use the "undo the exponent function" and remember that we retain the equality whenever we do the same thing to both sides.

Method 1: $N_o e^{kt} = R \Rightarrow \ln [N_o e^{kt}] = \ln [R] \Rightarrow \ln [N_o] + \ln [e^{kt}] = \ln [R]$

$\Rightarrow \ln [e^{kt}] = \ln [R] - \ln [N_o] = \ln [R/N_o]$

$\Rightarrow kt \ln [e] = \ln [R/N_o] \Rightarrow$ $\quad \boldsymbol{t = k^{-1} \ln [R/N_o]}$

Method 2: $N_o e^{kt} = R \Rightarrow e^{kt} = R/N_o \Rightarrow \ln [e^{kt}] = \ln [R/N_o] \Rightarrow kt = \ln [R/N_o] \ldots$

d) How did you do it? I rewrote it as: $\log_a b \log_b N = \log_a N$. As a memory aid, notice that the b's, in a sense, cancel. Let's see why it's true.

$$\log_a b \log_b N = \log_a N \Rightarrow N = a^{(\log_a b)(\log_b N)} = \left(a^{\log_a b}\right)^{\log_b N} = b^{\log_b N} = N!!$$

EX. B.6: Make a quick sketch if you need to. Then notice if the sign of the coordinates of the end of the radial line. As you know: If $x > 0$, then $\cos\theta > 0$. If $y > 0$, then $\sin\theta > 0$. If x and y have the same sign, $\tan\theta > 0$. If x and y have opposite signs, $\tan\theta > 0$.

EX. B.7:

a) 220° is in the third quadrant, so the x part (cos 220°) and the y part (sin 220°) are negative. Since right triangle angles give positive trig values, we'll need to put in the negative sign for each part. The result should be evident from a quick sketch:

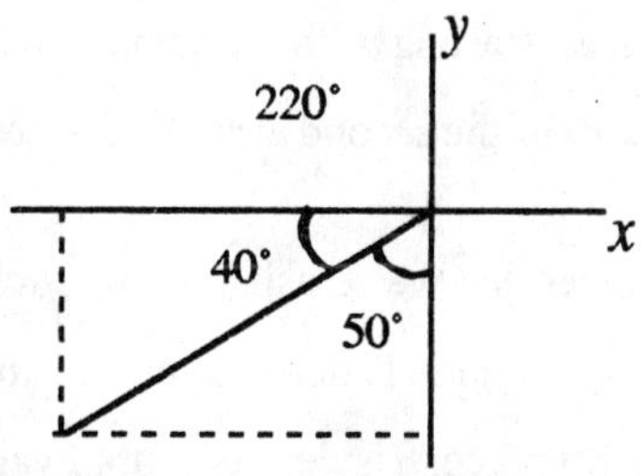

b) In the fourth quadrant, $x > 0$, $y < 0$, so cos 340° > 0, and sin 340° < 0. The x length is opposite the 80° angle, so it is + sin 80°, and adjacent to the 20° angle, so it's cos 20°. The y length is opposite the 20° angle and negative, so it is – sin 20°, and it is adjacent to the 80° angle, so it's – cos 80°. Can you find other 80° and 20° angles that are not marked in the triangles? Sure, I knew you could.

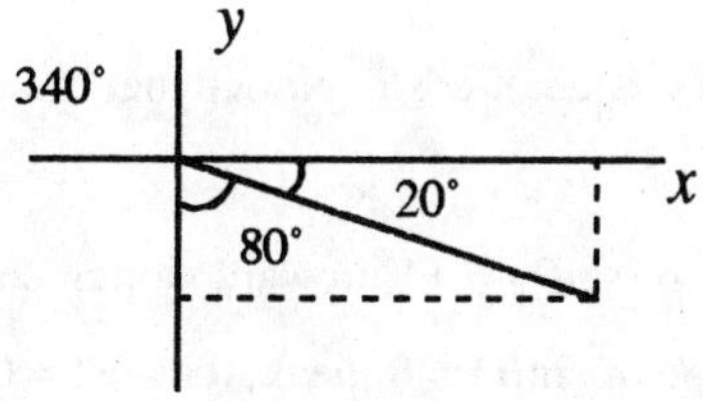

c) Let's place the unit radial arm in two locations – at any angle θ and then at the angle $\theta + 90°$. The statement that $\cos\theta = \sin(\theta + 90°)$ should be reflected in the unit circle by the fact that the x component when the arm is at θ is equal in length to the y component when the arm is at $\theta + 90°$. Let's make a sketch that clearly illustrates this. (It's generally good to avoid using 45°; do you see why?)

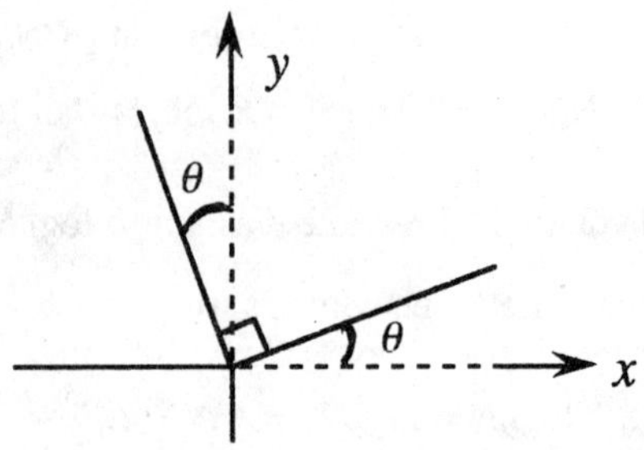

Now imagine the two arms rotating around to different angles, but keeping the relation of θ and $\theta + 90°$. Can you visualize this? When x is long for the first, y is long for the second. Furthermore, when x is + (or –) for the first, y is + (or –) for the second. Challenge: can you see how $\sin\theta$ and $\cos(\theta + 90°)$ are related? If so, you're smokin'!! [Hint: in the challenge, one of them needs a minus sign. Which?]

d) Now we draw one arm at θ and the other back 90° behind the first. Which components of each arm are we looking at? The sine of the first (so y of that one), and the cosine of the second (so x of that one). Are they equal in length? Did I forget a minus sign somewhere? You make the sketch and decide. I suggest starting with the same radial arm that we started with above and then going back 90° to 4th quadrant for the second arm. Again, see if you can now relate $\cos\theta$ and $\sin(\theta + 90°)$.

EX. B.8: Remember that we're using a unit circle so that $r = 1$ and $y = \sin\theta$ (not $r\sin\theta$).

a) $-1 \le y \le +1$, and $\sin\theta$ is defined for all values of θ.

b) $-1 \le x \le +1$, and $\cos\theta$ is defined for all values of θ.

c) $-\infty \le y/x \le +\infty$, so $\tan\theta$ is undefined when $x = 0$, i.e. when $\theta = \pi/2,\ 3\pi/2,\ 5\pi/2$, etc.

d) $-\infty \le x/y \le +\infty$, so $\cot\theta$ is undefined when $y = 0$, i.e. when $\theta = 0,\ \pi,\ 2\pi,\ 3\pi$, etc.

e & f) I know you can take it from here!

EX. B.9: I really hope that you did this one ... physically.

EX. B.10:

a) Slope of $\sin\theta$ vs value of $\cos\theta$. Notice that the slope of $\sin\theta = +\cos\theta$. This implies that the derivative of $\sin\theta = +\cos\theta$!

$\theta = 0$: (slope of $\sin\theta$) = +1, upward slope; $\cos 0 = +1$

$\theta = \pi/2$: (slope of $\sin\theta$) = 0, peak; $\cos \pi/2 = 0$

$\theta = \pi$: (slope of $\sin\theta$) = –1, downward slope; $\cos \pi = -1$

$\theta = 3\pi/2$: (slope of $\sin\theta$) = 0, trough; $\cos 3\pi/2 = 0$

b) Slope of $\cos\theta$ vs value of $\sin\theta$. Notice that the slope of $\cos\theta = -\sin\theta$. This implies that the derivative of $\cos\theta = -\sin\theta$!

$\theta = 0$: (slope of $\cos\theta$) = 0, peak; $\sin 0 = 0$

$\theta = \pi/2$: (slope of $\cos\theta$) = −1, downward slope; $\sin \pi/2 = +1$

$\theta = \pi$: (slope of $\cos\theta$) = 0, trough; in $\pi = 0$

$\theta = 3\pi/2$: (slope of $\cos\theta$) = +1, upward slope; $\sin 3\pi/2 = -1$

EX. B.11: There are two ways to look at this. It's important to understand both. You may want to come back to this when it shows up later on. At least preview it now.

METHOD 1: Pick any value for θ. Look at one curve at $\theta \pm \pi/2$ ($+\pi/2$ means that you look one quarter cycle to the right, while $-\pi/2$ means that you look one quarter cycle to the left), then look at the other curve at θ. Personally, I think $\theta = 0$ or $\pi/2$ are the easiest choices, but anything will work.

a) $\sin(0 + \pi/2) = +1 = +\cos 0 \Rightarrow \sin(\theta + \pi/2) = +\cos\theta$

b) $\sin(0 - \pi/2) = -1 = -\cos 0 \Rightarrow \sin(\theta - \pi/2) = -\cos\theta$

c) Since $\sin 0 = 0$, we won't be able to tell the sign if we use $\theta = 0$, so let's go with $\pi/2$ here.

$\cos(\pi/2 + \pi/2) = -1 = -\sin \pi/2 \Rightarrow \cos(\theta + \pi/2) = -\sin\theta$

d) $\cos(\pi/2 - \pi/2) = +1 = +\sin \pi/2 \Rightarrow \cos(\theta - \pi/2) = +\sin\theta$

METHOD 2: This is kind of the reverse of method 1. Rather than shifting where we look on the horizontal axis, we'll shift the curve instead. Here's a simple exercise that will help you see what's going on. Quickly sketch the lines $y = x$, $y = x + 1$, $y = x - 1$. (Simply plot two points. First, use $x = 0$, then, use $y = 0$.) Notice that adding 1 shifts the curve *left*, not right! Why? The +1 adds to the value, so you don't need as big an x to obtain the same y as for either of the other two lines. Think about it until you get this point! If you see this, you can see that …

a) Adding $\pi/2$ to the "argument" of the sine, $\sin(\theta + \pi/2)$, shifts the curve LEFT. If you look at the curves and imagine shifting the sine curve back a quarter of a cycle, you'll see that it will lay on top of the cosine curve. Thus: $\sin(\theta + \pi/2) = +\cos\theta$

b) Now we subtract $\pi/2$ from the argument of the sine and shift it RIGHT a quarter cycle. Do you see that this puts the peak of the sine where the trough of the cosine is? Then you see that: $\sin(\theta - \pi/2) = -\cos\theta$

c & d) You check 'em!

Be sure to notice that adding or subtracting a set amount from the argument shifts the curves one way or the other. ADDITION to the argument shifts the curve to the LEFT, while SUBTRACTION from the argument shifts the curve to the RIGHT.

EX. B.12: Only by grinding away at this stuff will you get any good. Note that parts (a) and (b) reflect what we did in exercise B.11.

a) Notice in the angle addition formulae that we have two things to get right, the sign between the terms and whether it's sin · sin (or cos · cos) or sin · cos (or cos · sin). In each of the four cases, one thing is the same while the other is different. That is, if the sign on the left is the same as that on the right, we multiply different functions together, and vice-versa. See the pattern? ... (OK, we do have to worry about where the α and β go too, but you can see that pattern too, no?)

$$\sin(\theta + \pi/2) = \sin\theta \cos \pi/2 + \sin \pi/2 \cos\theta = \sin\theta \cdot (0) + (+1) \cdot \cos\theta = \cos\theta$$

b) Hmmm, cosine → "complimentary" (opposite) sign, so must be the same functions multiplied together. It's a cosine, so start with the cosine term first.

$$\cos(\theta - \pi/2) = \cos\theta \cos \pi/2 + \sin \pi/2 \sin\theta = \cos\theta \cdot (0) + (+1) \cdot \sin\theta = \sin\theta$$

c) If you want a hint, multiply the top and bottom by $(\cos\alpha \cos\beta)^{-1}$. Now go give it a shot and don't come back until you're done! :-)

$$\frac{\sin(\alpha \pm \beta)}{\cos(\alpha \pm \beta)} = \frac{\sin\alpha \cos\beta \pm \sin\beta \cos\alpha}{\cos\alpha \cos\beta \mp \sin\alpha \sin\beta}$$

$$= \frac{\sin\alpha \cos\beta \pm \sin\beta \cos\alpha}{\cos\alpha \cos\beta \mp \sin\alpha \sin\beta} \cdot \frac{\frac{1}{\cos\alpha \cos\beta}}{\frac{1}{\cos\alpha \cos\beta}}$$

$$= \frac{\frac{\sin\alpha}{\cos\alpha} \pm \frac{\sin\beta}{\cos\beta}}{1 \mp \frac{\sin\alpha \sin\beta}{\cos\alpha \cos\beta}} = \text{The answer! See it?}$$

d) I'll prove one, then you can do the other:

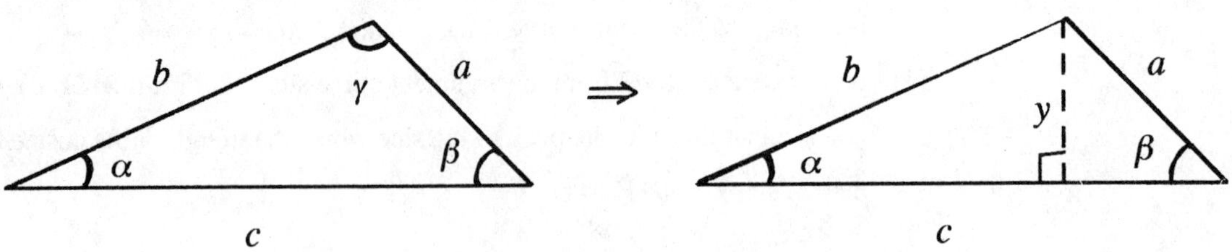

Notice that $\sin\alpha = y/b$ so $y = b \sin\alpha$. Similarly, $y = a \sin\beta$.

Setting these equal gives $b \sin\alpha = a \sin\beta \Rightarrow \dfrac{\sin\alpha}{a} = \dfrac{\sin\beta}{b}$ ... Now you finish it up.

EX. B.13:

Vectors: position, velocity, force, momentum, magnetic field (all are defined with an associated direction)

Non vectors: time, distance, speed, # of people, energy (no direction to point in for these)

EX. B.14: The vectors are drawn and the coordinates and axes are chosen. What's the next step? (Memorize the steps! It'll really save you time.) Below I sketch the grids and notice the triangles (when there are triangles). Then I use trig and be careful to note the signs.

a)

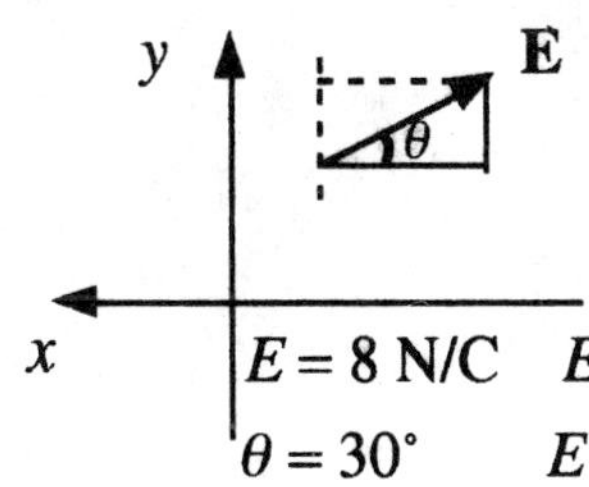

$E = 8$ N/C $E_x =$ ____

$\theta = 30°$ $E_y =$ ____

- $Ex = (8\text{ N/C})\cos 30° = 7$ N/C
- $Ey = (8\text{ N/C})\sin 30° = 4$ N/C

b)

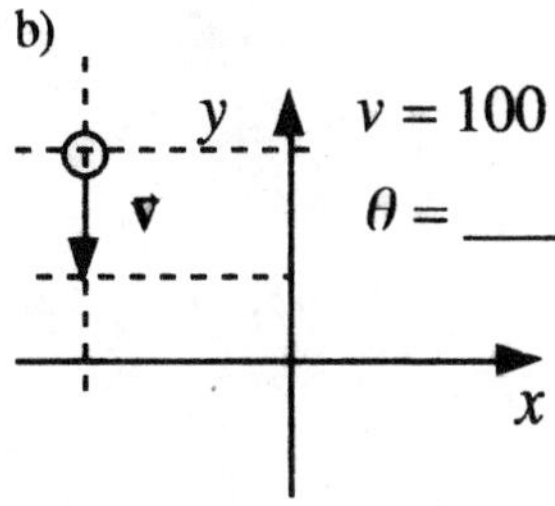

$v = 100$ m/s $v_x = 0$

$\theta =$ ____ $v_y = -100$ m/s

So what's the angle? It doesn't really matter if see the components. You could say that = + 270° or — 90° from the +x axis, 180° from the + y axis, or choose another reference direction.

c)

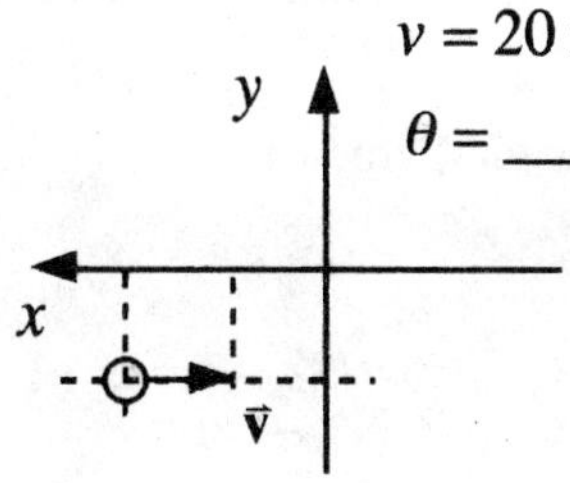

$v = 20$ m/s $v_x = -20$ m/s

$\theta =$ ____ $v_y = 0$ m/s

d)

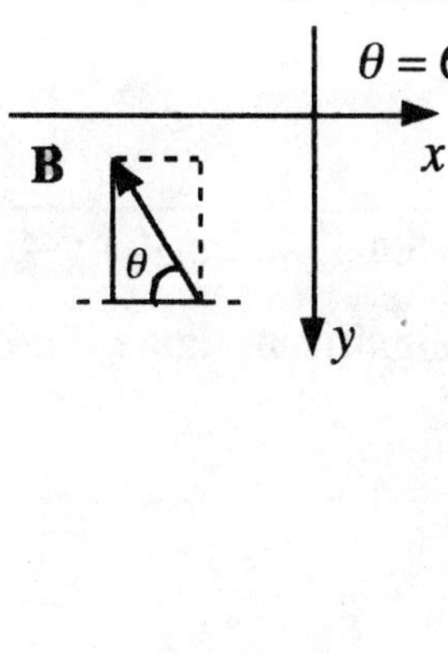

$B = 0.5$ gauss $B_x =$ ____

$\theta = 60°$ $B_y =$ ____

- $Bx = -(0.5\text{ gauss})\cos 60° = -0.25$ gauss
- $By = -(0.5\text{ gauss})\sin 60° = -0.43$ gauss

e)

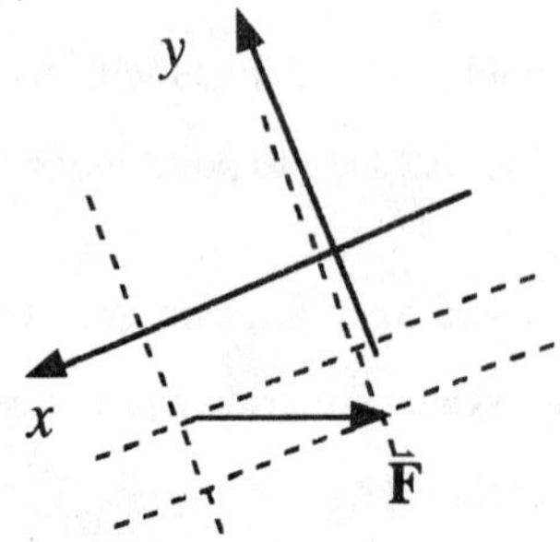

It's easier to see if you expand the diagram!

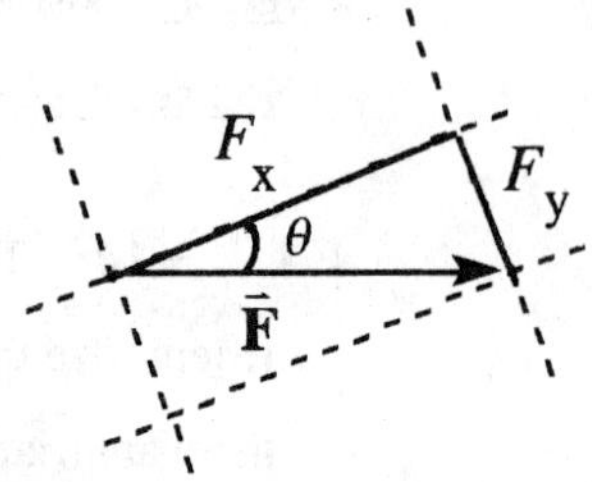

Make sure you get both the magnitudes (via the trig functions on the right triangle) and the directions (via +/–) correct. In unit vector notation, we would write:

$$\vec{\mathbf{F}} = F_x\,\hat{\mathbf{i}} + F_y\,\hat{\mathbf{j}} = [(222\text{ N})(\sin 80^\circ)]\,(-\hat{\mathbf{i}}) + [(222\text{ N})(\cos 80^\circ)]\,(-\hat{\mathbf{j}})$$

$$= (-219\text{ N})\,\hat{\mathbf{i}} + (-39.5\text{ N})\hat{\mathbf{j}}$$

I used the 80° angle just to show you that you can. The result will be the same with the 20° angle, you'll just reverse the trig functions that are used.

EX. B.15: Piece of cake!

a) (23 m/s, east) b) (22 m/s, east) c) Yes, if you walk at the same speed, opposite direction.

EX. B.16: Notice: no direction in the result!

a) With the magnitudes and angle between the vectors, the choice is obvious:

$$\vec{\mathbf{F}}\cdot\vec{\mathbf{d}} = |\vec{\mathbf{F}}||\vec{\mathbf{d}}|\cos\theta_{\text{between them}} = (40\text{ N})(2\text{ m})\cos 60^\circ = 40\text{ N}\cdot\text{m}$$

b) With the components given, it's easy to find the scalar (dot) product!

$$\vec{\mathbf{F}}\cdot\vec{\mathbf{d}} = F_x d_x + F_y d_y + F_z d_z$$

$$= (2\text{ N})(-5\text{ m}) + (10\text{ N})(1\text{ m}) + (-3\text{ N})(-4\text{ m})$$

$$= 12\text{ N}\cdot\text{m} \quad (\text{or } 10\text{ N}\cdot\text{m, to only 1 significant figure})$$

c) We want the angle, so we may as well start with the relation that includes it. Notice that I use the Pythag. th'm to find the magnitudes:

$$\vec{\mathbf{F}}\cdot\vec{\mathbf{d}} = |\vec{\mathbf{F}}||\vec{\mathbf{d}}|\cos\theta_{\text{between them}}$$

$$= \left(\sqrt{(2\text{ N})^2 + (10\text{ N})^2 + (-3\text{ N})^2}\right)\left(\sqrt{(-5\text{ m})^2 + (1\text{ m})^2 + (-4\text{ m})^2}\right)\cos\theta$$

$$= (68.89\ldots\text{N}\cdot\text{m})\cos\theta \quad (\text{keeping the significant figures until the end})$$

But we still don't know θ. How can we get it? Use the other method for finding the scalar product! That is, multiply parallel components and add them up. ... Hey, it just so happens that we already did that in part b with the same vectors. I'll use 12 N·m as the result, since 1 sig. fig. is really quite low. Then

$$\vec{\mathbf{F}}\cdot\vec{\mathbf{d}} = (68.89\ldots\text{N}\cdot\text{m})\cos\theta$$

$$\overbrace{12\ \text{N}\cdot\text{m}} = (68.89\ldots\text{N}\cdot\text{m})\cos\theta \Rightarrow \theta = \cos^{-1}\left(\frac{12\ \text{N}\cdot\text{m}}{68.89\ldots\text{N}\cdot\text{m}}\right) = \mathbf{80^\circ}$$

EX. B.17: Notice that the direction *is* included in the vector (cross) product...unless all we want is the magnitude. Also, be careful not to reverse the order of the product. It matters!

a) With the magnitudes and angle between the vectors, the choice is obvious:

$$|\vec{\mathbf{d}}\times\vec{\mathbf{F}}| = |\vec{\mathbf{d}}||\vec{\mathbf{F}}|\sin\theta_{\text{between them}} = (2\text{ m})(40\text{ N})\sin 60^\circ = 69\text{ N}\cdot\text{m}\ \ (70,\text{ to 1 sig. fig})$$

WAIT! Are we done? No. We only found the magnitude. What about the direction? Use a right hand rule. This is tough, since I didn't draw the vectors for you. Let's try two cases:

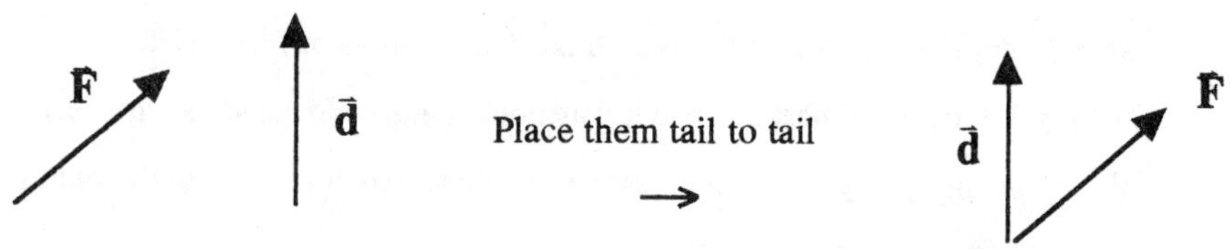

The plane of the vectors is the plane of the page. The result is either into or out of the page. Try both ways and discover that placing your thumb **into** the page allows you to push (curl) $\vec{\mathbf{d}}$ onto $\vec{\mathbf{F}}$. Let's try another arrangement of vectors:

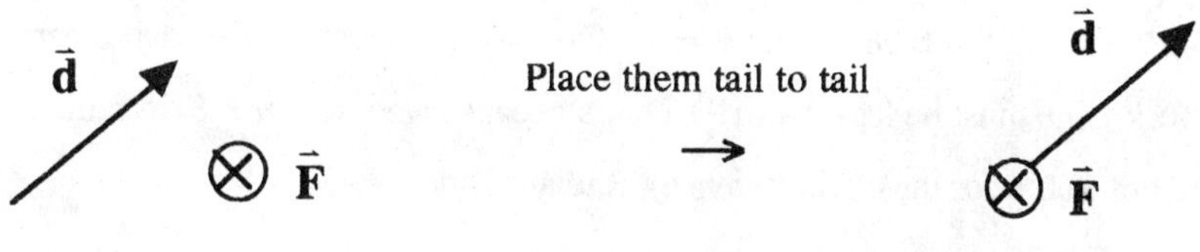

Where's the plane formed by the vectors? It's *not* horizontal, vertical, or the plane of the page. If you were to stand a piece of paper on this page so that the edge is along $\vec{\mathbf{d}}$, and imagine that plane extends into the page, you would see the plane. (A karate chop along $\vec{\mathbf{d}}$ would show it too!) The two directions perpendicular to this plane are up and to the left, and down and to the right. Now with the palm of your right hand, push $\vec{\mathbf{d}}$ onto $\vec{\mathbf{F}}$ (i.e. into the page) to discover that your thumb is **up and to the left** giving the direction of "$\vec{\mathbf{d}}$ cross $\vec{\mathbf{F}}$."

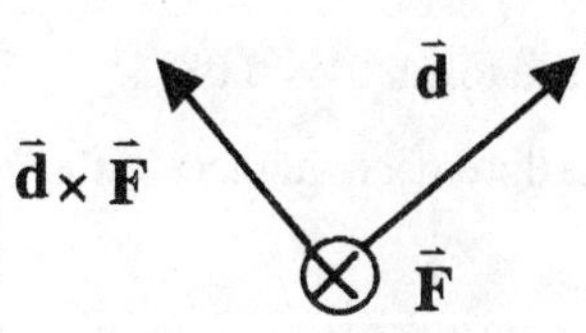

b) Use the component method! Make sure that you have gotten the hang of writing this down quickly.

$$\vec{\mathbf{F}} \times \vec{\mathbf{d}} = \hat{\mathbf{i}}\left(F_y d_z + F_z d_y\right) + \hat{\mathbf{j}}\left(F_z d_x + F_x d_z\right) + \hat{\mathbf{k}}\left(F_x d_y + F_y d_x\right)$$
$$= \hat{\mathbf{i}}((10\text{ N})(-4\text{ m}) + (-3\text{ N})(1\text{ m})) + \hat{\mathbf{j}}((-3\text{ N})(-5\text{ m}) + (2\text{ N})(-4\text{ m}))$$
$$+ \hat{\mathbf{k}}((2\text{ N})(1\text{ m}) + (10\text{ N})(-5\text{ m}))$$
$$= (-43\text{ N}\cdot\text{m})\hat{\mathbf{i}} + (7\text{ N}\cdot\text{m})\hat{\mathbf{j}} + (-48\text{ N}\cdot\text{m})\hat{\mathbf{k}}$$

That is the torque vector. Or IS IT?! I've gone and multiplied the vectors in the wrong order! Oh my, what can I do? Must I repeat the process? While that may be good practice, I can also just use the fact that reversing the order gives me the negative of the vector I want. Hence I just need to reverse the signs of the components and my answer will match yours. Then it's got the components right there, and you know how to get the magnitude and direction of any vector (Pythag. and inverse trig....here you could specify the angle made with the axes or whatever you like. Just make a sketch so that you can see the angles. You could also sketch the vectors to try and see the directions.)

c) Hey, did I say to equate the two expressions for the "dot" product?! Yes, the dot product. If we had not already done part b, we might as well use the component form of the scalar product, since it's easier. (Check Exercise 16, part c, since it was done there.) We *could* use the vector (cross) product by simply using Pythag to get the magnitude of each vector and then the magnitude of the result found in part b from the component method. Then,

$$\left|\vec{\mathbf{d}} \times \vec{\mathbf{F}}\right| = \left|\vec{\mathbf{d}}\right|\left|\vec{\mathbf{F}}\right| \sin\theta_{\text{between them}} \Rightarrow \sin\theta = \frac{\left|\vec{\mathbf{d}} \times \vec{\mathbf{F}}\right|}{\left|\vec{\mathbf{d}}\right|\left|\vec{\mathbf{F}}\right|} \Rightarrow \theta = \sin^{-1}\left(\frac{\left|\vec{\mathbf{d}} \times \vec{\mathbf{F}}\right|}{\left|\vec{\mathbf{d}}\right|\left|\vec{\mathbf{F}}\right|}\right)$$

EX. B.18:

a) Let's see, $\sin 5^\circ = 0.087\ldots$ and $\theta = 5^\circ$. The two are nowhere near each other! (Of course it wouldn't work, $\sin\theta$ must be less than 1!!) That's because we *must* use θ in radians (just like we must use $2\pi r$, not $360\, r$, for the circumference). Radians Rule!

b) Here we go. It doesn't matter if we use $\theta = 5^\circ$ or the equivalent radians for $\sin\theta$, since you'll get the same result (UNLESS YOU FORGET TO TELL YOUR CALCULATOR THAT YOU ARE USING RADIANS! WHEN YOU PUNCH IN A NUMBER AND PRESS "SIN," THE CALCULATOR INTERPRETS THE NUMBER AS DEGREES OR RADIANS, DEPENDING ON HOW YOU HAVE IT SET.) OK, OK, let's convert to radians.

$$5^\circ = 5^\circ \cdot \left(\frac{\pi\text{ rad}}{180^\circ}\right) = \frac{\pi}{36}\text{ rad} = 0.08726\ldots$$

Say, that is close to $\sin 5^\circ = 0.08715\ldots$ We're only off by one in the fourth decimal place. At what angle does the third decimal place differ? What about the second? How well do 15° and 30° work? Note that at $\theta = \pi/2$ rad $\approx$ 1.5 rad, the sine is 1.0, which is a pretty big error. The error gets bigger as the angle increases.

c) Let's tabulate this

$x =$	$e^x =$	$e^x \approx 1$	$e^x \approx 1 + x$	$e^x \approx 1 + x + x^2/2$
0	1	1	1	1
0.5	1.649	1	1.5	1.625 (try one more term!)
1	e = 2.718	1	2.0	2.50
–2	0.1353	1	–1	+1 (keep going!)
10	22026	1	11	61 (more terms!)

d) Use the last expansion listed. At $x = 10$, the exact function gives $(1 - x)^{-0.5} = (1 - 10)^{-0.5} = [(-9)^{0.5}]^{-1}$ ... hey, you can take the square root of a negative number! Let's see what the expansion gives: we get 1 after the first term, $1 - 5 = -4$ after the second term, $1 - 5 + \frac{3}{8}(1000) = 371$, and gibberish, gibberish, gibberish. It simply doesn't work outside the prescribed domain.

Now, we set $x = (v/c)^2$, so at half the speed of light, $v = 0.5$ c, we get $x = (0.5)^2$ and

$$\text{Exact solution:}\quad \frac{1}{\sqrt{1-x}} = \frac{1}{\sqrt{1-(v^2/c^2)}} = \frac{1}{\sqrt{1-(0.5^2)}} = 1.15$$

$$\text{Approx. solution:}\quad \frac{1}{\sqrt{1-x}} \approx 1 + \frac{1}{2}x = 1 + \frac{1}{2}\left(\frac{v^2}{c^2}\right) = 1 + \frac{1}{2}\left(0.5^2\right) = 1.125$$

Wow! Not too bad for only two terms. And look how much easier the approx. expression, $1 + \frac{1}{2}x$, is to work with. Do you think this could be useful?

EX. B.19: Using these expansions turns taking the derivative of a function into taking the derivative of terms like x^n, which is easy.

a) I'll do the first one for you, then you check the derivative of the sine.

$$\frac{d(\cos x)}{dx} = \frac{d}{dx}\left(1 - \frac{x^2}{2!} + \frac{x^4}{4!} - \frac{x^6}{6!} + \ldots\right)$$

$$= \frac{d}{dx}(1) + \frac{d}{dx}\left(-\frac{x^2}{2!}\right) + \frac{d}{dx}\left(\frac{x^4}{4!}\right) + \frac{d}{dx}\left(-\frac{x^6}{6!}\right) + \ldots$$

$$= 0 - x + \frac{x^3}{3!} - \frac{x^5}{5!}\ldots = -\left(x - \frac{x^3}{3!} + \frac{x^5}{5!}\right)$$

$$= -\sin x$$

b) I'll take the derivative of e^{9x} and leave e^x to you.

$$\frac{d(e^{9x})}{dx} = \frac{d}{dx}\left(1 + (9x) + \frac{(9x)^2}{2!} + \frac{(9x)^3}{3!} + \frac{(9x)^4}{4!} \ldots\right)$$

$$= \frac{d}{dx}(1) + \frac{d}{dx}(9x) + \frac{d}{dx}\left(\frac{(9x)^2}{2!}\right) + \frac{d}{dx}\left(\frac{(9x)^3}{3!}\right) + \frac{d}{dx}\left(\frac{(9x)^4}{4!}\right) + \ldots$$

pulling out the constants gives …

$$= 0 + 9\frac{d}{dx}(x) + \frac{(9)^2}{2!}\frac{d}{dx}(x^2) + \frac{(9)^3}{3!}\frac{d}{dx}(x^3) + \frac{(9)^4}{4!}\frac{d}{dx}(x^4) + \ldots$$

$$= 0 + 9(1) + \frac{(9)^2}{2!}(2x) + \frac{(9)^3}{3!}(3x^2) + \frac{(9)^4}{4!}(4x^3) + \ldots$$

$$= 0 + 9(1) + (9)^2(x) + \frac{(9)^3}{2!}(x^2) + \frac{(9)^4}{3!}(x^3) + \ldots$$

$$= 9\left[1 + 9x + \frac{(9x)^2}{2!} + \frac{(9x)^3}{3!} + \ldots\right]$$

$$= 9e^{9x}$$

Certainly the way to differentiate e^{9x} that you already know is easier, but you are taking a glimpse, a preview, of some powerful stuff right there.

Congratulations!

PART I . . .

I. Newtonian Mechanics

"Only with quantitative observations can one arrive at quantitative relationships, which are the heart of physics." - Feynman

Overview of Part I

The 4 "pillars" of physics create a foundation for most of the physical sciences and engineering. In the 1st semester of a physics course, you study these topics in detail. Later on, you use these ideas to study any type of wave motion (light, earthquakes, and so on), thermodynamics, electricity and magnetism, relativity, quantum mechanics, etc. Without further ado, I present to you ...

~ THE BIG-4 PILLARS OF PHYSICS ~

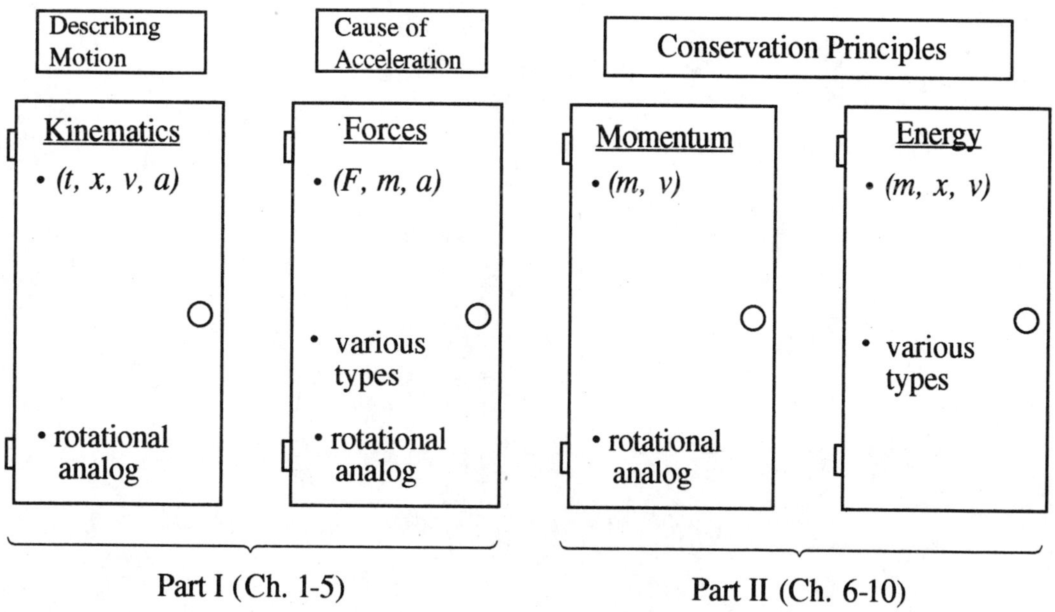

I represented these "pillars" more as doors to *tool sheds* than as pillars (support columns) because the concepts, relationships, and principles within each provide the *tools* for detailed analyses of physical systems.

When you begin a problem, you need to decide which approach you are going to attempt first. In the spirit of "top-down thinking," you need to be very clear on your decision *before* you begin. This will be easy in chapters 2 & 3, since you won't have studied what's behind door number 2 yet. But as you acquire more and more tools, you are presented with the task of selecting the right tool for the job. In many cases there is more than one way to solve a problem. Sometimes one method is easier than others. For more advanced problems you will need to pull out tools from 2 or 3 of the tool sheds in order to fully analyze the problem. In any case, always be conscious of what approach you have chosen to take at each stage of a problem.

Now let's get to work installing those pillars, tool sheds, or whatever you want to call them, in your mind. It truly is an exciting and admirable task.

Ch. 0: The Roots of Science

Overview

As twentieth and twenty-first century students of science, we spend most of our time trying to remember facts, learning about experimental techniques and equipment, learning to use the language of mathematics to make predictions from mathematical models of various systems, and … getting the best grades we can! Something is often missing in the rush to get the next assignment done. A truly remarkable, and frankly rather enchanting, breakthrough is waiting to be had, but you'll need to step back, have an espresso or herb tea, and wonder, and notice, and challenge what you take for granted. In Ch. 0, the authors share with you some of the ideas that came from all those caffè lattes throughout history. The very birth and evolution of "science," Greek for "to know," is highlighted there. It is a story of people trying to understand the *true* nature of the universe, and in the process engaging in the tug-of-war between their preconceptions (their "world view") and their desire to know what is *actually* so, even if it means throwing out cherished beliefs. In this course *you* have the opportunity to radically shift how you view the world as well as develop powerful tools that can be used to enhance our lives. I suggest that you create a brief summary of the questions and ideas that are presented in Ch. 0 and every so often go back and think about the enormous conceptual leaps that have been made and that you are making. Perhaps one night you will look up at the sky and see that the "stars" are really holes in an outer sphere through which the fire sphere shines.

Conceptual

I offer the following as an example of how you might summarize a chapter for yourself. The process of organizing the information on paper, in your own way, and in your own words (or paraphrasing), is often very helpful. See if you can find anything that I've left out. (Note that I've quoted and paraphrased the text liberally without using quotations.)

Questions

- What is the *fundamental stuff* that things are made of? (Water alone? Earth, air, fire, & water? Tiny, indivisible building blocks (called "atoms" by Democritus - hey, I took humanities courses too!)? Consciousness? Ravioli?)
- What *rules* does the physical universe follow? What are the rules on earth? What are the rules in the heavens? Are they the same or different rules?
- What are the *causes* of various occurrences?
- How can we take that land from those people? Oops, wrong subject. Although technology often plays a part.

- How or why do objects fall? What happens when an object falls?
- What's the deal with projectiles (things thrown into the air)?
- Can we predict what will happen in the future (at least with inanimate objects)?
- Why are humans so afraid of letting go of antiquated beliefs?
- If a chicken and a half could lay an egg and a half in a day and a half, how long would it take a grasshopper with a wooden leg to kick all of the seeds out of a dill pickle?

People

- **100,000 years ago** people recorded the phases of the moon (according to archaeologists).
- **Stonehenge**, apparently an early astronomical observatory, suggests a recognition and study of the natural order of the heavens. (The stones are believed to date from **1800 to 1500 BC** ... I looked it up in another source.)
- **600 BC**, Ionian Philosophers (in northern Greece and the Mediterranean coast of Turkey) thought that fixed laws might hold on Earth too. A huge conceptual leap? Cool. Also, **Thales** suggested that water was the universal substance. Clearly problematic.
- **500 BC**-ish, **Pythagoras** really got into numbers. He suggested that the properties of matter arise from the mathematical structure of the arrangement of the universal substance, not simply from the substance itself. His discovery of the simple mathematical relationships among harmonic notes certainly supports this. (Do we use math as a language to understand and describe quantitative relationships, or does the math exist in and of itself, **Plato**?)
- **350 BC**-ish, **Aristotle** (student of Plato, who was a student of Socrates)
  - gave one of the first careful discussions of the nature of *causes*
  - was first to give a complete, coherent view of the universe
  - thought the *heavens* (the realm of the gods) consisted of an ideal substance and uniformly rotating spherical shapes (perfect symmetry, befitting the gods)
  - thought *intricate spheres* carried the stars, the Sun, the Moon, and the planets

- thought that there were *four basic substances* - fire, air, water, and earth - from which all things within the Moon's sphere were composed in differing proportion
- thought that there were 4 spheres between the Moon and the Earth (surrounding the Earth), one for each of the 4 elements. The *proper place* for fire would be in the fire sphere, etc. The proper place for a composite object would depend on which element was predominant.
- thought that there were 2 types of motion. *Natural change* is when an object moves to return to its proper place. Objects do this because of their internal *potential state* (predisposition). *Unnatural/violent change* is any motion other than natural and is a result of *external causes*.
- Two big problems with Aristotle's cosmology: it couldn't make numerical predictions of the planets' motions, and couldn't explain projectile motion (it's not natural motion, but then what's the external cause that keeps pushing the thing through the air?)

Of course, this summary doesn't really do Aristotle service. If you were the first to formally try and explain everything from politics, to psychology, to physics, where would you begin? Would you realize that the physical world obeyed certain rules? What rules would you come up with to explain the planets? What about cannon balls? And microwave ovens?

- **100 AD**-ish, **Ptolemy** of Alexandria came up with a model to quantitatively predict the planetary motions. His model kept the idea of planets moving on larger spheres, but had the centers of those spheres not be at the center of the Earth, but rather move around the earth in circles. This method isn't bad, but becomes noticeably inaccurate after about 200 years.

- **1500 AD**-ish, **Copernicus** noticed that the simpler model of the planets, including the Earth, moving around the sun works well. There were objections: "But wouldn't there be enormous winds from the motion of the Earth? And if you dropped something, wouldn't it fall behind you if you were facing in the direction of the Earth's motion? Besides, this would mean that Aristotle's theories would have to be tossed out. No, we *are* the center of the universe!"

- **1600 AD**-ish, Tycho **Brahe**, an interesting character, made highly reliable observations of Mars (without a telescope, by the way!).

- **1600 AD**-ish, Johannes **Kepler** used the Copernican model to determine Mars' orbit (matched with Brahe's data) 100 times more accurately than any prior attempt, but noticed that there were still discrepancies. Came up with 3 laws that describe the motions. (It took 7 years to come up with the first 2!)
  1) Throw out the circles! Planets move on elliptical paths with the Sun at one foci.
  2) The line from the Sun to the planet sweeps out equal areas in equal times. (This effectively says that the planets move faster when they get closer to the Sun. Check it out with a simple diagram!)
  3) The time to go around once, the "period, T," squared is proportional to the average distance to the sun, "a", cubed: $T^2 \propto a^3$. (Mnemonic: think of "**tee two**" to remind you of T^2. T has the hard t sound like "two", not "three.")

  Kepler didn't know *why* the planets moved this way, just that observations show that they do. It took Newton to give more of an explanation. But we mustn't forget ...

- **1600 AD**-ish, **Galileo**, now that's a Pisa! He filled the need for a new physics required by Copernicus, Kepler, and their new astronomy. His resume reads: math prof.; inventor of precision instruments; expert in theories of fortification and the use of artillery; turned the telescope to the sky to see sun spots, lunar mountains, and the satellites of Jupiter, thereby strongly supporting the Copernican model; house sitter. Galileo
  - realized that projectile motion may be thought of as a constant velocity motion in the horizontal direction and an *independent* accelerated motion in the vertical direction. Do you see what an enormous shift in perception this is?
  - established the principle of inertia. The motion of an object requires no cause. Only the change in motion requires an external cause.
  - used a thought experiment as well as inclined planes to discover and support these things.
  - thought in terms of a **quantity of motion** which wasn't particularly well defined, but was said to be related to the height from which an object falls.
  - mistakenly thought that ideal unforced motion was in a circle around the earth. This conflicts with Kepler's elliptical orbits. We now know that unforced motion is along a straight line at a constant speed (or at rest).

Methods

- Assume that the entire universe is rational and follows *consistent rules*.
- Assume that *causality* holds.
- *Experiment*: measurement and observation to develop or test a theory, or to discover something entirely new.
- *Theory*: explanation (always quantitative in physics) of the presumed rules/principles behind a large class of phenomena. A theory may not be absolutely true, but it may be useful. Also, the theory may not apply to all phenomena.
- *Scientific process*: observe and measure phenomena, put forth a theory, test the theory and apply it to more and more phenomena, ask new and more detailed questions, possibly make new observations and discoveries in the process, come up with a new theory when phenomena is found that doesn't fit the old theory (this may require major conceptual breakthroughs).

Other Notes

- The idea that reliable rules exist had to be *invented*!
- Historical figures overcame major conceptual barriers and led the way for future development.
- Galileo's experiments and Kepler's astronomy proved Aristotle's theories wrong.
- A **physical theory** is an attempt to understand the structure and substance of the universe. It explains physical phenomena in terms of *causes*. Good theories are consistent with experiment, useful in understanding nature, and increase our ability to make quantitative predictions.
- That events have causes is (usually) a basic human assumption.
- Numerical inaccuracy isn't just inconvenient, it means something is wrong.
- Belief is not proof. While experiments cannot prove the exact truth of any physical concept, they do put ever tighter constraints on the ideas that deserve belief.
- In physics, one tries to remain flexible in choosing a point of view for describing a physical system. Often you can simplify the description of a physical system by imagining yourself moving in some particular way, looking at a system from some particular place, or focusing your attention on some particular size scale.
- Mathematical models of ideal situations are used to model the actual processes. Striving for more extreme accuracy is often time consuming, difficult, and costly. Hence, simpler models that give results that are accurate enough for our needs are the best choice. These models may leave out the more subtle effects such as the size and shape of an object.

EXERCISE 0.1: You've heard the story about an apple falling on Newton's head and stimulating a great revelation. Now you try it. Drop an object and watch it fall. What do you see? Try to describe what's happening. Try to see/notice something that you've never noticed before. From this simple experiment, can you determine anything about the nature of physical process in general or the nature of this particular process? So, do you see the world from an Aristotelian, Galilean/Newtonian, or some other world view? [Check the "Exercise Solution Plans" to see if your approach matches mine.]

EXERCISE 0.2: Is it valid to extrapolate from what we observe about objects falling on earth to the motion of the sun, moon, and stars? That is, do the same laws apply on earth as in the "heavens"? Justify your response. (Have fun with this. Be creative. Consider "yes." Consider "no." Is your inclination a belief or is it knowledge? What is *knowledge*? [Oooh, my head hurts!]

EXERCISE 0.3: Can our analysis of falling objects assist us in understanding other mechanical processes, such as a spinning top, a rolling bowling ball, or the motion of an athlete? Can it help in understanding heat (transfer of thermal energy), electricity (neurology, computer circuitry, home wiring), magnetism (the cause of flirtation), atomic, and nuclear phenomena, radiation, and so on?

EXERCISE 0.4: Most humans want to know "why"? Why do you think we often tend to fixate on "why"?

EXERCISE 0.5: How can we quantitatively define something akin to Galileo's "quantity of motion"? In parts I and II you will see that there are 3 useful ways: speed (v) = distance (d)/time (t); kinetic (moving) energy = $^1/2$ (mass) (speed)^2 = $^1/2$ m v^2; and momentum = (mass) (speed) = m v. If a person in a car (total mass of 2000 kg) reads their unusual metric speedometer as 30 $^m/s$ (roughly 60 mph), what three quantities of motion can you give?

Advanced

Why are things the way they are? What is the ultimate reason or purpose of everything? (Any assumptions in those questions?)

Quiz

1. Who introduced the idea of "quantity of motion"?

2. What were the two physical processes that clearly point out problems in Aristotle's cosmology (explanation of the universe)?

3. True or false? "Any assertion can be proved correct or incorrect by argumentation alone."

4. True or false? "Mathematical equations afford exact descriptions of physical processes."

5. True or false? "Science makes no assumptions."

6. Okay, okay, what are some of the assumptions that are made in the scientific method?

Quiz Solutions

1. The big G. ... Galileo, that is.

2. Quantitative predictions of the planets, and projectiles.

3. False! Sorry, A., it helps to do a reality check to see what's actually so. Enter experimentation.

4. We wish! Actually, there are some fundamental equations/relations that we believe to be exactly correct. Pretty amazing accomplishment! But many equations are *models* of physical processes. So we'd have to say "false" here, since the statement is by no means always true.

5. False! It's not "bad" to make assumptions. Just be clear and forthright about them.

6.
   - Assume that the entire universe is rational and follows *consistent rules.*
   - Assume that *causality* holds.

   If you want some real fun, throw these assumptions out the window and see what you can come up with. But be careful. The minute you start giving reasons, you're contradicting yourself. This is not to say that there aren't other ways of viewing the world that are enlightening. But who can deny the power and successes of science?

Exercise Hints & Solution Plans

EX. 0.1: This is not about mathematical problem solving ... yet. Actually look and notice. Be a Natural Philosopher. Observe the process. Observe your thinking about the process. What assumptions do you make? Are you thinking or are you waiting for the "answer" to be given to you? Is this just too obvious to consider? (Often the "obvious" is merely familiar but not at all fully comprehended. Furthermore, when we assume that we know all there is to know, we shut down and stop questioning. ... Dogma-city.)

EX. 0.2: You're the budding scientist, how are you approaching the question? What's your method? What is the structure of your thinking? How would we humans go about answering this if we weren't spoon fed the answers in school? We all stand on the shoulders of past individuals involved in the discovery process. How did they make their discoveries? How can we make ours?

EX. 0.3: See the note for Ex. 0.2 above.

EX. 0.4: My suggestion is to consider the apparent difference between human actions and the motion of "inanimate" objects.

EX. 0.5: This is really a "basic" exercise since it involves plugging in numbers into simple relationships in order to get a feel for those definitions.

Exercise Solutions

EX. 0.1: Aristotle thought that the impetus for an object to fall was within the object itself. Galileo started to look outside the object for the cause of the fall. And Newton gave us lucid, coherent rules for the causes of change in motion and the cause of falling objects: objects fall because of "gravity" between the object and the earth - an *external* impetus! You will study this process in detail soon enough (Ch. 2), for now enjoy your wonderment. Try to "see" falling objects from both points of view. Can you do it? What a difference a shift in world view makes!

EX. 0.2: By looking at falling objects and "heavenly" bodies, we can only guess. We can come up with elaborate arguments, woven with flawless logic, to give reason to our assertions. But if we are honest with ourselves we will see that we really don't know absolutely. All logic is based on starting premises ("axioms"). If these premises come from the way we would *like* the world to be, how much stake can we put in these arguments? Scientists can, and have gone beyond logical analysis alone by introducing quantitative inquiry: measure (experiment), predict quantitative results (theory), measure again to test the theory (experiment), and so on. It is the *verifiability* and *repeatability* of the finding that lead us to believe more and more in our theories.

EX. 0.3: As it turns out, the study of falling objects can be very ... fruitful (get it, Isaac?). The concepts and equations used are directly related to most mechanical analysis and offer a springboard into fields such as electromagnetism, Einstein's relativity, and quantum mechanics. Who'd've known?

EX. 0.4: We always have a reason for *why* we are late, and *why* we couldn't get home on time. So *why* do things fall the way they do? And *why* is the world as it is? *Why* are we here? I'd sure like to know! But perhaps there is no answer to why. Perhaps *why?* comes from the way of the human cerebrum. Perhaps *why?* isn't an appropriate question. Perhaps there is no reason. Perhaps *why?* isn't answerable. Then again, maybe it is. How would you go about answering it? Many people have an answer to *why?* based on hearsay or personal preference. Is that okay? Why not?!

It is powerful to acknowledge the strengths and limitations of science. The scientific method is impressively good at investigating what is going on and how things work. And while why things are the way they are is any body's guess, a deeper understanding of the way in which things work turns out to be exceptionally intriguing ... especially when coupled with a nice beverage and a few good friends!

EX 0.5:

$v = 30\ ^{m}/_{s}$, as read from the speedometer

kinetic energy (K) = $^1/_2\ (2000\ kg)\ (30\ ^{m}/_{s})^2 = 9 \times 10^5\ ^{kg \cdot m^2}/_{s^2}$

momentum (p) = $(2000\ kg)\ (30\ ^{m}/_{s}) = 6 \times 10^4\ ^{kg \cdot m}/_{s}$

Ch. 1: Introducing the Language of Physics

Overview

To model physical processes, we will not only need to define quantities of importance, but we will need to cut up space and time, so to speak, in a useful manner. The task is to choose the following:

- reference point (place)
- reference lines (directions)
- standard increments (units) of space, time, and any other quantity of interest
- reference motion (a person on a train might say that they are at rest or that a person on the ground outside is at rest, depending on what simplifies the analysis.)

Furthermore, it will be extremely important to distinguish those quantities which are specified by amount alone (for example: size, number of objects, price, distance traveled, speed, number of seconds left in a game, length of a stick, and so on) and those that also require a *direction* to specify them (position relative to your navel, a push, velocity, etc.). Quantities defined solely in terms of a magnitude (amount) are called "**scalar**" quantities. Those which require specification of a direction in addition to magnitude are called "**vector**" quantities. The rules for doing mathematics with scalars are the same rules of mathematics that you are familiar with. The rules for doing mathematics with vectors are similar in some respects, but quite different in others. You must BE CAREFUL to note whether you are working with scalar or vector quantities! This is one of the most common and frequent sources of error.

Among the essential math skills that you will use throughout this course and that are discussed in this chapter are:

- Check units of an equation to find errors
- Determine the units (or dimensions) of an unknown variable
- Use dimensional analysis to guesstimate the relationship between variables
- Convert from one system of units to another
- Keep correct significant figures in a calculation to reflect the accuracy with which the numbers are known
- Be ever aware of which variables are scalars and which are vectors.
- Do standard (scalar) mathematics.
- Do vector mathematics, employing basic trigonometry and the Pythagorean theorem.

Make sure that you soon become proficient, efficient, and down right flame-throwingly fast with units and vectors. Then, and only then, you can get on with the task of learning and applying physics!

Conceptual

Reference Frames

Take a clean sheet of paper and put a dot on it anywhere. Where is the dot? How far away is it? In order to give numerical answers, you need to create a grid that has a set reference point (the "origin") and reference directions (the "axes" and lines parallel to the axes). So draw a grid on the paper. Pick 6 different places (or more) that could be the origin. Can you come up with 6 different grids? Make them wild; make them silly. Make them rectangular, circular, or elliptical. Cut up the space in any way that serves you best.

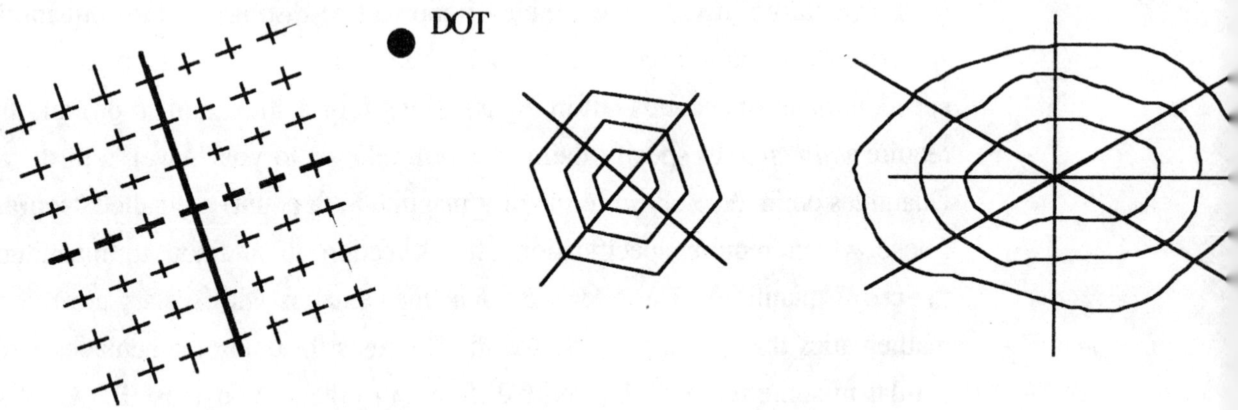

Two of the most useful grids are the rectangular ("Cartesian") grid and the polar grid. As you well know, the former uses parallel and perpendicular lines that are evenly spaced, while the latter uses concentric circles and radial lines.

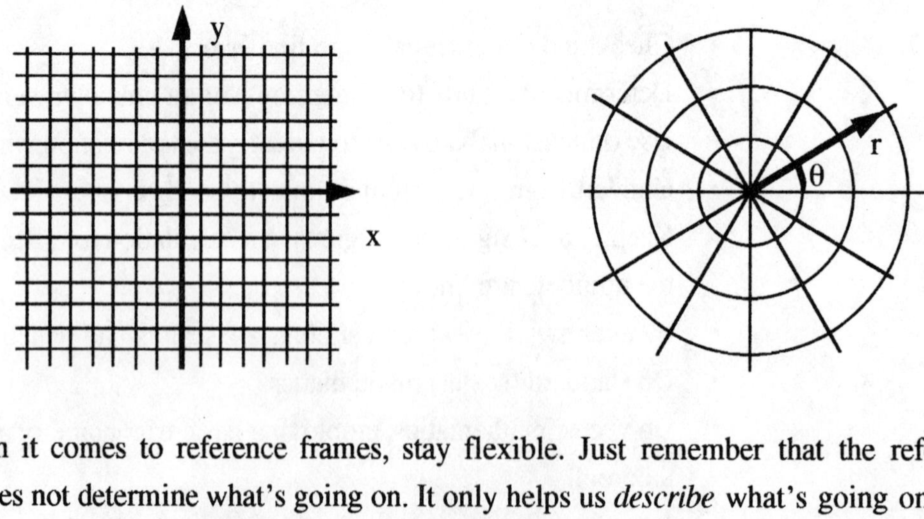

When it comes to reference frames, stay flexible. Just remember that the reference frame does not determine what's going on. It only helps us *describe* what's going on. And some systems make describing certain processes easier than others.

Dimensions & Units

- Read A.5.c of this study guide before proceeding.

If I say that an object is 100 meters away, can you visualize that distance? If not, then you really don't know what I'm saying. If I say that a car is traveling at 200 $^{m}/_{s}$, can you tell me if this is reasonable or preposterous? Is it possible to walk 1m in 1 second? Is this a fast walk, a snail's pace, a jog, or what? How many miles per hour is 1 $^{m}/_{s}$? If you rent a car in Europe and the speedometer is in $^{km}/_{h}$, can you make a quick, rough conversion to $^{mi}/_{h}$? It's actually quite easy!

Unless you have a *feel* for the physical meaning of the numbers that you are working with, you will probably view this course as a boring, academic exercise in doing word problems. When you can visualize the magnitudes (even very roughly), the physics begins to come to life and you will begin to *see* more of the physics all around you. Train yourself to develop an intuition for the magnitudes by visualizing the relationships and doing the exercises below.

Dimension of time, T

S.I. unit of second

1 s = $^{1}/_{60}$ min = $^{1}/_{60}$ ($^{1}/_{60}$ h) = $^{1}/_{60}$ ($^{1}/_{60}$) ($^{1}/_{24}$ day) = $^{1}/_{60}$ ($^{1}/_{60}$) ($^{1}/_{24}$) ($^{1}/_{360}$ y)

1 s ≈ time to say "one thousand one" at normal speaking speed

$^{1}/_{2}$ s ≈ time to say "one thou"

Dimension of length, L

S.I. unit of meter

1 m = 3.28 ft ≈ 3.25 ft = 3 $^{1}/_{4}$ ft = 3 ft 3 in (since $^{1}/_{4}$ ft = $^{1}/_{4}$ x 12 in = 3 in)

1 m ≈ 1 yd (+ 3 in)

unit of kilometer

1 km = 1 (10^3 m) = 1000 m ≈ 1000 (3.25 ft) = 3250 ft = 0.62 mi ≈ 0.5 mi

1 km ≈ 0.5 mi (+ 0.1 mi)

Measurement of angle

S.I. unit of radian - (reread B.6.c of this guide!)

2π radians = 360 degrees

π rad = 180° ; $^{\pi}/_{2}$ rad = 90° ; $^{\pi}/_{3}$ rad = 60° ; $^{\pi}/_{4}$ rad = 45° ; $^{\pi}/_{6}$ rad = 30°

1 rad = $(^{180}/_{\pi})°$ = 57.3°

1 rad ≈ 60° (for a very rough visualization of the angle, it is useful to think of $\pi \approx 3$)

Dimension of mass, M

S.I. unit of kilogram

1 kg = 1 (10^3 g) = 1000 g (where g means gram in this context)

1 kg ≈ 2 lb (+0.2 lb) [Ahhhh! Be careful here! Mass is NOT weight. I should really say that an object with mass of 1 kg weighs 2.2 lb on earth. More on this distinction later.]

EXERCISE 1.1: [Count it out!] Roughly how long does it take for an object to fall from your navel to the ground? How long from the top of your head? How about out of a second floor window? [Watch out below!]

EXERCISE 1.2: Roughly how many meters tall are you? Roughly how many meters wide, shoulder to shoulder, are you? Roughly how many meters is your arm span? Roughly how many meters long is a car? Very roughly, what are the width, length, and height of ceiling of the room that you're in now? If you look around you and do this for a while, you will become much more in tune with the lengths and distances around you.

EXERCISE 1.3: Distinguish *length*, *distance*, and *displacement*.

EXERCISE 1.4: [No calculators allowed!] A football field is 100 yd, not including the end zones. Each end zone is 10 yd. So how can you use this to help you visualize 100 m very roughly? What image roughly corrects for the error in your first image? Roughly how many football fields are in 1 km? Very roughly, how many football fields are in a mile? [Make a sketch!]

EXERCISE 1.5: [No calculators allowed!] Can you walk 1 m in 1 s? Is 200 $^{km}/_{h}$ a reasonable or preposterous speed for a car?

EXERCISE 1.6: [No calculators allowed!] Physically get involved with physics. I encourage you to do this one physically, not just mentally. Straighten both of your arms, flatten your hands, and make an angle of π radians between your arms. Make an angle of $^{\pi}/_{2}$ rad. Now make an angle of $^{\pi}/_{4}$ rad. Roughly make an angle of 1 rad. Roughly make an angle of $^{1}/_{2}$ rad. Roughly make an angle of 2 rad, now 3 rad, now 4 rad, and if you can make angles of 5 and 6 rads, join the circus!

EXERCISE 1.7: [No calculators allowed!] Very roughly, what's your mass? What's the mass of a 10 lb sack of potatoes? 1 liter is defined as the volume of 1 kg of water. About how much does a full 2 liter (plastic) bottle of soda weigh? [HINT: Not surprisingly, soda weighs about the same as water.]

Significant Figures

In the section above, I was rather sloppy with significant figures because I simply wanted you to get a feel for magnitudes. If you want to actually do any science or engineering, your numbers should reflect the degree of precision that you know and can defend.

EXERCISE 1.8: Trying to decide which contractor to hire to work on your home, you offer the following friendly challenge: "Measure the square footage of my home." From the results you will select your contractor. Contractor A reports a measurement of 1607.4 square feet, while contractor B reports 1606.932546 square feet. What are your thoughts on the matter? What questions would you ask? Who would you hire?

Vectors

- Read B.8 of this study guide for more discussion on vectors.

I've placed much of the discussion in B.8 so that you can use it as a reference as you work through the course. Since vectors are so important in physics, and since they are such a stumbling block for many students, the topic is certainly worthy of a few guided exercises. To be sure, you'll have plenty of practice as the course progresses. But let's get you primed for the task now. You are likely to have many questions arise about vectors. Be sure to ask them and get clear ASAP!

Types of Vectors

How many pairs of shoes do you have? How much money is in your bank account? How tall are you? How long have you been on this planet? All of these questions can be answered with a number. How many yachts do you have? A number was made up to answer that question too. Now get in a car, or better yet use public transportation, and ask yourself how you can describe your motion at any instant. You can give a speed by using a number, but you should also say which direction you're traveling.

EXERCISE 1.9: Come up with 5 ways to specify your direction of travel.

EXERCISE 1.10: Try to list 5 concepts that have both magnitude (amount) and direction. [I won't tell if you have to peek. These concepts are called "vector concepts," or just "vectors" for short. Vector concepts are intrinsically different than scalar concepts such as your age and height which require only a number to specify them, these.

Notation for Vectors

If I say "write down a variable that stands for the amount of time I've been working on this chapter," you would probably write "*t*.." If I asked you to choose a variable to represent the speed of an electron, you would probably write "*s*" for speed, or the commonly used "*v*." But how would you choose to denote a vector? An arrow above the variable says "I'm more than my numerical amount; I have a direction which must also be specified." Thus to represent the speed *and* the direction of motion, i.e. the velocity, we put an arrow above the variable: $\vec{\mathbf{v}}$. (Some texts use only bold letters because it's a bit of a pain to get the arrows in.)

So, the notation for vectors, complete with magnitude *and* direction, is a bold variable with an arrow above it, $\vec{\mathbf{A}}$. Sometimes you will want to refer to a vector's magnitude and direction separately. The magnitude can be denoted one of two ways. One, using the absolute value sign $|\vec{\mathbf{A}}|$, clearly says "I am the magnitude of the vector $\vec{\mathbf{A}}$." Always, always, remember that this thing represents a number (a scalar), and is treated as such. There is no direction carried with this symbol. The second way of denoting a vector's magnitude is simply to leave off the arrow and "unbold" the variable, A (frequently used because it is easier to write). The danger is in not realizing that it is the magnitude of a vector. If you treat it algebraically as the scalar it is, you will always be aware of the meaning of your variables, as laid out in the secret initiation rites of the Royal Order of People Who Have Taken a Physics Class and it won't present a problem. **So use the first notation, $|\vec{\mathbf{A}}|$, to emphasize the meaning and the second, A, for expedience.**

<u>Vector Notation Recap</u>:

- vector: an arrow above a bold variable, $\vec{\mathbf{A}}$
- magnitude of a vector : absolute value signs around the vector symbol, $|\vec{\mathbf{A}}|$,

  or no arrow above the variable, A (not bold and italicized here and in text).
- direction of a vector: somehow designate the orientation (draw an arrow; use north, south, etc. as a reference; use vertical and horizontal as a reference; use "unit orientation vectors," any other ideas).

Never, ever, ever, ever, ever write v when you mean $\vec{\mathbf{v}}$, or vice-versa. And never, ever, ever, ever, ever *read* $\vec{\mathbf{v}}$ as v, or vice-versa. The distinction is very important: one says to worry about the direction, the other doesn't–the mathematics is generally very different for vector and non vector quantities.

Basic

Reference Frames

- See section B.7 of this guide.

Now let's make use of your basic trigonometry, including inverse trig. functions, and the Pythagorean theorem. These are fundamental tools when working with the geometry of space, be it physical space (3-D) space or an abstract space.

EXERCISE 1.11: Using the two rectangular ("Cartesian") coordinate systems shown ($x-y$ and $x'-y'$), what is the position of the center, C, of the object shown? That is, what is (x,y) and (x',y') of C? How far away (distance) is the center from the origin, O, of the x-y frame? How far from O$'$? If you draw a straight line from O to C, what angle would it make with the x-axis? What about a straight line from O$'$ to C? How long (length) is the object in each frame? How wide? [See Solution Plan section for help.]

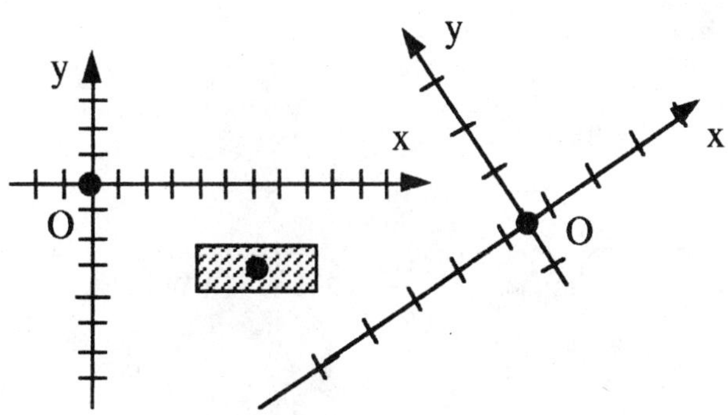

EXERCISE 1.12: Give the position coordinates of C using the polar coordinate grid shown below: (r, θ) = (__, __). [Notice that this immediately tells us how far away C is from the origin and the angle that the line from O to C makes with the $\theta = 0$ radian line. However, it isn't easy to see the length nor the width of the object.]

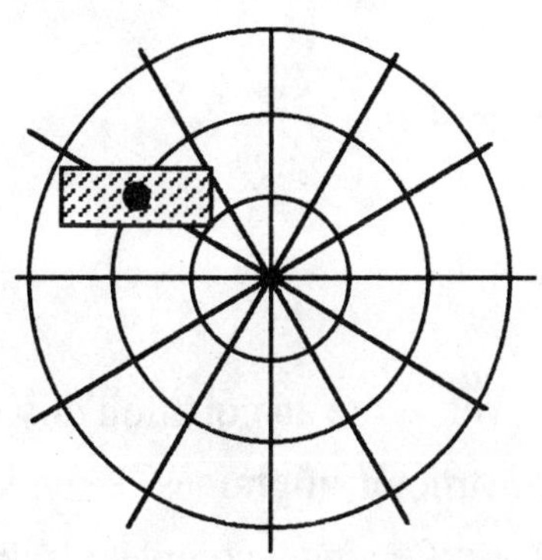

Dimensions & Units

Algebra with units

2 Rules to live or die by:

1. You can add or subtract numbers *only* if the units of each quantity is the *same.*

2. You *can* multiply or divide any quantity by *any* other quantity, regardless of the units! You'll just get something with different (combined) units.

EXERCISE 1.13: Calculate the following. Give the units of the final answer. [HINT: The answer to some of these is "nonsense."]

a) $1 \text{ hr} + 30 \text{ min} =$

b) $(10 \text{ N})(3 \text{ m} - 5 \text{m}) =$

c) $4.0 \times 10^6 \text{ J} / 2 \text{ s} =$

d) Trick question: $\dfrac{11 \text{ N}\cdot\text{m}}{10 \text{ s}} + 2.2\dfrac{\text{J}}{\text{s}} + 1.1 \text{ W} =$

e) $3.00 \text{ m} - 2.50\dfrac{\text{m}}{\text{s}}(2.00 \text{ s}) + \tfrac{1}{2}\left(2.00\dfrac{\text{m}}{\text{s}^2}\right)(2.00 \text{ s})^2 =$

f) $\left(3.0\dfrac{\text{m}}{\text{s}}\right)^2 - 2\left(9.8\dfrac{\text{m}}{\text{s}^2}\right)(1.01 \text{ s}) =$

EXERCISE 1.14: Which of these equations is absolutely invalid? By looking at the units, can you determine which is definitely correct? [x has units of meters, m; t has units of seconds, s; v has units of m/s; and a has units of m/s²; see Solution Plan section for help.]

a) $x = x_o + v_o t + \tfrac{1}{2}at^2$

b) $x = x_o + v_o t + at^2$

c) $v = v_o + at$

d) $v^2 = v_o^2 + 2a\Delta x$, where $\Delta x = x_f - x_o$

e) $v = v_o + ax$

f) $a \equiv \dfrac{dx}{dt}$

Never, ever, write an equation down that is dimensionally absurd...unless you want to give your instructor migraines. Even so, it's *your* grade! These sorts of errors are often interpreted as reflecting a complete lack of understanding, even if they are merely errors of haste.

Finding the units of an unknown variable

Suppose there is a variable in an equation whose units you either don't know or have forgotten. If you know the units of the other variables in the equation, it is a simple matter of algebra to determine the unknown units. Here are two very similar approaches that you can try out on the exercise that follows.

1. Solve for the variable whose units you are looking for. Replace the other variables with their units. Reduce the units until you're satisfied. (Note: when two terms are added or subtracted, you may drop one of them. After all, meters + meters are meters.)
2. Replace the variables with the corresponding units first. Then do the algebra to find the units of the variable of interest.

EXERCISE 1.15: Use the units given below for the variables in this exercise. Be careful to distinguish a letter that stands for a variable and one that stands for a unit. Our convention is to use italics for variables and non-italicized letters for units. For example, "*m*" is the variable for mass which has units of kg, and "m" stands for meters, a unit of length. Spatial variables (length, position, displacement) are denoted by such variables as *x*, *y*, *z*, *d*, *r*, etc.

- *x* and *y* have units m (meters)
- *r* has units m (meters)
- *t* has units s (seconds)
- *m* has units kg (kilograms)
- *v* has units of m/s
- *q* has units of C (coulombs)
- *V* has units of V (volts)
- *F* has units of N (newtons)

a) Find the units of *G*: $F = G\frac{m_1 m_2}{r^2}$

b) Find the units of *k*: $F = -kx$

c) Find the units of *g*: $y = y_o + v_o t + \frac{1}{2}gt^2$

d) Find the units of τ: $\tau = rF\sin\theta$

e) Find the units of *k*: $F = k\frac{q_1 q_2}{r^2}$

f) Find the units of *E*: $F = qE$

g) Find the units of *E*: $V = E \cdot d$

h) Find the units of *B*: $F = qvB\sin\theta$

Guessing a relationship

Suppose that you want a relationship between some variables, but you can't find it anywhere. It is always a splendid achievement to derive an equation based on physical reasoning and sound mathematics. But before you begin, why not try to get a feel for the sort of relationship you might expect? In simple cases, you can often guess (or recall) the exact equation just by looking at the units (or alternatively, the dimensions).

EXERCISE 1.16: Guess the relationships between the variables listed in each part by paying attention to the units. Use the following table of units. [See Solution Plan section for help.]

- x and y have units m (meters)
- t has units s (seconds)
- v has units m/s
- a, g have units m/s^2
- m has units kg (kilograms)
- F has units N (newtons)
- U has units N·m (newton·meter)
- U has units kg·m^2/s^2 (equivalent to N·m above)

a) Relate v, x, t: $v \approx$?; $x \approx$?; $t \approx$?

b) Relate y, g, t: $y \approx$?

c) Relate v, a, x: [Hint: v has an s, while a has an s^2.]

d) Relate F, U, x: $F \approx$?; $U \approx$?

e) Relate U, m, g, y: $U \approx$?

f) Relate U, m, v: $U \approx$?

Units conversion - 2 methods

Method 1: Conversion Factors

You know that 1 = 1 and 2 = 2, but did you know that 1 = 12? And if that's true, surely $^1/_{12} = 1$. What I'm getting at here is the difference between numerical equality and physical equality. Physically, you know that a length of 1 ft is the same as a length of 12 in, so 1 ft = 12 in. Consequently the ratio $^{1\ \text{ft}}/_{12\ \text{in}}$ is a **physical unity**. The punch line is that, just as multiplying a number by 1 doesn't change the numerical value, multiplying a physical quantity (with units) by a physical unity ratio doesn't change the physical amount. What does change is the number and the units. Which is why we call the physical unity ratio a "conversion factor."

To construct a conversion factor, simply answer "How many of these units are in how many of those units? For example, there are 60 min in 1 h, 24 h in 1 day, π radians in 180°, 2π radians in 1 revolution, and 1000 m in 1 km. Use these physical equivalencies as follows:

Multiplication by $(^{60\ \text{min}}/_{1\ \text{h}})$ converts from h to min:

$$4\ \text{h} = 4\ \text{h}\ (^{60\ \text{min}}/_{1\ \text{h}}) = 4\ (^{60}/_{1})\ \text{min} = 240\ \text{min}$$

Multiplication by $(^{1\ \text{h}}/_{60\ \text{min}})$ converts from min to h :

$$4\ \text{min} = 4\ \text{min}\ (^{1\ \text{h}}/_{60\ \text{min}}) = 4\ (^{1}/_{60})\ \text{h} = {}^{1}/_{15}\ \text{h}$$

Multiplication by $(^{\pi\ \text{rad}}/_{180^\circ})$ converts from ° to rad :

$$30^\circ = 30^\circ\ (^{\pi\ \text{rad}}/_{180^\circ}) = 30\ (^{\pi}/_{180})\ \text{rad} = {}^{\pi}/_{6}\ \text{rad}$$

Multiplication by $(^{180^\circ}/_{\pi\ \text{rad}})$ converts from rad to ° :

$$3\ \text{rad} = 3\ \text{rad}\ (^{180^\circ}/_{\pi\ \text{rad}}) = 3\ (^{180}/_{\pi})^\circ = 172^\circ$$

... and so on. Notice how the units cancel. Thus if I have 4 dogs and I want to convert that to pigs, I write

$$4 \text{ dogs} = 4 \text{ dogs} \left(\frac{\text{---pigs}}{\text{\_\_dogs}}\right) = 4 \text{ dogs} \left(\frac{3 \text{ pigs}}{2 \text{ dogs}}\right) = 6 \text{ pigs}$$

where I've used the well known fact that every house with 2 dogs has 3 pigs. That second step sets up the conversion. It is guided by what units I want to get rid of, i.e. cancel (dogs), and what units I want to end up with (pigs). I then insert the correct numbers for the ratio. To convert back to dogs, I'd simply use the reciprocal ratio.

Method 2: Rapid Replacement Therapy

I use this method whenever I can. (I like to think that I made this up, but surely others use it without giving it a name. As an example, how many miles are there in 250 km? Since there are 0.62 mi. in 1 km, I simply replace the "km" with 0.62 mi, and I've got it! 250 km = 250 (0.62 mi) = 160 mi. (or 155 mi, if you aren't paying too much attention to sig. fig.'s). For another example, if an object spins around 32 complete revolutions, how many radians has it rotated around? Since there are 2π radians in 1 rev., I replace "rev" with "2π rad" to obtain 32 rev = 32 (2π rad) = 64 π rad. This works particularly well with the metric system.

The metric system is wonderful because of its use of powers of ten. This helps us get a feel for magnitudes more easily, and allows us to convert between units with simple factors of 10. We may, however, not want to write out those factors of 10^n all of the time. In that case we can use the prefix notation. A table of prefixes that you must commit to memory follows. (Texts give more complete tables.)

NAME	PREFIX		POWER OF TEN EQUIVALENT
giga	G	=	10^9
mega	M	=	10^6
kilo	k	=	10^3
centi	c	=	10^{-2}
milli	m	=	10^{-3}
micro	μ ("myou")	=	10^{-6}
nano	n	=	10^{-9}

Thus, any time that you see μ as a unit prefix you may replace it with 10^{-6}. This works "backwards" as well. If there is no prefix and you want the unit in terms of a prefix, insert some form of unity. Since k = 10^3, 10^{-3} x k = 1. Consequently, 5.6 m = 5.6 (1) m = 5.6 (10^{-3} x k) m = 5.6 x 10^{-3} km. Now convert from W (watts) to MW (megawatts): 3.21 x 10^8 W = 3.21 x 10^8 (1) W = 3.21 x 10^8 (10^{-6} x M) W = 3.21 x 10^2 MW.

EXERCISE 1.17: Convert 4 km to cm first using conversion factors, then using replacement therapy. Even if you can see the answer without writing it down, go ahead and write it out formally so that you are proficient with the methods for less transparent cases.

Significant Figures

For quick reference when you do the short exercises below, I include the 3 rules and a few pointers for keeping track of significant figures.

Rule 1: Unless uncertainty (AKA the *precision*) in a number is explicitly stated, it is assumed to be ½ of 1 in the last decimal place shown.

Rule 2: The number of significant figures (AKA the *accuracy*) in a **product or quotient** is equal to the smallest number of significant figures in any of the numbers being multiplied or divided. (Nothing is said about the uncertainty, so defer to rule 1.)

Rule 3: The uncertainty (AKA *precision*) of a **sum or difference** is the same as the *greatest* uncertainty of the numbers added or subtracted. This will also tell you how many significant figures to keep–only keep the first uncertain figure.

These will take some practice and rereading before they make much sense to you.

Notes:

- *Accuracy* is synonymous with # of sig. fig.'s. ("It's accurate to 3 sig. fig.'s.")
- *Precision* is synonymous with uncertainty. It is the ± amount for the number.
- Round up if the first non significant figure is 5 or greater; round down if it's less.
- Round-off errors build up more slowly if you keep extra decimal places until the end.
- Zeros at the end of a number may or may not be significant. They are if they are to the right of the decimal place (e.g. 2.30 has 3 sig. fig.'s.); they are not, by convention, if they are to the left (e.g. 230 is assumed to have only 2 sig. fig.'s, even though that zero may in fact be significant).

EXERCISE 1.18: Write the number "230" with the implied precision written explicitly. That is, write it as "# ± precision." Next, assume that the zero is significant and write it with the precision. Finally, use scientific notation, that is, "(0 < # < 10) x 10^n," to distinguish 230 with 2 sig. fig.'s and with 3 sig. fig.'s.

EXERCISE 1.19: State the number of significant figures, accuracy, uncertainty, and precision in the numbers below. In c - f, perform the calculations using rules 1 - 3 and the associated notes.

a) 1040

b) 620.30 ± 0.002

c) $\dfrac{\left(2.2\times 10^{2}\right)(30)^{2}}{3.9876}$

d) $(2.9 \pm 0.06) - (2.9 \times 10^{2})$

e) 3300 + 11

f) $(3.84 \pm 0.009) + \left(\dfrac{(1.234 \pm 0.0001)\times 10^{2}}{2.4}\right)$

EXERCISE 1.20: Evaluate the following expression in two ways. First use rules 1 - 3 at every step. Second, keep all significant figures on your calculator at each step and, at the end, shorten your result to the same number of significant figures as your first calculation. Would you say that the rules for sig. fig.'s are rigid and exact, or are they reasonable guidelines?

$$x = (2.0 + 2.378)(21.28) = ?$$

Vectors

Addition & Subtraction of Vectors

Do you know what else an arrow above a variable says? It says "I can't be added or subtracted like those simpleton variables that have no direction. Oh sure, there are similarities in the way you add us. For example, $a + b = c \Rightarrow a = c - b$ would represent a scalar addition and subtraction, while $\vec{\mathbf{a}} + \vec{\mathbf{b}} = \vec{\mathbf{c}} \Rightarrow \vec{\mathbf{a}} = \vec{\mathbf{c}} - \vec{\mathbf{b}}$ would represent a vector addition and subtraction. So you can move the symbols around in the same manner as with non vector variables, but you can *never* ever ever add or subtract vectors to non vectors. When you take into account the directions in a computation, the results are usually vastly different from when you are dealing only with numerical amounts. So be careful! We vectors are refined, and expect to be treated in the proper manner!"

EXERCISE 1.21: Evaluate the following either numerically or symbolically. [HINT: The answers to some of these are "gibberish!"]

a) What does 3 feet forward + 2 feet back result in (i.e. "add up to")?

b) What does (3 ft) + (2 feet back) result in (i.e. "add up to")?

c) Solve for $\vec{\mathbf{C}}$ in the equation: $\vec{\mathbf{A}} + \vec{\mathbf{B}} + \vec{\mathbf{C}} + \vec{\mathbf{D}} = \vec{\mathbf{E}}$.

d) Solve for $\vec{\mathbf{F}}_1$ in the equation: $\vec{\mathbf{T}} + F_2 - \vec{\mathbf{f}} = -\vec{\mathbf{F}}_1 + \vec{\mathbf{N}}$.

e) Solve for $\vec{\mathbf{F}}_1$ in the equation: $\vec{\mathbf{T}} + \vec{\mathbf{F}}_2 - \vec{\mathbf{f}} = -\vec{\mathbf{F}}_1 + \vec{\mathbf{N}}$

f) What is 33 + 22 - 1? = ?

g) Solve for x in the equation: $3x + mgh - c - 5x = 62$.

h) What is: $\vec{\mathbf{A}} + \vec{\mathbf{A}} + \vec{\mathbf{A}} + \vec{\mathbf{A}}$?

i) What is: $2\vec{\mathbf{A}} + 3\vec{\mathbf{A}}$?.

EXERCISE 1.22: Michael Jordan is about 2 m tall (6' 6" to 6' 7"). In his hand he holds a 1 m stick at one end. For this exercise, let's say that he keeps this hand on the top of his head. At the other end of the stick, a measuring tape is attached. Michael holds the other end of the tape measure with his big ol' feet so that the tape measure stretches from his feet to the end of the meter stick. Got the picture? If not, sketch it! The question is, how far is the tape measure stretched? Can you imagine moving the meter stick around so that the tape is stretched from his feet to the end of the stick more or less? [See Hint section.]

a) In what direction should he point the stick so that the tape is stretched as far as possible?
b) In what direction should he point the stick so that the tape is as short as possible?
c) What if he holds the stick horizontally so that it makes a right angle with his body?
d) Is the numerical sum or difference of his height and the length of the stick the only meaningful way to add these lengths? How might you figure out the length of the tape measure in part c?

We'll return to this problem in the intermediate section after we have seen how to use vector arithmetic. Then we'll find out just how far that tape is stretched no matter how the stick is held!

Multiplication of Vectors

There are 3 types of multiplication appropriate to vectors: Why and when we use each of these is certainly a matter to be discussed. For now just know what the three types of multiplication are. We will layer in the knowledge in like manner to how Monet layered paint, no layered light, onto the canvas ... or something. Have another cappuccino.

1. Multiplication by a Scalar ($-2\vec{\mathbf{A}}$) - Multiply a vector by a number (positive or negative). (For now, "scalar" is just another word for number. It comes from "scale," which has to do with size. No big whoop.) The number changes the magnitude (amount); a minus sign reverses the direction.

2. Scalar (AKA Dot) Product ($\vec{\mathbf{A}} \cdot \vec{\mathbf{B}} = C$) - Multiply a vector by a vector and get a number (sans direction). It is called a *scalar* product because the result is a scalar. It is sometimes called a *dot* product because we use a dot as the symbol for this type of multiplication.

3. Vector (AKA Cross) Product ($\vec{\mathbf{A}} \times \vec{\mathbf{B}} = \vec{\mathbf{C}}$) - Multiply a vector by a vector and get a vector (with direction). It is called a *vector* product because the result is a vector. It is sometimes called a *cross* product because we use a cross as the symbol for this type of multiplication.

   But what's it all mean, Dorothy?

- $2\vec{\mathbf{A}}$ means twice the amount, same direction, same units as $\vec{\mathbf{A}}$.
- $-\vec{\mathbf{A}}$ means the same magnitude, opposite direction, same units as $\vec{\mathbf{A}}$.
- $-5\vec{\mathbf{A}}$ means 5 times the amount, opposite direction, same units as $\vec{\mathbf{A}}$.

- The meaning of $\vec{\mathbf{A}} \cdot \vec{\mathbf{B}} = C$ is not quite as transparent. For now, just notice that you end up with a number, not a vector. The value of C not only depends on the magnitudes of $\vec{\mathbf{A}}$ and $\vec{\mathbf{B}}$, but also on the angle between them. (More on this later.) What should be clear is that C's units are the product of the units of $\vec{\mathbf{A}}$ and $\vec{\mathbf{B}}$.

- Similarly, the meaning of $\vec{\mathbf{A}} \times \vec{\mathbf{B}} = \vec{\mathbf{C}}$ is not very transparent. Here, you end up with a vector, instead of a number. Again, the magnitude of $\vec{\mathbf{C}}$ not only depends on the magnitudes of $\vec{\mathbf{A}}$ and $\vec{\mathbf{B}}$, but also on the angle between them. Furthermore, the *direction* of $\vec{\mathbf{C}}$ depends on the directions of $\vec{\mathbf{A}}$ and $\vec{\mathbf{B}}$. (More on this later.) It should be clear that $\vec{\mathbf{C}}$'s units are the product of the units of $\vec{\mathbf{A}}$ and $\vec{\mathbf{B}}$.

With just 2 caveats, you can move the vector variables around algebraically just as you do with scalar variables. Always remember that:

1. You cannot divide by a vector. (You *can* divide a vector by a number, after all, this is the same as multiplying by a fraction.)

2. $\vec{\mathbf{A}} \times \vec{\mathbf{B}} \neq \vec{\mathbf{B}} \times \vec{\mathbf{A}}$!! You'll see why later, but $\vec{\mathbf{A}} \times \vec{\mathbf{B}} = -\vec{\mathbf{B}} \times \vec{\mathbf{A}}$ (opposite directions). So the order matters for the vector (AKA cross) product. The order does NOT matter for the scalar (AKA dot) product: $\vec{\mathbf{A}} \cdot \vec{\mathbf{B}} = \vec{\mathbf{B}} \cdot \vec{\mathbf{A}}$.

3. There's no place like home.

Ready to try some algebraic maneuvers with vector variables? Piece of cake!

EXERCISE 1.23: Use standard algebraic maneuvers, keeping the two exceptions noted above in mind, to perform the algebra requested. Some of these may be "utter nonsense!" [See Hint section for help.]

a) Rewrite the following expression by distributing the $\vec{\mathbf{A}}$. Is it a vector or a scalar?

$$\vec{\mathbf{A}} \cdot \left(\vec{\mathbf{B}} + \vec{\mathbf{C}}\right) =$$

b) Rewrite the following expression by distributing the $\vec{\mathbf{A}}$. Is it a vector or a scalar?

$$\vec{\mathbf{A}} \times \left(2\vec{\mathbf{B}} - b\vec{\mathbf{C}}\right) =$$

c) Solve for $\vec{\mathbf{D}}$: $\left(\vec{\mathbf{B}} \cdot \vec{\mathbf{C}}\right) + \left(\vec{\mathbf{A}} \times \vec{\mathbf{B}}\right) = \vec{\mathbf{D}} + \vec{\mathbf{E}}$

d) Solve for $\vec{\mathbf{F}}$: $2\vec{\mathbf{F}} - 3\left(\vec{\mathbf{A}} \times \vec{\mathbf{B}}\right) = 5\left(\vec{\mathbf{A}} \times \vec{\mathbf{B}}\right) - 2\vec{\mathbf{F}}$

e) Solve for $\vec{\mathbf{r}}$: $\vec{\tau} = \left(\vec{\mathbf{r}} \times \vec{\mathbf{F}}_1\right) + \left(\vec{\mathbf{r}} \times \vec{\mathbf{F}}_2\right)$

f) Solve for E : $5\vec{\mathbf{A}} \cdot \left(\vec{\mathbf{B}} \times \vec{\mathbf{C}}\right) = \dfrac{4D}{E}$

g) Solve for $\vec{\mathbf{E}}$: $D\vec{\mathbf{A}} = \dfrac{4D}{\vec{\mathbf{E}}}$

EXERCISE 1.24: Determine the units for the variable on the left side of the equality. Note that $\vec{\mathbf{r}}$ has units of meters (m) and $\vec{\mathbf{F}}$ has units of newtons (N). [Hint: Simply multiply the units together. There is an interesting lesson here that I'll share with you in the solutions.]

a) The scalar quantity "work" is defined as: $W = \vec{\mathbf{r}} \cdot \vec{\mathbf{F}}$

b) The vector quantity "torque" is defined as: $\vec{\tau} = \vec{\mathbf{r}} \times \vec{\mathbf{F}}$

Okay, so why no division by a vector? (Remember, you *can* divide a vector by a number. What would it mean to divide by a *direction*? I don't know either. If and when you study "linear algebra," you will come upon a sort of division associated with "inverse matrices." But that'll have to wait.

Intermediate

Reference Frames

Just a few more things before you're ready to start doing some physics. (I know you're champing at the bit! Let's take a quick look at the 3-D rectangular coordinate system and the 3-D Pythagorean theorem, moving reference frames, and the introduction of time as a coordinate.

3-D Pythagorean Theorem

EXERCISE 1.25: [As best you can, try to visualize and draw this on your own. It will be tough at first, but try hard. If you're totally stuck, take a peak at the solution.] Draw a 3-D rectangular coordinate system with the z-axis up, y-axis to the right, and x- axis "out of the page," i.e. down and to the left. Place a point in the quadrant where x, y, z are all positive, and call this point "P." Draw a line from the origin to the point. Draw a line down from P parallel to the z-axis and ending at the x-y plane. Now draw a line back to the origin (this should form a diagonal line in the x-y plane). The coordinates of P are (x,y,z), so the length of the vertical line is z. Use the Pythagorean theorem to determine the length of the diagonal in the x-y plane. [HINT: Draw a right triangle in the x-y plane with x as one side, y as the other side, and that diagonal as the hypotenuse. Finally, use the Pythagorean theorem to find how far away P is from the origin. Your answer should be in terms of x, y, and z, and should look like a 3-D Pythagorean theorem!

Moving Coordinate Systems

If you're in a plane that's traveling at 500 mi/h and a person throws a pillow from behind you at 10 mi/h, aren't you the least bit concerned that the pillow is traveling over the earth at 510 mi/h? Of course not. Since it is only approaching you at 10 mi/h, there's no real worry. For getting up to go to the rest room, working on your laptop computer, or passing the peanuts, the natural frame of reference is that of the plane, not that of the Earth. As far as the control tower is concerned, the Earth is the reference frame of choice.

It turns out to be very valuable to physicists and engineers to be able to take data from one reference frame and deduce what that data would be relative to another reference frame. For the example given above, that would mean that the control tower could measure the plane speed as 500 mi/h and if the captain said "Control tower we have a problem. Someone threw a pillow and it's traveling at 10 mi/h toward me." Control tower could deduce that the pillow is traveling at 510 mi/h over the ground. If the captain threw the pillow *back* at the passenger with a speed of 15 mi/h, it would be traveling at (500 - 15) mi/h relative to the Earth. And if she threw it sideways, at her copilot, ... well, we'd need to use the triangle relationships for vectors.

We can draw two reference frames that are moving relative to one another as moving coordinate axes. We usually assume one of them is not moving to draw the diagram, but who's to say who is moving and who is not? You are!

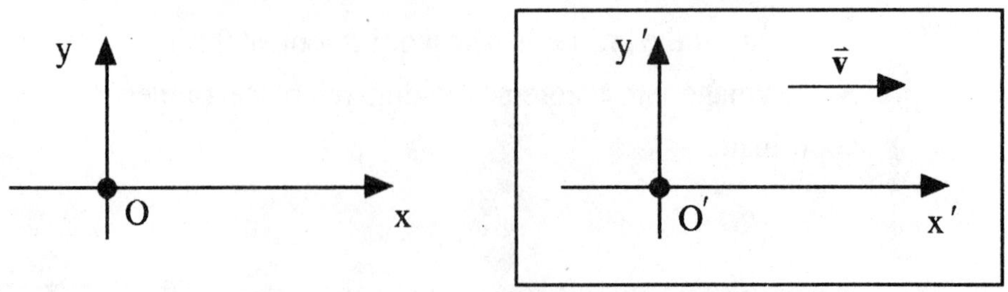

"Events": Introducing Time as a Coordinate

If you want to specify where an object is, you might give the spatial coordinates (x,y,z). If you want to say where an object is at a particular time, you might give the space-time coordinates (t,x,y,z). In physics we call this an "event." I guess everyone else calls it that too, except that here we are talking about an instant in time rather than an interval of time, as with a "sporting event." An event could be a light bulb flashing, you sitting there now, or now, or now, or me typing this Q. Anything at some place at some time is an event. Now the weird thing is that you can graph time on a spatial axis. But maybe that's not too weird, since we can graph any variable we want on spatial axes, like exam score vs. number of students, profits vs. cost, or position vs. time. Just because we use a spatial representation doesn't mean the variable is a spatial dimension, sheesh! For more on this, see Ch. 2!

Vectors

- See section B.6 of this guide for additional ... well, guidance.

Types of Vectors

Since *any* concept that has an associated amount (magnitude) and direction is a vector concept, there are lots of different types of vectors. For instance, there are: displacement vectors in units of m, km, ft, etc.; velocity vectors in units of m/s, km/h, mi/h, ft/s, etc., force vectors in units of newtons (N), or pounds (lb); and many, many more. In this chapter, your text focuses on three types. You will learn about the rest in time.

- <u>Displacement Vectors</u>, $\vec{\mathbf{D}}$

  Concept: the displacement between 2 locations, the beginning (initial) position and the end (final) position.

  Magnitude: the straight line distance between the 2 places. This is only equivalent to the distance traveled if the path is a straight line from start to finish. Otherwise, the magnitude of the displacement is less than the actual distance along the path traveled.

  Direction: *from* the initial position *to* the final position.

- <u>Position Vectors</u>, $\vec{\mathbf{r}}$

  Concept: points to the location of an object relative to some chosen reference position ("origin").

  Magnitude: the straight line distance between the chosen origin and the location in question. Note that for some particular location, say that of the tip of your nose now, the position vector is different for different origins, say your right knee vs. your left elbow.

  Direction: *from* the chosen origin *to* the position of interest.

- <u>Unit Orientation Vectors</u>, $\hat{\mathbf{i}}, \hat{\mathbf{j}}, \hat{\mathbf{k}}$, or $\hat{\mathbf{a}}$

  Concept: used to designate a direction only. They say "That'a way!"

  Magnitude: These vectors have a magnitude of 1 ("unity"), but are dimensionless and so have no units in the sense of meters, seconds, and so on. They are usually just called "unit vectors."

  Direction: in general, a unit vector $\hat{\mathbf{a}}$ points in the direction of the vector $\vec{\mathbf{a}}$. $\hat{\mathbf{B}}$ points in the direction of $\vec{\mathbf{B}}$. The directions of the chosen coordinate system are often used as reference directions. For a 3-D rectangular coordinate system, we say that $\hat{\mathbf{i}}, \hat{\mathbf{j}}, \hat{\mathbf{k}}$ point in the direction of the positive x, y, and z axes, respectively.

- <u>Other Vectors</u>, $\vec{\mathbf{v}}$, $\vec{\mathbf{F}}$

  I couldn't resist mentioning these two favorites. You'll see them and more in the weeks to come.

  Concepts: velocity and force, respectively.

  Magnitude: speed and umph, respectively.

  Direction: that of the motion and push (or pull), respectively.

Vector Representation

- See section B.8.a & B.8.b of this guide.

There are 4 ways that we will commonly represent a vector concept.

- **Vector Variable Symbols:** $\vec{\mathbf{A}}, \vec{\mathbf{B}}, \vec{\mathbf{D}}, \vec{\mathbf{r}}, \vec{\mathbf{v}}, \vec{\mathbf{a}}, \vec{\mathbf{F}}, \vec{\mathbf{p}}$, need I go on?

- **Arrows:** →, ↓, ←, provide a graphical representation. The orientation of the arrow gives the orientation of the vector quantity. The length of the arrow may be proportional to the magnitude of the vector (if we have chosen a scale and drawn the vector with a ruler), may give a rough sense of the magnitude (in a rough sketch), or may not even be taken into consideration in a very rough diagram. You'll also see that sometimes the positions of the tail and head (tip) matter, sometimes they don't. But you can always "pick them up" and move them around to perform graphical addition and subtraction. Stay tuned!

- **(Magnitude, Direction):** (11 miles, 22° south of east), (65 m/s, 15° above horizontal), (100 lb, down). To point in the direction "22° south of east," point east and then rotate 22° toward south. It's 22° southward *from* the direction of east. In short, 22° S of E.

- **Unit Vector Representation:** $\vec{\mathbf{a}} = \left(9\frac{\mathrm{m}}{\mathrm{s}^2}\right)\hat{\mathbf{a}}$; $\vec{\mathbf{v}} = \left(3\frac{\mathrm{m}}{\mathrm{s}}\right)\hat{\mathbf{i}} - \left(4\frac{\mathrm{m}}{\mathrm{s}}\right)\hat{\mathbf{j}}$; $\vec{\mathbf{F}} = F_x\hat{\mathbf{i}} + F_y\hat{\mathbf{j}} + F_z\hat{\mathbf{k}}$ This is also known as the component representation, or component form, since it shows how much of the vector is in each direction. The amount in the x-direction is called the "x-component," and so on. Note that "component" simply means "part."

EXERCISE 1.26: Would you say that the following vectors have the same *orientation*? Would you say that they point in the same *direction*? Justify your response.

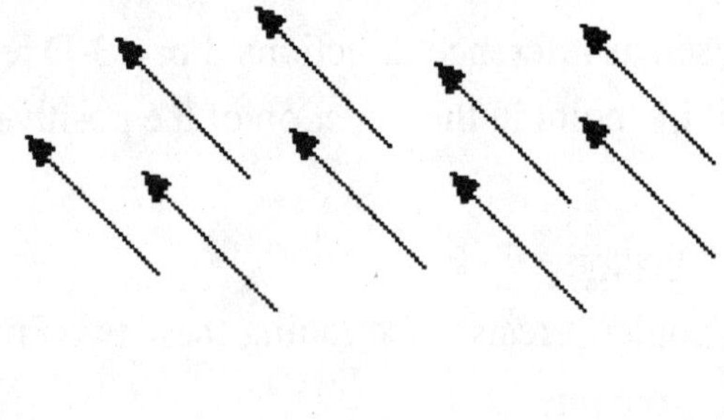

EXERCISE 1.27: "33° north of west," "northwest," "20° east of south," "20° south of east," what the heck does it all mean?! Draw north-south-east-west directional axes. [Notice that this is a *top view* of the horizontal directions.] Now draw vectors pointing in each of the directions cited above. [HINT: Try reading this as "33° north *from* west," or "starting from west, rotate around 33° toward the north axis."]

EXERCISE 1.28: If you were to get out a ruler and draw your vectors to scale, you would first need to decide on a scale. There's an interesting point to made here. I'll let you in on it after you choose some scales to graphically represent your vectors. Choose scales for the magnitudes of vectors with: a distance of 1 mi; a distance of 100 m; a speed of 50 $^{mi}/_{h}$; a speed of 10 $^{m}/_{s}$; a force of 25 lb; and a force of 10 N.

Seeing a Vector's Components

Extremely Important!!

Follow these steps very methodically each time that you look for the components of a vector. (Commit these steps to memory as soon as possible and practice!)

1. Sketch the vector.
2. Choose reference directions (coordinate axes).
3. Draw a GRID through the vector's tip and tail. Make the grid lines parallel to the reference directions.
4. "Walk" ***from*** the tail ***to*** the tip along either triangle to see how much the vector points in each direction. Notice the direction of the components!

EXAMPLE:

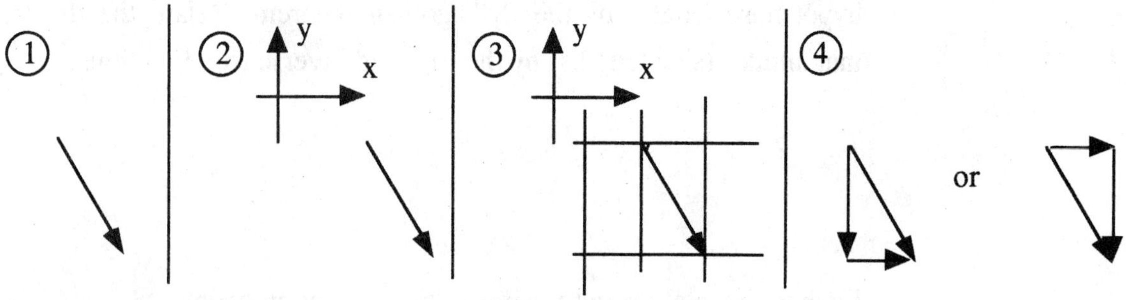

EXERCISE 1.29: For each vector and reference directions (coordinate axes) shown, draw in the grid and show the two possible component triangles. State whether the x-component is in the positive or negative x-direction. Likewise, determine the sign of the y-component.

(a)

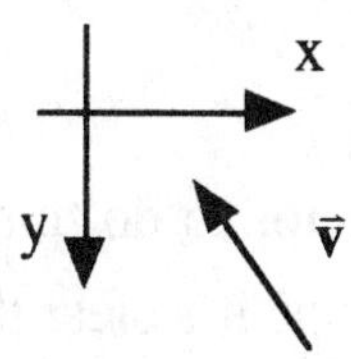

(b)

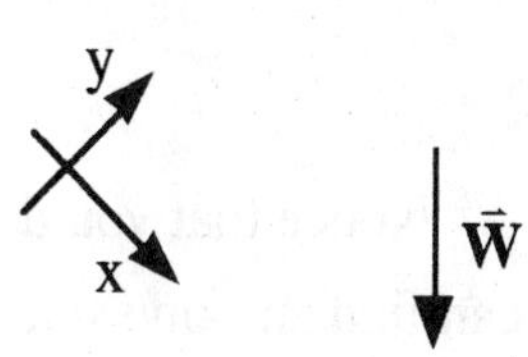

(c)

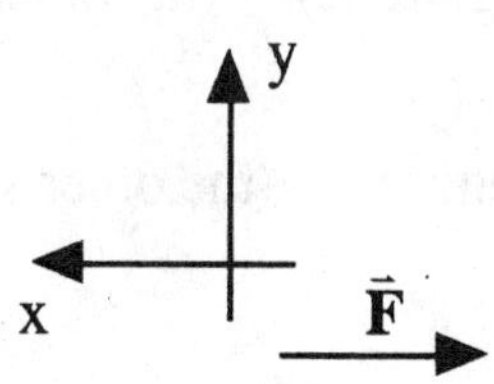

(d)

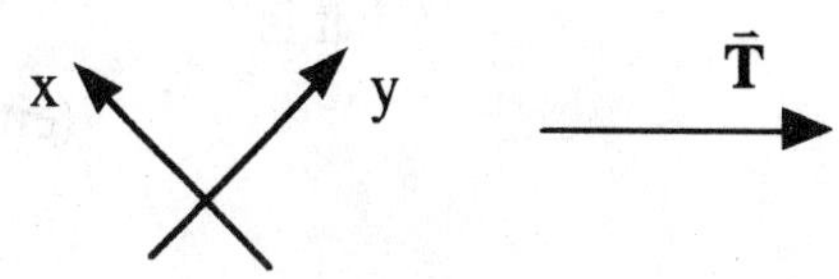

Everything You Always Wanted to Know About a Vector (... but were afraid to ask)

- See section B.6.a and B.8.c of this guide.

This is big. No, enormous. Get this, and you'll cut your aspirin intake by half. And that's a money-back guarantee. (Well, figuratively.) The method is always the same. Here's a quick outline of the procedure, then it's time for you to get out a pencil and your workbook.

MAY NEED: the magnitude, the direction (relative to reference directions), and the components (in the reference directions).

MAY HAVE: any of the above (i.e. magnitude, direction, components)

TOOLS: the basic tools for analyzing any right triangle, including the component triangle of a vector, are the Pythagorean theorem and the basic trig. relations. That's it!

METHOD: sketch the vectors and components as shown above. Relate the component magnitudes (represented by the side lengths) to the total magnitude (represented by the hypotenuse length) by the Pythagorean theorem. Relate the directions (angles) and magnitudes (side lengths) by the trig. and inverse trig. functions.

Finding the total magnitude if you know the components

Quick and simple - apply the Pythagorean theorem with the components as the lengths of the sides:

$$v \equiv |\vec{\mathbf{v}}| = \sqrt{v_x^2 + v_y^2}$$

$$F \equiv |\vec{\mathbf{F}}| = \sqrt{F_x^2 + F_y^2 + F_z^2}$$

Notice that you don't even need to sketch the vector to do this. Further notice that you can find any one side if you know the other two. I hope it's clear that

$$(\text{either component})^2 = (\text{total magnitude})^2 - (\text{the other component})^2$$

that is,

$$(\text{either short side})^2 = (\text{hypotenuse})^2 - (\text{the other short side})^2.$$

Finding a component if you know a side and an angle

You *must* know this cold–with minimal thought and little hesitation. Get it now so that you aren't stumbling throughout the course. (See exercise that follows.)

Essential:

(side opposite the angle, θ) = (hyp.) sin θ ... since sin θ = opp./hyp.

(side adjacent to the angle, θ) = (hyp.) cos θ ... since cos θ = adj./hyp.

Other Possibilities:

(side opp. the angle, θ) = (adj.) tan θ ... since tan θ = opp./adj.

(side adj. to the angle, θ) = (opp.) cot θ ... since cot θ = adj./opp.

Finding an angle if you know two sides

First set up a trig. ratio with two known sides. Then simply use the inverse trig. functions on your calculator. For example,

$$\tan\theta = \frac{a}{b} \quad \Rightarrow \quad \theta = \tan^{-1}\left(\frac{a}{b}\right)$$

EXERCISE 1.30: (Really try to get this before checking the solutions.) You walk from the origin 50 m along a straight line. Your displacement vector makes an angle of 300° (counterclockwise from the $+x$-axis). How much have you moved in the x-direction? How much in the y-direction? Verify your results with the Pythagorean theorem. Further verify your results by using the inverse tangent of the ratio of the components. [See Hint section for help.]

EXERCISE 1:31: Write the displacement vector in exercise 1.30 in component form. That is, use the unit vectors $\hat{\mathbf{i}}$ and $\hat{\mathbf{j}}$.

Vector Addition and Subtraction

- See section B.8.d of this guide!

Aaah. The final preparatory topic before you embark on your study of physics. You probably know that there are basically two ways to add and subtract vectors.

Graphically, we draw arrows and move them end to end. We can either use the head to tail method or the parallelogram method (they are essentially the same). If we are drawing the vectors to scale by using a ruler and protractor, we can measure the resulting ("resultant") vector's length and angle. Otherwise we use the basic trig. relations, the law of cosines, and/or the law of sines to find the angles and magnitudes of interest.

Analytically, we break vectors down into components, add or subtract components that are along the same line, and thereby obtain the components of the resulting ("resultant") vector. ... Now it's your turn. Practice makes perfect.

******

Essential Training!!

******

EXERCISE 1.32: You're walking along a city street where each block is 100 m. You start at point O, and walk 3 blocks to point A. You make a right turn and then walk 4 blocks from A to point B. a) Draw in the displacement vectors for your walk from O to A, $\vec{\mathbf{D}}_1$, your walk from A to B, $\vec{\mathbf{D}}_2$, and your net displacement from O to B, $\vec{\mathbf{D}}_{\text{total}}$. b) What is the magnitude of each displacement vector? c) What distance did you walk? d) What is the angle between $\vec{\mathbf{D}}_1$ and $\vec{\mathbf{D}}_{\text{total}}$?

EXERCISE 1.33: [Exercise 1.22 revisited.] Michael Jordan is about 2.00 m tall (6' 6" to 6' 7"). In his hand he holds a 1.00 m stick at one end. For this exercise, let's say that he keeps this hand on the top of his head. At the other end of the stick, a measuring tape is attached. Michael holds the other end of the tape measure with his big ol' feet so that the tape measure stretches from his feet to the end of the meter stick. Got the picture? The question is, how far is the tape measure stretched? Can you imagine moving the meter stick around so that the tape is stretched from his feet to the end of the stick more or less?

a) What is the *greatest* distance that he can stretch the tape measure? How should he do this?
b) What is the *shortest* distance that he can stretch the tape measure? How should he do this?
c) If he holds the stick horizontally so that it makes a right angle with his body, what is the length of the tape measure? What angle does it make with his legs?
d) Now suppose that he moves the meter stick so that the tape makes an angle of 16.2° with his legs. How far is the tape stretched? Make a sketch. What is the angle between his body and the meter stick? [HINT: Use the law of cosines!]

How about some practice using unit vector notation? When you get the hang of it, you realize that it's a great way to deal with vectors!

EXERCISE 1.34: Lets go for a walk in the neighborhood. We begin at Mr. Rogers' house. For this walk, let's choose East to be the positive *x*-direction, and North to be the positive *y*-direction. For the first part of our walk, we have a displacement $\vec{\mathbf{D}}_1 = (20.1\text{ m})\hat{\mathbf{i}} - (3.15\text{ m})\hat{\mathbf{j}}$. For the second part, we have a displacement $\vec{\mathbf{D}}_2 = (51.3\text{ m})\hat{\mathbf{i}} + (40.1\text{ m})\hat{\mathbf{j}}$. And for the third and forth parts of our walk, we have displacements $\vec{\mathbf{D}}_3 = -(11.7\text{ m})\hat{\mathbf{i}} + (62.2\text{ m})\hat{\mathbf{j}}$ and $\vec{\mathbf{D}}_4 = (9.73\text{ m})\hat{\mathbf{i}} + (33.3\text{ m})\hat{\mathbf{j}}$, respectively. (a) How far are we away from the house (our starting point)? (b) In what direction should Mr. Rogers shout in order for us to come home? (In other words, what is the direction of our total displacement vector?) (c) How far and in what direction do we need to walk to return, assuming that we can walk in a straight line?

EXERCISE 1.35: Relative to some fixed point (origin), say Mr. Rogers' house, your position vector is $\vec{\mathbf{r}}_{\text{you}} = (311\text{ m})\hat{\mathbf{i}} + (212\text{ m})\hat{\mathbf{j}}$. My position vector is $\vec{\mathbf{r}}_{\text{me}} = -(113\text{ m})\hat{\mathbf{i}} + (314\text{ m})\hat{\mathbf{j}}$. (a) What is the displacement from me to you? (b) What is the displacement from you to me? (c) What is the angle that the displacement vector makes with the +*y*-axis (north) in each case?

Quiz

1. You can add a velocity vector to a force vector

a) anytime it is useful to do so.
b) only when the units are the same.
c) ... never! Velocity and force are entirely different concepts and so they can never have the same units.

2. True or False? Your choice of reference frames (coordinates) does not affect what is physically taking place.

3. True or False? Polar coordinates are actually better than rectangular coordinates.

4. The displacement vector is defined as

a) the sum of the initial and final position vectors.
b) the distance along the path taken.
c) the final position vector minus the initial position vector.
d) being relative to north-south-east-west axes.

5. Very roughly, 1 km is ___ mi, so 30 km/h is very roughly ___ mi/h.

a) 2; 60
b) 1/2; 60
d) 1; 30
c) 1/2; 15
e) 2; 15
f) none of these choices is correct

6. $3\hat{i} + 4\hat{j}$ is equivalent to

a) $7\hat{k}$
b) $5\hat{k}$
c) 5
d) ... is not reducible
e) ... is nonsense!

7. 72.0° is equivalent to ____ radians.

a) 5
b) 3/4
c) 1.26
d) 3
e) 57.3

8. 1 μm = _____ km.

a) 10^{-9}
b) 10^{9}
c) 10^{-3}
d) 10^{3}
e) 10^{-15}

9. An angle of 2π radians (roughly 6 1/4 rad, actually 6.28 rad) is physically equivalent to 360°. So to convert *from* degrees *to* radians we may use the conversion factor:

a) π rad/30
b) 30°/π rad
c) π rad/180°
d) 180°/π rad
e) π rad/90°
f) 90°/π rad
g) 3 rad/180°
h) 180°/3 rad

10. In the equations $x = 0.5\ C_1\ t^2$ and $v^2 = 2\ C_2\ x$, the SI units of the constants C_1 and C_2 must be, respectively:

a) m/s & m/s^2
b) m s^2 & m
c) m/s^2 & m/s^2
d) m s^2 & m/s

Quiz Solutions

1. c

2. True!

3. False. The best choice of coordinates depends on the situation.

4. c

5. c. (Actually, 1 km = 0.62 mi, so the approximation is pretty rough.)

6. d. The only way to add these is either graphically or by getting the magnitude with the Pythag. Th'm and the angle by an inverse trig. function.

7. c (1.26 rad). You really shouldn't even have to make the conversion to guess the correct answer here. Since 1 rad ≈ 60° (57.3° = $^{180}/_{\pi}$°), and 3 rad ≈ π rad = 180°, 72.0° must be smaller than 3 radians and larger than 1 radian. I purposely gave you only one choice that fits. (Also note that π/2 = 90°, and π/2 ≈ 3/2 = 1.5. The answer is evident from this.)

8. a (10^{-9} km). You should be able to get this three ways.

   First, think. 1 μm is very small compared to 1 km, so it should be a very small number of kilometers, certainly much, much less than one. That leaves 10^{-3}, 10^{-9}, and 10^{-15} as candidates. Since 1 μm is 10^{-6} m, and 1 km is more than 1 m, 10^{-3} is out. Down to two choices. To think it out further, μ is 10^{-6}, and k is 10^3, and since we add or subtract exponents when multiplying or dividing by powers of 10, 10^{-9} is a better guess than 10^{-15} (remember we ruled out 10^{-3}).

   Second, use the replacement technique. Put in 10^{-6} for μ to get 10^{-6} m = ? km. Now 1 m is how many km? … 10^{-3}, so 1 μm = 10^{-6} m = 10^{-6} (10^{-3} km) = 10^{-9} km. (Remember that 10^{-3} k = $10^{-3} \times 10^3 = 1$.)

   Third, use conversion factors. $1\ \mu\text{m} = 1\ \mu\text{m}\left(\frac{10^{-6}\,\text{m}}{1\,\mu\text{m}}\right)\left(\frac{10^{-3}\text{km}}{1\,\text{m}}\right) = 10^{-9}\ \text{km}$. Note that we could also write $1\ \mu\text{m} = 1\ \mu\text{m}\left(\frac{1\ \text{m}}{10^{6}\ \mu\text{m}}\right)\left(\frac{1\ \text{km}}{10^{3}\ \text{m}}\right) = 10^{-9}\ \text{km}$, since 1 μm = 10^{-6} m (1 μm is a fraction of 1 m) ⇒ 10^6 μm = 1 m (there are lots of micrometers in one meter, a million to be exact!), and so on.

9. c, π rad/180°. 90° = 90° · (π rad/180°) = (90°/180°) · (π rad) = (π/2) rad. Notice that this is a physical equality, not a numeric one.

10. c, m/s^2 & m/s^2. Make sure that you can verify this. If you can't, get some help soon.

Exercise Hints & Solution Plans

EX. 1.11: [Sol'n Plan]

Coordinates: Extend lines from C back to the coordinate axes being used. Make sure that the lines are parallel to the those axes. Then just read off the coordinates on the axes.

Distance: Use the Pythagorean theorem.

Angle: Draw it. Notice that you can use either of the two angles. Further notice that the tangent of the angle is a ratio of the coordinates. Use the inverse tangent to find the angle.

Length & Width: Do these depend on the axes that you use? Make it as easy as possible.

EX. 1.14: [Sol'n Plan] Check to see if all of the additive terms in the equation have the same units. Do this by replacing the variables with the units.

EX. 1.16: [Sol'n Plan] There are three methods that I can think of.

Method 1: Tinker. Just play around with the variables until you have a relationship that is at least correct in the units. "Tinker evers chance."

Method 2: Write down the units for the variable sought and then replace the units with other variables. For example, v has units of $^{m}/_{s}$. Replace the units of meters and seconds with the variables x and t, respectively, to obtain $v = {}^{x}/_{t}$ for a guessed equation. Of course this only works for simple cases.

Method 3: Solve for unknown exponents. Here we break out the big guns. This one will always work. Suppose that you want to find a dimensionally correct relation between A, B, C, and D. First write

$[A] = [B]^{n}[C]^{m}[D]^{p}$, where n, m, p are any exponents (+, -, 0, fractions, etc.)

Now replace the variables with the units and choose n, m, and p so that it works out. Remember that a negative exponent will invert units. After a few slow tries on your own, you'll start to get the hang of it.

EX. 1.22: [Hint] Sketches are almost always a good idea. One end of the stick must stay on the top of his head while the other end can rotate. As you rotate the stick, the tape measure gets longer or shorter:

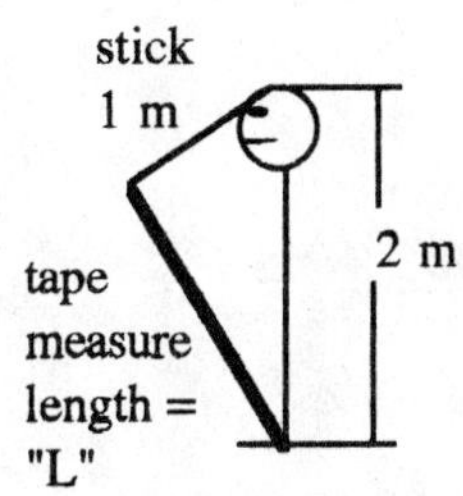

EX. 1.23: [Hint] Don't worry that you don't know the meaning of these symbols. (I don't either, I just made them up.) This is an exercise in distinguishing vectors and scalars. In preparation, answer these questions: When you add (or subtract) vectors, what do you end up with?______ Can you add a vector to a scalar?_____ What do you get when you multiply a vector times a scalar?_______ What do you get when you multiply a vector times a vector?__________ Can you divide by a scalar?______ Can you divide by a vector?_______ – That's all you need to know!

Answers: A vector; No, never!; Another vector; It depends on which of the two types of vector multiplication that you're using; Yes, always; No, never.

EX. 1.26: [Hint] What if they were all compass needles, would they have the same orientation? What if you walked exactly in the direction of each of the vectors, would you get to the same place? When dealing with vectors, do we say that these have the same direction?

EX. 1.30: [Hint] Recall that if you can form a right (90°) triangle, and you know any two pieces of information (2 sides, 2 angles, 1 side and 1 angle), then you can find out everything else from the basic trig. identities and the Pythagorean theorem:

$$\sin\alpha \ (\text{or } \cos\beta) = \frac{a}{c} \Rightarrow a = c\sin\alpha \ (\text{or } \cos\beta), \text{ and so on}$$

$$c^2 = a^2 + b^2 \quad (\text{or } a^2 = c^2 - b^2)$$

EX. 1.33: [Hint} In part (d) there are two ways to achieve the desired result...above and below his head.

EX. 1.35: [Hint] Remember that displacement ($\vec{\mathbf{D}}$) = final position ($\vec{\mathbf{r}}_f$) - initial position ($\vec{\mathbf{r}}_i$) . A very rough sketch will help immensely.

Exercise Solutions

EX. 1.1: What's going on? I count...

Dropped from	*Time counted*
my navel	"one thou" ≈ $^1/_2$ second
top of my head	"one thous" ≈ $^1/_2$ second (twice as high but not twice as long! What's up with that?)
2nd floor window (12 ft ≈ 4 m)	"one thousand" ... still not quite 1 s!!

EX. 1.2: Me? Well, *very* roughly I guess: height ≈ 2 m (ok, more like 1 $^3/_4$ m); shoulder width ≈ $^1/_2$ m; arm span ≈ 2 m; since 1 step is a bit less than 1 m, I guess a car might be ≈ 4 m long; room widths and lengths are about 3 to 6 m; room height ≈ 2 $^1/_2$ m. When you get a chance, measure and check your guesses. A yard stick will give you an estimate if that's all you have.

EX. 1.3: I'll leave this for you to consider as you study.

EX. 1.4: A sketch is helpful when trying to visualize:

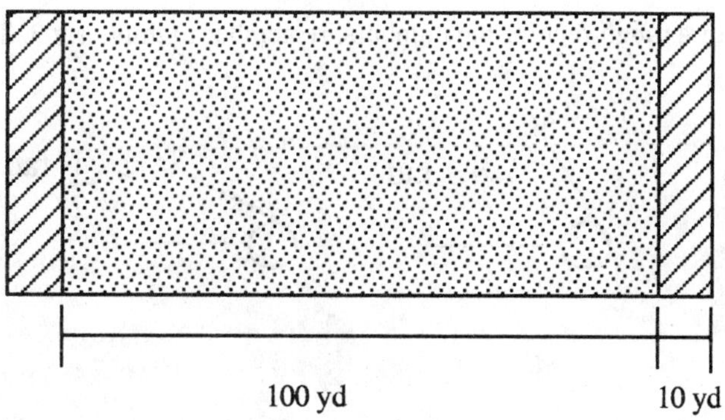

- 100 m ≈ 100 yd = length of field between the end zones.
- Better yet, 100 m ≈ 110 yd = length of playing field + 1 end zone
- Now, 1 km = 1000 m = 10 x 100 m ≈ 10 football fields
- Furthermore, 1 km = 0.62 mi ≈ $^6/_{10}$ mi (or use $^1/_2$ mi for an easier, albeit less accurate, estimate).

  So, 1 mi ≈ $^{10}/_6$ km ≈ 1.7 km (or 2 km if you used $^1/_2$ mi) ≈ 1.7 · (10 football fields) ...

  we see that 1 mi ≈ 17 football fields (or 20 if you used 1 mi ≈ 2 km). Does this makes sense when you realize that 4 times around the oval track that is around a high school or college field is about a mile?

Try to visualize the physical meaning of your numbers!

EX. 1.5: See it! Do it!

Walk it!

1 m = 1 to 2 steps and 1 second ≈ "one thousand one." You can certainly take more than one step in one second. Thus 1 m/s is a slow walk (2.2 mi/h) while 1.5 m/s is probably about an average pace (3 mi/h).

Drive it! ... Or convert.

For reasons that you may have already ascertained, I'll convert rather than drive this one. Using 1 km ≈ 0.5 mi (very rough), we find 200 km/h ≈ 100 mi/h. (124 mi/h is more accurate, but either way it'll probably get you a speeding ticket.)

EX. 1.6: Give me a "p"! Give me an "i"! What's that spell?! What's that spell?!

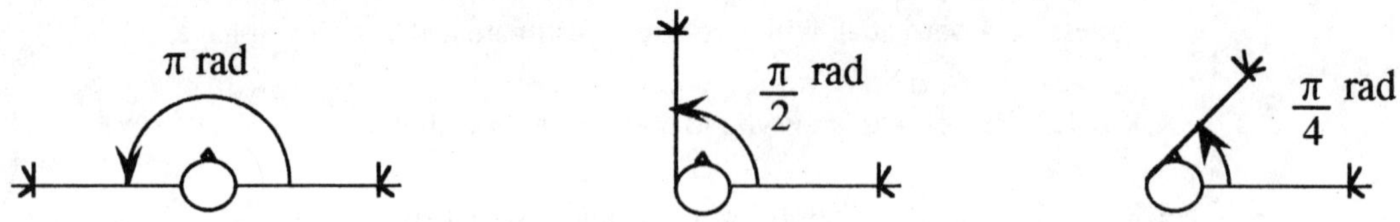

For a rough feel for the numbers, remember to think $\pi \approx 3$. Then 1 rad ≈ (π/3) rad ≈ (1/3) 180° = 60°. So,

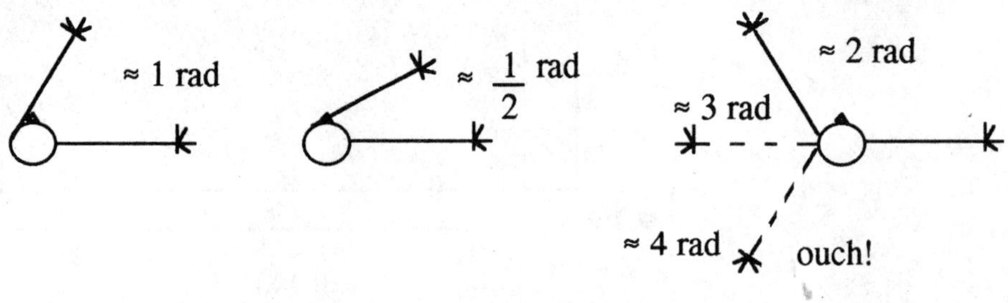

EX. 1.7: Use the fact that 1 kg weighs 2.2 lb ≈ 2 lb on Earth. Some very rough

<u>weight</u>	<u>*mass*</u>
baby's wt ≈ 7 lb	baby's mass ≈ 3.5 kg (really 3.2)
sack of potatoes ≈ 10 lb	sack of potatoes mass ≈ 5 kg (really 4.5)
child's wt ≈ 60 lb	child's mass ≈ 30 kg (really 27)
your wt ≈ ___ lb	your mass ≈ ___ kg (really ___)

Thus, a 2 liter container of water (or soda), with a mass of 2 kg, weighs about 4 lb (really 4.4 lb).

EX. 1.8: Here are my thoughts:

- What tools did B use to get that kind of accuracy, a laser?!
- B is either lying to try to impress me or is very unclear about basic measurement and the significance of the numbers obtained. Either way, it doesn't bode well. Thus, I question B's honesty and/or competence.
- A's figure seems to have reasonable accuracy. The last digit, i.e. tenths of a square inch, is duly uncertain.
- If B's figures are justifiable, say by laser technology, then how much is it going to cost me for that kind of accuracy and do I really require that kind of accuracy for this job? Furthermore, does it take B more or less time than A to obtain the measurement? Money and time are important factors when determining the desired accuracy in any project.
- In sum, if B's figures are true and B can do the job at a cost and in a time that beats A, I'd go with B. Otherwise, if A's numbers are accurate to the degree claimed by the number reported–which is an accuracy that is certainly adequate for the job–and the cost and quotes are acceptable, I'd go with A.

EX. 1.9: What did you come up with? Here is my list:

1. Use the *horizontal* reference directions: North-South-East-West. A compass would be handy for this method. You could also assign a horizontal x-y coordinate system to these directions.

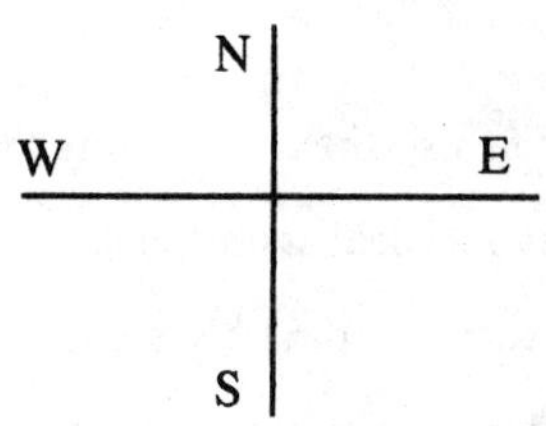

2. Use reference directions: forward-backward-right-left-up-down.
3. Physically point in that direction! Use your finger or an arrow as your pointer.
4. Draw a map and place an arrow on the map.
5. Can you come up with more?

EX. 1.10: I'll contrast the vector concepts with some scalar concepts.

<u>*Scalar Concepts*</u>	<u>*Vector Concepts*</u>
time ($1^1/2$ hours)	displacement (3 m, North)
distance (2 m)	position (2 m, in front of me)
speed (5 $^{mi}/_{h}$)	velocity (5 $^{mi}/_{h}$, SE)
volume (3 ft^3)	force (20.9 lb, down)
(your choice)	(Can you think of another? There are many.)

EX. 1.11: Sketch it! See the triangles!

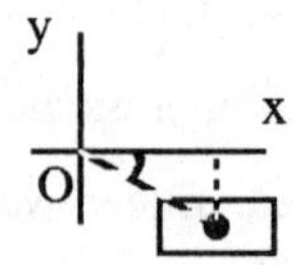

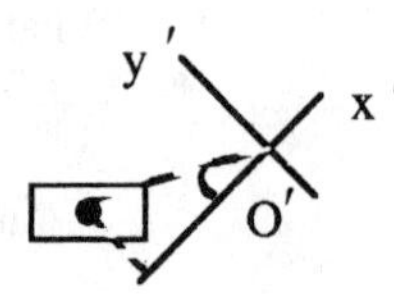

	x-y system	*x'-y' system*
position of C:	$(x,y) = (6, -3)$	$(x',y') = (-5, 2)$
distance from origin to C:	$\sqrt{(6)^2 + (-3)^2} = \sqrt{27}$	$\sqrt{(-5)^2 + (2)^2} = \sqrt{29}$
angle of line from origin to C (measured as shown):	$\tan\theta = \dfrac{y}{x} \Rightarrow$ $\theta = \tan^{-1}\left(\dfrac{3}{6}\right) = 26.6°$	$\tan\theta' = \dfrac{y'}{x'} \Rightarrow$ $\theta' = \tan^{-1}\left(\dfrac{2}{5}\right) = 21.8°$
length of object:	4 units	same as in the *x-y* system
width of object:	1 unit	same as in the *x-y* system

Notes:

- It is much easier to use the *x-y* system to find the length and width of the object. Once we've done that, we don't have to deal with the more difficult process using the *x'-y'* system. This technique of switching to an easier coordinate system becomes extremely important in later work.
- For finding angles using inverse trig. functions, I like to sketch the triangle so that I can see what angle I am calculating. Notice that I use *positive* lengths for the sides of the triangles even though the coordinates may be negative. Try keeping the negative signs in. The answer is correct, you just need to be able to see what angle it is that you have found. For now, I suggest the method that I've used.

EX. 1.12: Look at either of the axes and notice that the circle that goes through the center of the object has a radius of 4 units; hence, $r = 4$. Now notice that the radial lines divide up the circle evenly. A quarter circle is divided into three parts, so each segment is 30°. Thus the angle from the $+x$-axis is 5 segments, or 150° (30° from the $-x$-axis).

$$(r, \theta) = (4 \text{ units}, 150°)$$

EX. 1.13: You need to be able to do this sort of thing quickly. Get help if you have *any* questions.

a) Nonsense! ... unless you convert so that both terms have the same units. Then 1 h + ½ h = 1½ h, 60 min + 30 min = 90 min, or 3600 s + 1800 s = 5400 s. Any way you look at it, it sure isn't 31!

b) (10 N)·(3 m - 5 m) = (10 N)·(-2 m) = **-20 N·m**

c) $\dfrac{4.0 \times 10^6\ \text{J}}{2\ \text{s}} = \dfrac{4.0 \times 10^6}{2} \dfrac{\text{J}}{\text{s}} = \mathbf{2 \times 10^6 \dfrac{J}{s}}$

d) OK, you didn't know this, but the three units shown are different ways of expressing the same unit (it's a unit of power). Because they are the same, we can go ahead and add them.

$$\left(\frac{11}{10} + 2.2 + 1.1\right)\left(\begin{array}{c}\text{in the units expressed}\\ \text{any of the three ways}\end{array}\right) = (1.1 + 2.2 + 1.1)\left(\frac{\text{N}\cdot\text{m}}{\text{s}}, \frac{\text{J}}{\text{s}}, \text{or W}\right) = \mathbf{4.4\ W}$$

e) Since $\text{m/s} \cdot \text{s} = \text{m}$, and $\text{m/s}^2 \cdot \text{s}^2 = \text{m}$, all terms have the same units and we can add them to obtain:

$$[3.00 - (2.50)(2.00) + \tfrac{1}{2}(2.00)(2.00)^2]\ \text{m} = [3.00 - 5.00 + 4.00]\ \text{m} = \mathbf{2.00\ \ m}$$

f) Same units for each additive term? No, so ... nonsense!

EX. 1.14: In fact, we cannot determine if an equation is correct just by looking at the units (compare exercises a and b). We can only determine if the equation is dimensionally correct (using either dimensions–time, T, length, L, and mass, M– or units). Still, this alerts us to conceptual, algebraic, or transcription errors. It also familiarizes us with equations that we check.

a) x and x_o have units of m, v_ot has units of $(\text{m/s})\cdot\text{s} = \text{m}$, and $\tfrac{1}{2}at^2$ has units of $(\text{m/s}^2)\cdot\text{s}^2 = \text{m}$. Looks OK.

b) Everything is the same as in (a) above except that there is a $\tfrac{1}{2}$ missing in front of the at^2 term. Since the $\tfrac{1}{2}$ doesn't carry units, each term still has the same units and may be added. So which one is correct? Only theoretical derivation and/or experiment can determine that. All we can tell is that they aren't ridiculous. (Psst. It turns out that you do need that $\tfrac{1}{2}$.)

c) v and v_o have units of m/s, but at has units of ... $(\text{m/s}^2)\cdot\text{s} = \text{m/s}$. Can't trick you. This one passes too.

d) OK, you could guess that this one is wrong, right? Actually, it's OK too, since $2a\Delta x$ has units of $(\text{m/s}^2)\cdot\text{m} = (\text{m/s})^2$.

e) No way! The v's aren't squared, whereas the ax term has units of $(\text{m/s})^2$.

f) Interestingly enough, since dx can be thought of as an infinitesimal piece of x (a small Δx), it has units of m. Likewise, dt has units of s. Thus dx/dt has units of m/s, which is not the unit for acceleration! This equation is ABSURD! [On the other hand, $a = d^2x/dt^2$ does pass the unit check.]

¡¡For your own benefit, check the units of every new equation that you run into or derive!!

EX. 1.15: This is just a matter of careful algebra. You don't even have to know what the heck the equations mean! Verify that:

a) Units of $G = \text{N}\cdot\text{m}^2/\text{kg}^2$.

b) Units of k = N/m.

c) Units of $g = \text{m/s}^2$. Notice that you only really need to work with the last term. It must have units of m in order to be added to the other terms.

d) Units of $G = \text{N}\cdot\text{m}^2/\text{kg}^2$.

e) Units of $k = \text{N}\cdot\text{m}^2/\text{C}^2$. Same letter as that used in (b), but the meaning is quite different!

f) Units of E = N/C. (The old electric field!)

g) Units of E = volts/m. Same letter as in (f) ... and actually the exact same meaning. These are equivalent units.

h) Units of $B = \text{N}/(\text{C}\cdot{}^{\text{m}}\!/\!_{\text{s}}) = (\text{N}\cdot\text{s})/(\text{C}\cdot\text{m})$. (The old magnetic field!) Notice that sin q doesn't have any units; it's just a number.

EX. 1.16: [See the hints section for three methods that you can use.] Not all of these are correct, but they are good ballpark guesses. You will encounter these relations later, so do try and work them out yourself as a preview.

a) $v \approx x/t,\ x \approx vt,\ t \approx x/v$

b) $y \approx gt^2$

c) $v^2 \approx ax$, or any algebraic variation such as $a \approx v^2/x$.

d) $F \approx U/x,\ U \approx F{\cdot}x$

e) $U \approx mgy$

f) $U \approx mv^2$

EX. 1.17:

<u>By conversion factors</u>: Since 1 km = 1000m and 1 m = 100 cm,

$$4\ \text{km} = 4\ \text{km}\left(\frac{1000\ \text{m}}{1\ \text{km}}\right)\left(\frac{100\ \text{cm}}{1\ \text{m}}\right) = 4\times 10^5\ \text{cm}$$

<u>By replacement</u>: 4 km = 4 (10^3 m) = 4 x 10^3 (m) = 4 x 10^3 (10^2 cm) = **4 x 10^5 cm**

EX. 1.18: By convention, we take "230" to mean "**230 ± 5**", which is to say that the last significant figure is the 3, not the 0. The precision is obtained from half of a unit in the last significant digit. If we break with convention and treat "230" as having three sig. fig.'s, we would then write "**230 ± 0.5**".

Scientific notation neatly handles the annoyance of not being able to treat the 0 as significant even if it is. 230 to two sig. fig.'s is **2.3 x 10^2**, while 230 to three sig. fig.'s is **2.30 x 10^2**. What a pain! But how else can we know if that last zero is to be taken as a significant figure?

EX. 1.19: Remember, # sig. fig.'s = accuracy and uncertainty = precision. Mnemonic: "P U!! ... That stinks having two words for the same thing." Get it? P and U stand for ...

a) 1040 → sig. fig.'s (AKA "accuracy") : **3 sig. fig.'s**

uncertainty (AKA "precision") : **± 5**

b) 620.30 ± 0.002 → sig. fig.'s (AKA "accuracy") : **5 sig. fig.'s**

uncertainty (AKA "precision") : ± **0.002**

c) $\dfrac{\left(2.2\times10^{2}\right)(30)^{2}}{3.9876} \rightarrow$

In any product and/or quotient, the number of significant figures ("accuracy") is determined by the factor with the fewest sig. fig.'s. Here, that is "30," with **1 sig. fig.**

The uncertainty ("precision") is taken as ½ of a unit of the last sig. fig., but what is the last sig. fig.? We'll have to calculate the number to see.

$$\frac{\left(2.2\times10^{2}\right)(30)^{2}}{3.9876} \xrightarrow{\text{rule \#2}} \frac{2\times10^{5}}{3.9876} \xrightarrow{\text{rule \#2}} 5\times10^{4}$$

Thus the uncertainty ("precision") is ± $\mathbf{0.5 \times 10^4}$.

d) $(2.9 \pm 0.06) - (2.9\times10^{2}) \rightarrow$

Both terms have **2 sig. fig.'s (accuracy)**, so no mystery there. The terms have a uncertainties (precisions) of ± 0.06 and ± 0.05 x 10^2 = ± 5, respectively. Clearly then, the second term is the least certain (has the greatest uncertainty, since 5 > 0.06). Thus the **uncertainty (precision) is ± 5**. From this we can see that the first term makes no difference. Want proof?

$$(2.9\pm0.06)-(2.9\times10^{2}) = 2.9+290 = 292.9 \xrightarrow{\text{to 2 s.f.'s}} 290$$

Alternatively,

$$(2.9\pm0.06)-(2.9\times10^{2}) = 0.029\times10^{2}+2.9\times10^{2} = 2.929\times10^{2} \xrightarrow{\text{to 2 s.f.'s}} 2.9\times10^{2}$$

e) 3300 + 11 →

Both have 2 sf's, and the uncertainty of each is ± 50 and ± 0.5, respectively. So 3300 + 11 = 3311 is, in the physical sciences, really only known to an accuracy of 2 sf's and a precision of ± 50 (largest uncertainty). Hence, 3300 + 11 = **3300 ± 50**!

f) First, follow the rules methodically. Then, check with your study partners! (See the next exercise!)

EX. 1.20:

Using the rules at each step:

$$(2.0+2.378)\cdot(21.28)\xrightarrow{\text{rule 3}}(4.4)\cdot(21.28)=93.63\xrightarrow{\text{rule 2}}94\pm0.5$$

Keeping all sig. fig.'s until the end (via a calculator)

$$(2.0+2.378)\cdot(21.28)\xrightarrow{\text{keeping all sf' s}}(4.378)\cdot(21.28)=93.16\xrightarrow{\text{reduce to 2 sf' s}}93\pm0.5$$

Which is "correct"? I'd use the latter to avoid round-off error – it's easier, too! Clearly, these rules are reasonable guidelines. It's a bit difficult to come up with a general rule that says when to round-off and when not to.

EX. 1.21: You can never add (or subtract) a vector to a scalar! Either direction matters, or it doesn't.

a) 1 foot forward, of course.

b) Gibberish! 3 ft what? Forward? Sideways? Up? 3 ft of string? What's this mean? Nothing!

c) Standard algebra rules apply: $\vec{\mathbf{C}}=\vec{\mathbf{E}}-\vec{\mathbf{D}}-\vec{\mathbf{B}}-\vec{\mathbf{A}}$

d) Gibberish! Because F_2 is not a vector, this equation makes absolutely no sense at all!

e) Phew! All is well again: $\vec{\mathbf{F}}_1=\vec{\mathbf{N}}-\left(\vec{\mathbf{T}}+\vec{\mathbf{F}}_2-\vec{\mathbf{f}}\right)=\vec{\mathbf{N}}-\vec{\mathbf{T}}-\vec{\mathbf{F}}_2+\vec{\mathbf{f}}$

f) Just checking. 54.

g) You must be fast with this basic algebra:

$$3x+mgh-c-5x=62\;\rightarrow\;-2x=62-mgh+c\;\rightarrow\;x=\frac{-62+mgh-c}{2}$$

h) Just $4\vec{\mathbf{A}}$. A vector that is in the direction of $\vec{\mathbf{A}}$ and has 4 times the magnitude.

i) $5\vec{\mathbf{A}}$

EX. 1.22: See sketch in the hints section.

a) He should hold the stick so that it goes straight above his head.

b) He should hold the stick so that it goes straight down from his head.

c) The tape will be stretched less than in (a), but more than in (b). Probably *not* halfway between the extreme lengths.

d) No! In (c) we need to use some trig. and/or the Pythag. th'm … or look at the tape measure!

EX. 1.23: Do make sure that you check the hints section if you need help.

a) Think of $\left(\vec{\mathbf{B}}+\vec{\mathbf{C}}\right)$ as the one vector that results from adding the two vectors. Then we certainly can multiply it by another vector. Using the scalar (dot) product results in a scalar. We can distribute the $\vec{\mathbf{A}}$ first if we like. Then we have $\left(\vec{\mathbf{A}}\cdot\vec{\mathbf{B}}\right)+\left(\vec{\mathbf{A}}\cdot\vec{\mathbf{C}}\right)$, the sum of two scalars.

b) $\vec{\mathbf{A}}\times\left(2\vec{\mathbf{B}}-b\vec{\mathbf{C}}\right)=\left[\vec{\mathbf{A}}\times\left(2\vec{\mathbf{B}}\right)\right]-\left[\vec{\mathbf{A}}\times\left(-b\vec{\mathbf{C}}\right)\right]=2\left[\vec{\mathbf{A}}\times\vec{\mathbf{B}}\right]+b\left[\vec{\mathbf{A}}\times\vec{\mathbf{C}}\right]$

Since $2\vec{\mathbf{B}}$, $b\vec{\mathbf{C}}$, and $\left(2\vec{\mathbf{B}}-b\vec{\mathbf{C}}\right)$ are all vectors, the vector (cross) product with $\vec{\mathbf{A}}$ is valid and results in a vector. Upon distribution of $\vec{\mathbf{A}}$, we end up with the addition of two vectors. Although the result is the same, this sort of algebraic manipulation is very useful.

c) Utter nonsense! While $\vec{\mathbf{D}}, \vec{\mathbf{E}}$, and $(\vec{\mathbf{A}} \times \vec{\mathbf{B}})$ are vectors, $(\vec{\mathbf{B}} \cdot \vec{\mathbf{C}})$ is a scalar. You know what happens to people that add scalars to vectors, don't you? It's not pretty.

d) OK, these are all vector terms, so just move 'em around to get: $\vec{\mathbf{F}} = 2(\vec{\mathbf{A}} \times \vec{\mathbf{B}})$

e) By reverse-distribution (i.e. factoring out the $\vec{\mathbf{r}}$), we can write this as:

$$\vec{\tau} = (\vec{\mathbf{r}} \times \vec{\mathbf{F}}_1) + (\vec{\mathbf{r}} \times \vec{\mathbf{F}}_2) = \vec{\mathbf{r}} \times (\vec{\mathbf{F}}_1 + \vec{\mathbf{F}}_2)$$

But we can't divide by $\vec{\mathbf{r}}$, so we're stuck ... for now.

f) Notice: $\underbrace{\underbrace{5\vec{\mathbf{A}}}_{\text{vector}} \cdot \underbrace{(\vec{\mathbf{B}} \times \vec{\mathbf{C}})}_{\text{vector}}}_{\text{scalar}} = \dfrac{4D}{\underbrace{E}_{\text{scalar}}}$ So this is OK! The answer is simply: $E = \dfrac{4D}{\left[5\vec{\mathbf{A}} \cdot (\vec{\mathbf{B}} \times \vec{\mathbf{C}})\right]}$

g) Utter nonsense! You can't divide by $\vec{\mathbf{E}}$.

EX. 1.24: As you probably surmised, the units for both are N•m, however the two concepts are totally different. So different, in fact, that in the case of work, W, we create another name for the unit: 1 N•m is also called 1 J (joule), for short. This is a unit of energy. However, when we are working with torques, the unit "N•m" is NOT equivalent to a joule (J). We simply leave it as N•m. The ideas that you are working with will make the distinction readily apparent. But we will get to that in time.

EX. 1.25: This is good training in being methodical, seeing in 3-D, and using the Pythag. th'm.

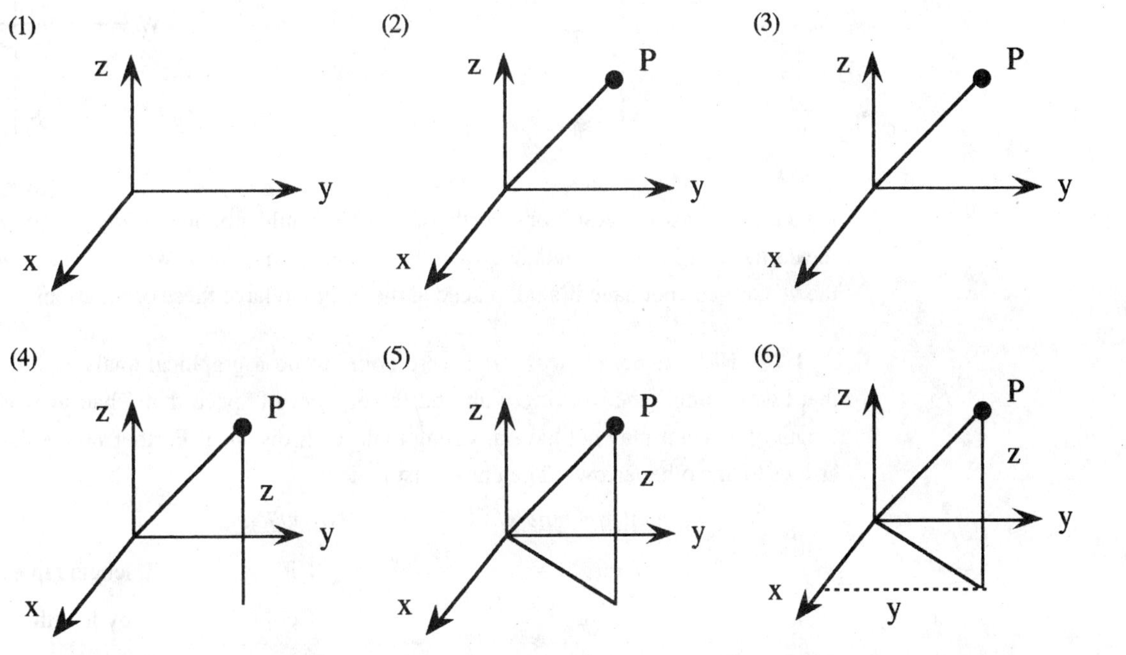

Notice in the final diagram that the Pythagorean theorem gives $\sqrt{x^2 + y^2}$ for the diagonal in the x-y plane. This side is now a side of the right triangle whose hypotenuse is from the origin to the point P, thereby giving a length of $\sqrt{\left(\sqrt{x^2 + y^2}\right)^2 + z^2} = \sqrt{x^2 + y^2 + z^2}$.

EX. 1.26: I think that it's clear that they all have the same orientation. – if they were compasses, they might all be pointing north. What may be a little confusing is that we say they have the same direction. It's confusing because if you were to walk along the line of one of the vectors, we might hit, say, a telephone pole, but while walking along a parallel vector's line we might not.

Try this: draw a vector that points to the right across the page. Did you do it? Why did you draw it there?! You could have drawn it anywhere to specify "to the right, across the page." Similarly, when we represent a vector quantity by an arrow, we can move it around on the paper and still consider it to be the *same* vector—as long as the length and orientation ("direction") stay the same. (See the next exercise.)

EX. 1.27: I think that the simplest way to get the hang of this terminology is to think of "33° N of E" as "33° N *from* E." What do you think?

1) 33° N of W = 57° W of N

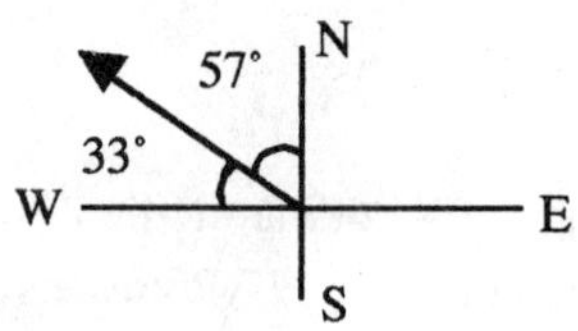

2) NW = 45° N of W = 45° W of N

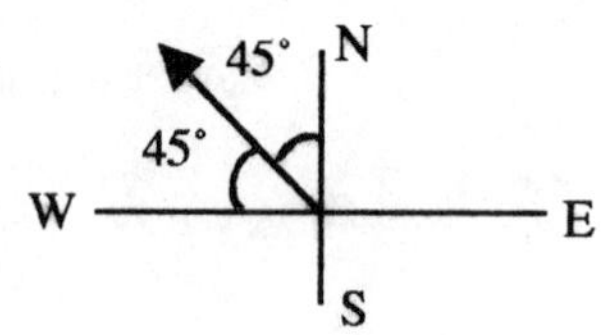

3) 20° E of S = 70° S of E

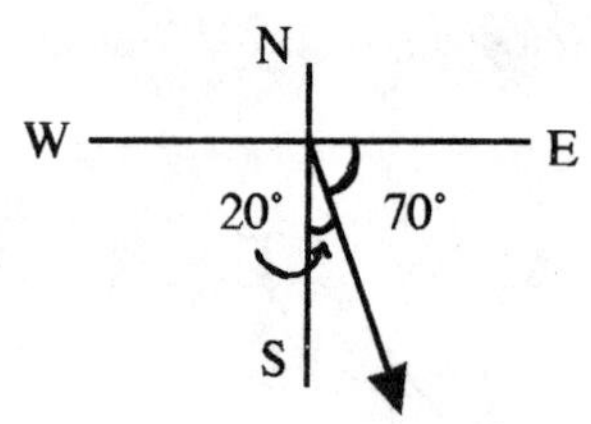

4) 20° S of E = 70° E of S

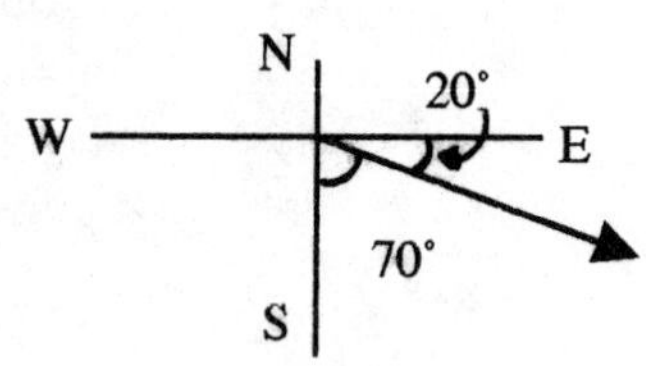

Notice that "east of west," or "north of south" would not make any sense. Also, "southwest, SW" means halfway between south and west (at 45°). By convention, we don't say "westsouth." And finally, the vector need not have it's tail placed at the origin to have these orientations.

EX. 1.28: Here are my choices. If I were going to do a graphical analysis, I'd look at the magnitudes that I was going to be working with and the amount of space that I had to work with. Notice that the length of a vector may not have anything to do with distance! Further notice that we can't just add any arrows to any other arrows. The units must match!

Actual Magnitude	*Chosen Scale*	
1 mi	1 in	length represented
100 m	1 cm	by length
50 mi/h	2 cm	speed represented
10 m/s	1 cm	by length
25 lb	5 cm	force magnitude
10 N	10 cm	represented by length

EX. 1.29: This is a very important exercise. Remember the grid and that components are *from* tail *to* head.

GRID	EQUIVALENT TRIANGLES	COMPONENT SIGNS
(a) 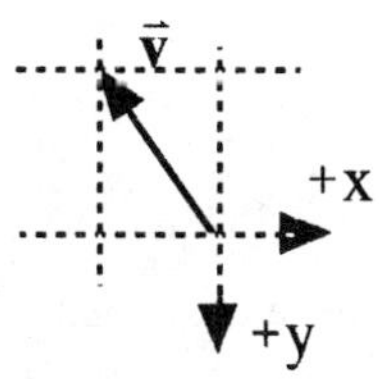	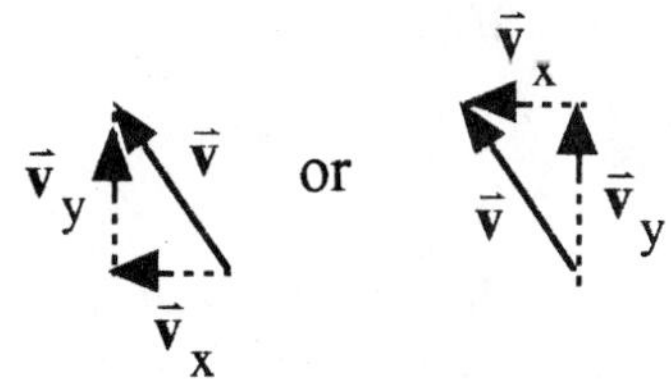	x-comp. : negative y-comp. : negative
(b) 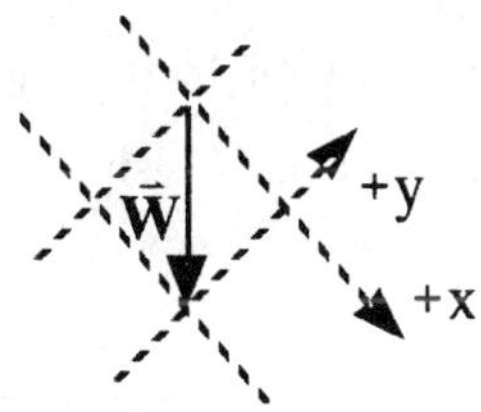	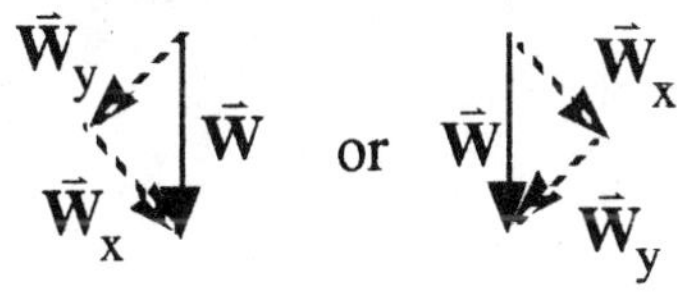	x-comp. : positive y-comp. : negative
(c) 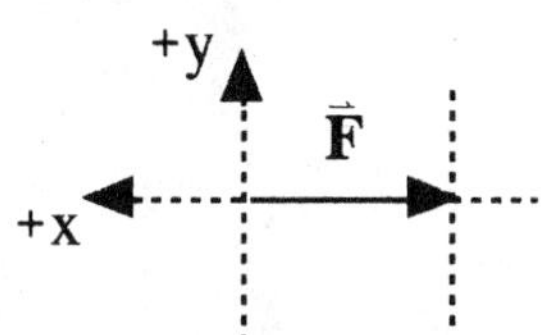	No triangles, since there is only one component. That is, the vector is entirely in one of the chosen coordinate directions.	x-comp. : negative y-comp. : 0
(d) 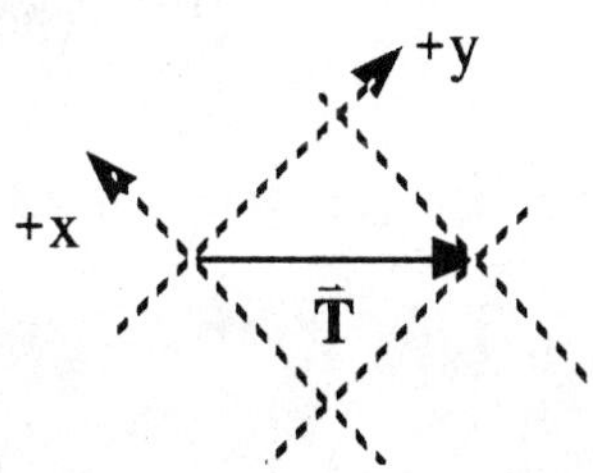	You draw these two. Start at the tail, move along either grid line, stop and put an arrow head at the corner, turn onto another grid line and move to the tip, draw the final component arrow head.	x-comp. : negative y-comp. : positive

EX. 1.30: Try to use the following to help you complete the solution before turning the page.

SKETCH	COMPONENT GRID	I'LL USE THIS TRIANGLE
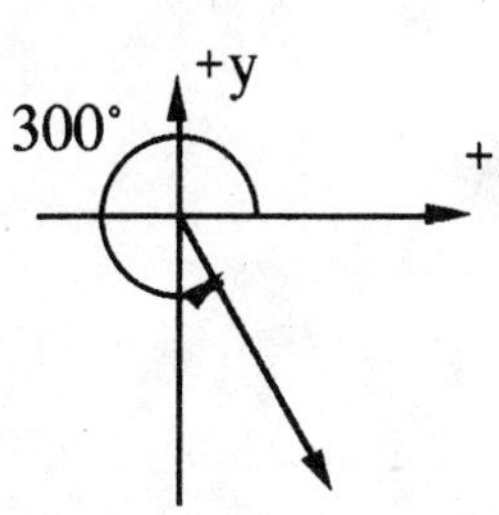	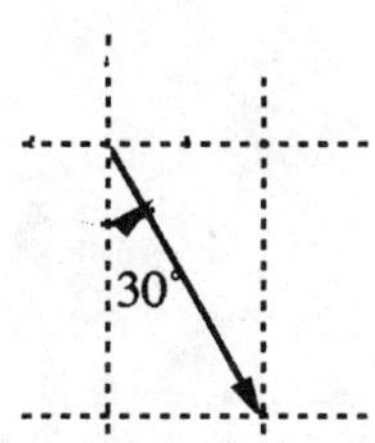	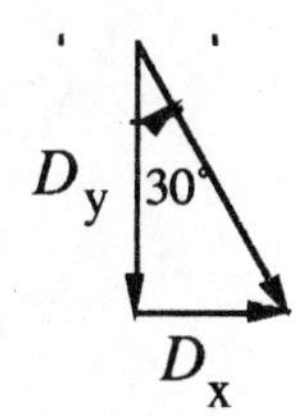

You can use either triangle. Using the one I chose, you need to be able to immediately see that:

$$\sin 30^\circ = \frac{|D_x|}{|D|} \Rightarrow |D_x| = |D|\sin 30^\circ = (50\text{ m})(\sin 30^\circ) \Rightarrow \boldsymbol{D_x = +25\text{ m}}$$

$$\cos 30^\circ = \frac{|D_y|}{|D|} \Rightarrow |D_y| = |D|\cos 30^\circ = (50\text{ m})(\cos 30^\circ) \Rightarrow \boldsymbol{D_y = -43\text{ m}}$$

Did you notice that I used positive numbers (absolute values) when I worked with the relations for the triangle that I chose? If you do this, you need to be very careful at the end to put in the correct signs for the components. From the sketch, you can see that you walked in the positive x direction and the negative y direction. The sketch will always keep things straight. (You might investigate what happens when you don't use absolute values. Just make sure that your numbers and +/- signs match your sketch!) Now let's check to see that our results match what we already know.

Pythagorean theorem verification: $D = \sqrt{D_x^2 + D_y^2} = \sqrt{(25)^2 + (-43)^2} = 50\text{ m} \rightarrow$ good!

Inverse tangent verification: $\tan^{-1}\left(\frac{|D_x|}{|D_y|}\right) = \tan^{-1}\left(\frac{25}{43}\right) = 30^\circ \rightarrow$ check the triangle!

Or, you can use:

$$\tan^{-1}\left(\frac{|D_y|}{|D_x|}\right) = \tan^{-1}\left(\frac{43}{25}\right) = 60^\circ \rightarrow \text{check the triangle!}$$

Consistency checks help you find out if any errors crept into your calculations.

EX. 1.31: $\vec{\mathbf{D}} = D_x\hat{\mathbf{i}} + D_y\hat{\mathbf{j}} = (25\text{ m})\hat{\mathbf{i}} + (-43\text{ m})\hat{\mathbf{j}}$

Note that this could also be written as follows. In practice, I would use the first form below, but I kind of like how the second form neatly separates the magnitude and the direction.

$$\vec{\mathbf{D}} = (25\text{ m})\hat{\mathbf{i}} - (43\text{ m})\hat{\mathbf{j}} \quad \text{or} \quad (25\text{ m})\hat{\mathbf{i}} + (43\text{ m})(-\hat{\mathbf{j}})$$

EX. 1.32: How are your basic trig. and Pythag. th'm coming along?

a) The all important sketch:

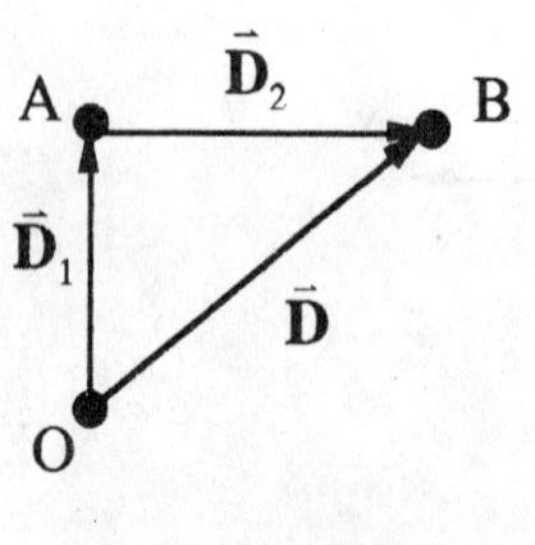

b) Since 1 block = 100 m: $\left|\vec{\mathbf{D}}_1\right| = D_1 = 300 \text{ m}, \quad \left|\vec{\mathbf{D}}_2\right| = D_2 = 400 \text{ m}$

So, $\left|\vec{\mathbf{D}}_{\text{total}}\right| = D_{\text{total}} = \sqrt{(300 \text{ m})^2 + (400 \text{ m})^2} = 500 \text{ m}$ (A 3 - 4 - 5 right triangle!)

c) Distance walked $= D_1 + D_2 = 300 \text{ m} + 400 \text{ m} = \mathbf{700 \text{ m}}$ (Add magnitudes without regard to direction.)

d) Refer to the sketch to see that we can write:

$$\tan\theta = \frac{400 \text{ m}}{300 \text{ m}} \Rightarrow \theta = \tan^{-1}\left(\frac{400}{300}\right) = \tan^{-1}\left(\frac{4}{3}\right) = \mathbf{53^\circ}$$

EX. 1.33: I'll leave the first two sketches to you. The lessons here are 1) You can't just add numbers anymore. If vectors are involved, the directions make a huge difference, and 2) the Pythag. th'm and basic trig. relations are the keys to analyzing vectors. (Note that 3 sig. fig.'s are implicit.)

a) He needs to place the stick so that it goes straight UP from the top of his head. Then the tape length (distance between his feet and the end of the meter stick) = 2 m + 1 m = **3 m** (no angles to deal with).

b) Now point the stick downward from the top of his head: shortest tape length = 2 m - 1 m = **1 m.**

c) The stick is horizontal and makes a right angle with his body. The tape measure, stretching from his feet to the end of the stick, forms the hypotenuse of a right triangle with sides of length 2 m and 1 m. Thus we expect a tape length somewhere between the extremes. By good old Pythag. and the definition of the tangent:

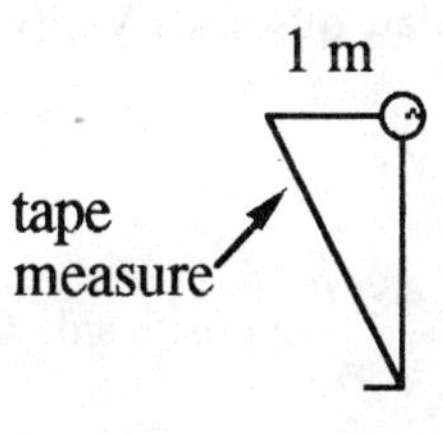

$$\text{tape length} = \sqrt{(1)^2 + (2)^2} = \mathbf{2.24 \text{ m}}$$

$$\theta_{\text{between tape and legs}} = \tan^{-1}\left(\frac{1}{2}\right) = \mathbf{26.6^\circ}$$

d) [This one is good!] You'll need clear sketches, clear application of the law of cosines, and the quadratic equation. But just take it as it comes. As you realized or read in the hints section, there are two ways to achieve this:

<u>(A) Short tape (below head)</u>

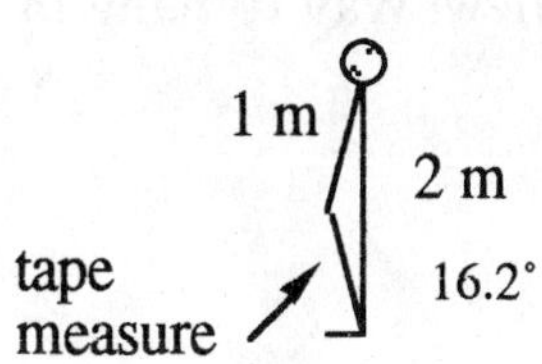

<u>(B) Long tape (above head)</u>

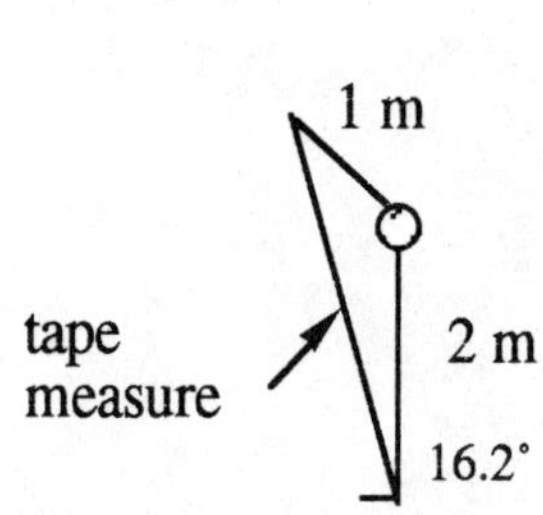

LAW OF COSINES: This applies to any triangle and is a primary tool when you don't have a right triangle. Biggest mistake = not paying attention to which angle is being used.

$$C^2 = A^2 + B^2 - 2AB\cos\theta_{\text{opposite side C}}$$

We only know the angle between his legs and the tape measure (16.2°), and the side opposite that angle has a length of 1 m. This is true in both cases sketched above, so either way we are forced to use: θ = 16.2° → C = side opposite θ = 1.00 m, either of the other variables can be 2.00 m, which leaves one unknown. Let's call the unknown "L," since it's the length of the tape.

$$1^2 = L^2 + 2^2 - 2L2\cos(16.2°) \xrightarrow{\text{quadratic form}} L^2 - (3.84)L + 3 = 0$$

There is no obvious way to factor this, so we turn to the quadratic equation:

$$L = \frac{-(-3.84) \pm \sqrt{(-3.84)^2 - 4(1)(3)}}{2(1)} \Rightarrow \begin{cases} L_- = 1.09 \text{ m} \\ L_+ = 2.75 \text{ m} \end{cases}$$

Two solutions! As expected, one is smaller than 2 m, the other is larger than 2 m, and they are both between the max and min possible values. Cool.

Now we just need to find the angle between his body and the meter stick. Again, you should expect two solutions – one small, the other larger than 90° (check the sketches!). We can't use the basic trig functions, so we'll go back to the law of cosines...carefully! (Actually, we could use the basic trig functions if we formed some right triangles and played around a bit. But guess what, we'd just derive the law of cosines and the law of sines!) Verify that a little algebraic manipulation of the law of cosines gives

$$\cos\theta_{\text{opposite side C}} = \frac{A^2 + B^2 - C^2}{2AB} \Rightarrow \theta_{\text{opposite side C}} = \cos^{-1}\left(\frac{A^2 + B^2 - C^2}{2AB}\right)$$

What side is opposite the angle in question? The tape measure, L, so use that as C. The other sides have lengths of 1 m and 2 m. Crank the wheel!

$C = L_{\text{short}} = 1.09$ m → θ = **17.6°**

$C = L_{\text{long}} = 2.75$ m → θ = **130°**

(I should write this as $(1.30 \times 10^2)°$, since there are 3 sig. fig.'s)

Phew! Way to hang in there!

EX. 1.34: You could sketch this out, and it would be good practice, but you don't have to. Simply add or subtract components.

a) The net displacement is the sum of each displacement:

$$
\begin{aligned}
\vec{\mathbf{D}} &= \vec{\mathbf{D}}_1 + \vec{\mathbf{D}}_2 + \vec{\mathbf{D}}_3 + \vec{\mathbf{D}}_4 \\
&= \left[(20.1\text{ m})\hat{\mathbf{i}} - (3.15\text{ m})\hat{\mathbf{j}}\right] + \left[(51.3\text{ m})\hat{\mathbf{i}} + (40.1\text{m})\hat{\mathbf{j}}\right] \\
&\qquad + \left[-(11.7\text{ m})\hat{\mathbf{i}} + (62.2\text{ m})\hat{\mathbf{j}}\right] + \left[(9.73\text{ m})\hat{\mathbf{i}} + (33.3\text{ m})\hat{\mathbf{j}}\right] \\
&= \left[(20.1\text{ m}) + (51.3\text{ m}) - (11.7\text{ m}) + (9.73\text{ m})\right]\hat{\mathbf{i}} \\
&\qquad + \left[-(3.15\text{ m}) + (40.1\text{m}) + (62.2\text{ m}) + (33.3\text{ m})\right]\hat{\mathbf{j}} \\
&= (69.4\text{ m})\hat{\mathbf{i}} + (132\text{ m})\hat{\mathbf{j}}
\end{aligned}
$$

But we wanted to know how far away (distance) we ended up from the start. We need the magnitude of the displacement vector. But that's easy!

$$\left|\vec{\mathbf{D}}\right| = D = \sqrt{(69.4\text{ m})^2 + (132\text{ m})^2} = \mathbf{149\text{ m}}$$

b) A rough, quick sketch always helps us get the orientation right. Simply notice that the displacement components are in the +x and +y directions (1st quadrant). Don't worry about drawing it to scale.

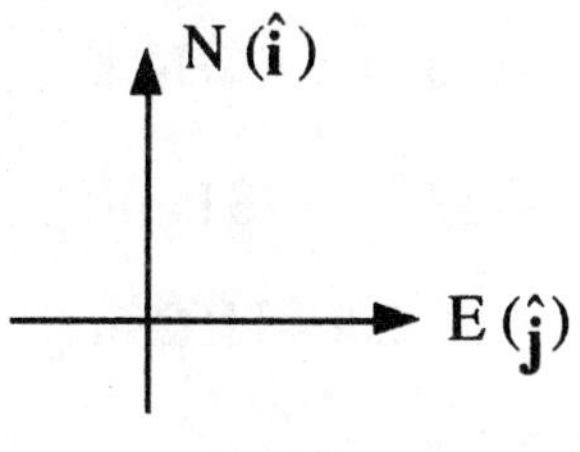

(Top view of horiz. directions.)

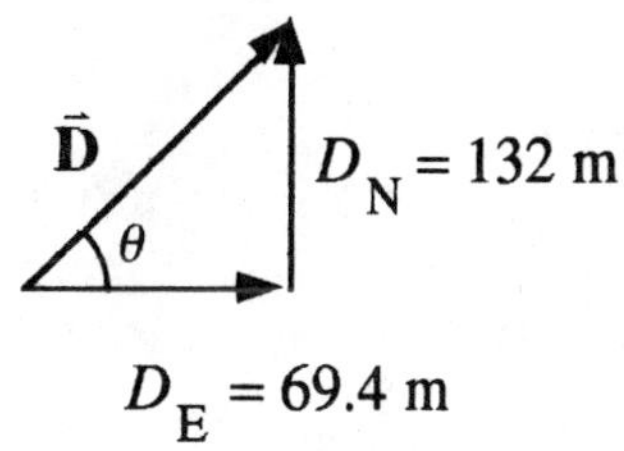

$$\theta = \tan^{-1}\left(\frac{132}{69.2}\right) = \mathbf{62.3^\circ\ N\ of\ E}$$

c) Without a sketch, you could easily get confused. A quick sketch is worth the time and space! (The distance is just D.)

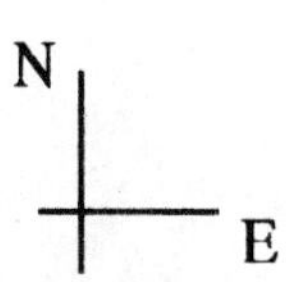

(Top View)

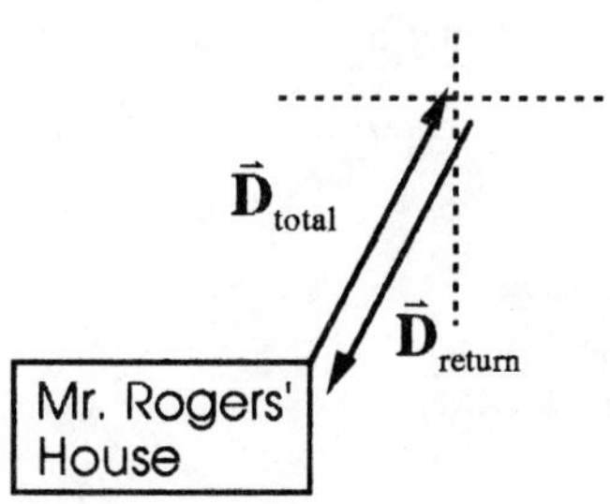

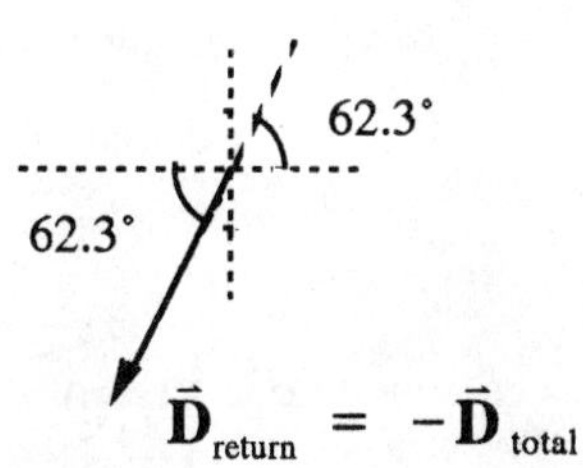

Head home at: **62.3° S of W**

EX. 1.35: Again, I'll just make a quick sketch. I won't worry about scale, but I will take note of the correct quadrants. I also might as well use a very rough scale (I can't resist) where my *y*-coordinate is about 3 to your 2, and my *x*-coordinate is like -1 to your 3. Get it?

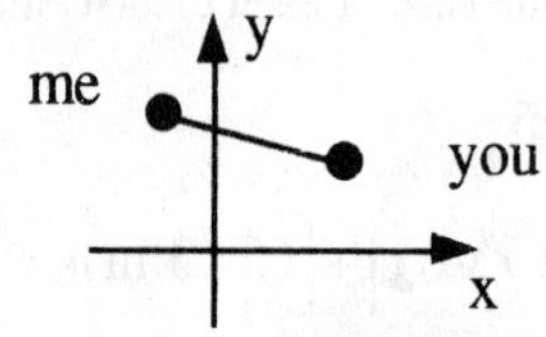

a) Notice that $\vec{\mathbf{r}}_{me}$ is the initial vector and $\vec{\mathbf{r}}_{you}$ is the final vector. The displacement is the final - initial position vector. The sketch helps us see which vector gets subtracted. Do you see that flipping $\vec{\mathbf{r}}_{me}$ gives a head-to-tail vector addition from me to you? Make sure that you see this.

$$\begin{aligned}\vec{\mathbf{D}}_{\text{from me to you}} &= \vec{\mathbf{r}}_{you} - \vec{\mathbf{r}}_{me} \\ &= \left[(311\text{ m})\hat{\mathbf{i}} + (212\text{ m})\hat{\mathbf{j}}\right] - \left[-(113\text{ m})\hat{\mathbf{i}} + (314\text{ m})\hat{\mathbf{j}}\right] \\ &= \left[(311\text{ m}) - (-113\text{ m})\right]\hat{\mathbf{i}} + \left[(212\text{ m}) - (314\text{ m})\right]\hat{\mathbf{j}} \\ &= (424\text{ m})\hat{\mathbf{i}} - (102\text{ m})\hat{\mathbf{j}}\end{aligned}$$

b) Now the initial vector is $\vec{\mathbf{r}}_{you}$ and the final vector is $\vec{\mathbf{r}}_{me}$. Graphically, flipping $\vec{\mathbf{r}}_{you}$ gives the proper head-to-tail result. (Remember: "final -initial, final - initial, fi…")

$$\begin{aligned}\vec{\mathbf{D}}_{\text{from you to me}} &= \vec{\mathbf{r}}_{me} - \vec{\mathbf{r}}_{you} \\ &= \left[-(113\text{ m})\hat{\mathbf{i}} + (314\text{ m})\hat{\mathbf{j}}\right] - \left[(311\text{ m})\hat{\mathbf{i}} + (212\text{ m})\hat{\mathbf{j}}\right] \\ &= \left[-(113\text{ m}) - (311\text{ m})\right]\hat{\mathbf{i}} + \left[(314\text{ m}) - (212\text{ m})\right]\hat{\mathbf{j}} \\ &= (-424\text{ m})\hat{\mathbf{i}} + (102\text{ m})\hat{\mathbf{j}}\end{aligned}$$

Hey! Do you notice something? We didn't have to go through all of that. This displacement is exactly the negative of the displacement in part (a). Save yourself some time by staying alert to such things.

c) [I hope you're getting the hang of this. It will make life sooo much easier in the coming weeks.] What's the key here? Yes, the sketch. We need to pick a triangle to use so that we get the correct sides and angle trig. relation. If you're clever, you only need to make one calculation. Using the magnitudes of the components of the $\vec{\mathbf{D}}$'s:

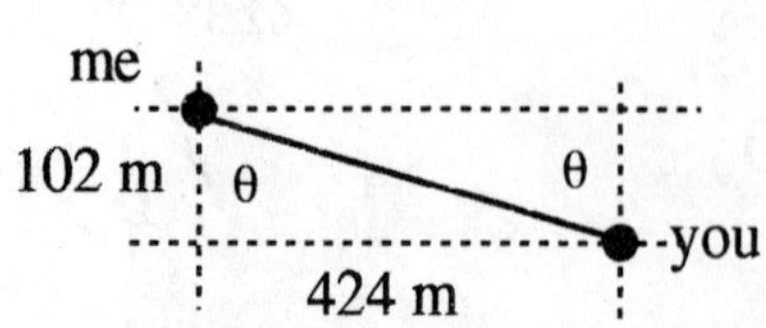

$$\theta = \tan^{-1}\left(\frac{424}{102}\right) = \mathbf{76.5^\circ \text{ in direction shown on sketches}}$$

Ch. 2: Kinematics

[I know that there are quite a few exercises in these early chapters. That's because you are just learning how to approach the subject and there are many lessons to be learned early on to prepare you for the rest of the course.] This is a very important chapter. The training that follows will work on many levels. I believe that you will gain lasting benefits from working through all of the exercises and the quiz. Especially make a point of getting to the intermediate section on 1–D uniformly accelerated motion–it's essential reading! Chapter 3 will be much easier having done this, as will the rest of the course.

Overview

Keep in mind that the task in kinematics is to *describe* the motion of an object. We are not yet describing what *causes* an object to move the way that it does. As you study kinematics, keep your eyes open and watch things move all around you. Actually get on a bike and notice your motion. What conceptual distinctions are useful in describing your motion? Later we will work on describing the *rotational* motion of the tires, pedals and gears, as well as motion while moving along a *circular* arc. In Ch. 2, we pay particular attention to the motion of an object moving back and forth along a *straight line* – 1–D motion. As you well know, it's easier to learn if you focus on the fundamental distinctions and gradually layer in the subtleties. So don't look down on this initially limited study. Kinematics is the groundwork for much more complex and interesting analysis. Herein, you will begin to develop the ability to **describe** and **predict** physical processes – a glorious achievement and an essential component of any field that uses science.

The following list categorizes some types of motion and will begin to give you a sense of the power and applicability of kinematics.

Some Types of Motion

0. **Rest:**
   - no motion, no speed (but there could be acceleration if the speed is different in the next moment!)

I. **Linear:**
   - also called "translational" motion–an object "translates" from one place to another, sort of like people "translate" from one language to another
   - motion along a straight line
   - the only change in direction is "forward" and "backward"
   - walking up and down a shopping aisle; a dropped object; etc.

II. Circular:

- motion along the arc of a circle
- ball on a string; a Ferris wheel ride is a vertical circle; satellites; roughly, the Earth around the sun (actually elliptical); etc.

III. Rotational:

- spinning around its own axis
- ice skater spinning; Earth; a compact disk (CD); etc.

IV. Oscillatory:

- back and forth motion
- pendula; springs; vibrations are small oscillations; etc.

V. Combinations:

- I + I ⇒ projectiles (surprising?)
- I + III ⇒ rolling (bowling ball, tires); yo–yo
- II + III ⇒ Earth (roughly)
- I + IV ⇒ waves! (of all types; you'll get to this later)

4 Primary Methods /Tools For Kinematics Analyses

Be sure to come back and read this overview often until it sinks in from doing problems!

- **Graphing of 1-D Motion:**
  - We often graph and analyze x vs. t, v vs. t, and a vs. t.
  - We may look at the points, the slopes, and/or how the slopes are changing
- **Application of the Definitions:**
  - While seemingly trivial, we must be careful with +/– signs and note whether or not the directions matter.
  - We often break problems down into *states* (instants) and *stages* (intervals of time). The states and stages should be clearly shown in your diagrams.
- **Constant Acceleration Equations:**
  - There are 3 (or 4) key equations that we can use to relate the time, position and velocity of an earlier instant (*state*) to the time, position and velocity of a later instant (*state*).
  - To use the constant acceleration equations, either the acceleration between the two states that we are relating must be constant, or we must use the average acceleration between the states.
- **Using Calculus:**
  - For functions that depend on time, we can move between position, velocity and acceleration functions by differentiation and integration.

Conceptual

Kinematics Concepts

There actually aren't that many kinematics definitions, but there are some subtleties that you'll need to be aware of. First, a simple exercise, then a brief list of concepts and distinctions, and then we'll get to the funny business.

EXERCISE 2.1: In describing motion, the four fundamental concepts are _____________, _____________, _____________, & _____________. [Method Tip: Simply look around and notice the essential ideas necessary to describe something's motion.]

Scalar (no direction in the definition)	**Vector** (includes direction)
	Position (+ or –): The location of an object specified relative to some chosen origin (reference place) and axes (reference directions). For instance, you may be to the left of the origin or to the right of it. In one case (either one), you'd have a positive position, and in the other, you'd have a negative position.
Distance (+): The amount traveled on the road, along the trail, etc. How far you actually walked or ran. The reading on the odometer of your car. All of these are positive numbers and do not depend on the direction.	**Displacement (+ or –)**: The net change in your position. "How far from your starting position are you (along a straight line, *not* along your path)?" The path taken doesn't matter, just the beginning and end positions. The *magnitude* of the displacement is simply the straight line distance between the initial and final positions. The *direction* is straight from the initial to the final position.

Scalar (no direction in the definition)	**Vector** (includes direction)
Ave. speed(+): The total *distance* traveled (independent of direction) divided by the total time elapsed. We needn't worry about what goes on between the initial and final times! Note that distance is always positive, so speed is too. Further note that your speedometer *always* reads a positive number. [Some sticky points will be brought out in the exercises.]	**Ave. velocity (+ or –)**: The total *displacement* divided by the time elapsed. You could drive all day and have an average speed of 40 mi/h, but if you end up at the place you started, your average velocity is zero. Why? Because the total displacement is zero. Direction most definitely matters when we're talking about velocity.
Instantaneous speed(+): What your speedometer…or the police radar…says at any instant. No direction specified.	**Instantaneous velocity (+ or –)**: The speed and direction at an instant. To find your instantaneous velocity while driving in your 4–wheeler, check your speedometer and your compass. The velocity vector is *always* tangent to the path–it is in the direction of travel at that instant.

Acceleration (+ or – ; vector): The rate of change of velocity. The more the velocity changes in, say, 1 second, the greater the acceleration. What are the only three ways that an object's velocity can change?

1. Change Speed ⇒ speed up
   ⇒ slow down
2. Change Direction

Of course ya got yer average and yer instantaneous accel.'s.

EXERCISE 2.2: You run a 1 mile race by making four complete circles around the track. You finish in 6 min. (a) What distance did you run? What is your displacement? (b) Is the ave. speed equal to the ave. velocity in this case?

EXERCISE 2.3: Give four ways to detect (notice) your own velocity. How would you detect (notice) your own acceleration?

EXERCISE 2.4: (a) Is it possible to have zero velocity and a non zero acceleration? If yes, give an example or two. If no, explain why not. (b) Is it possible to have zero acceleration and a non zero velocity? If yes, give an example or two. If no, explain why not.

EXERCISE 2.5: The *odometer* in a car registers the distance traveled, the *speedometer* registers the speed at any instant, and if you have a car *compass*, it tells you the direction you are heading at any instant.

(a) By looking at a *snapshot* of an odometer, can you tell anything about the speed of the car? What if the odometer reads 000000.0 miles?

(b) By looking at a *snapshot* of a speedometer, can you tell anything about the acceleration of the car? What if the speedometer is at 0 mi/h?

(c) If you look at a video of an odometer, how could you tell if the speed was non zero? Could this be seen even if the odometer starts out at 000000.0?

(d) If you look at a video of a speedometer, how could you tell if the acceleration was non zero? Could this be seen even if the speedometer starts out at 0 mi/h?

How much?!

In our continuing effort to have the numbers jump off of the page, let's get a feel for the magnitudes that you'll see when studying speeds and accelerations.

Speed

S.I. unit of meter/sec

1 m/s ≈ 2 mi/h (really 2.2 mi/h); a casual walk

10 m/s ≈ 20 mi/h (really 22 mi/h); world class sprinter, slow car

30 m/s ≈ 60 mi/h (really 66 mi/h); freeway speeds

unit of kilometer/hour

1 km/h ≈ 1/2 mi/h (really 0.62 mi/h); also about 28 cm/s (roughly 1 ft/s)–visualize it!

10 km/h ≈ 5 mi/h (really 6.2 mi/h); a slow jog

100 km/h ≈ 50 mi/h (really 62 mi/h); back on the freeway

Acceleration

S.I. unit of (meter/sec)/sec

$1\ \frac{m/s}{s} = 1\ m/s^2 \approx 2\ \frac{mi/h}{s}$ (really $2.2\ \frac{mi/h}{s}$); small acceleration

$10\ \frac{m/s}{s} = 10\ m/s^2 \approx 20\ \frac{mi/h}{s}$ (really $22\ \frac{mi/h}{s}$); can you do this in your car?

unit of (kilometer/hour)/second

$1\ \frac{km/h}{s} \approx 1/2\ \frac{mi/h}{s}$ (really $0.62\ \frac{mi/h}{s}$)

When you see a number, visualize what it means. Do you have a sense for the magnitudes in the list above? Well it's time to flex your physical intuition with the next two exercises.

EXERCISE 2.6: [No calculators allowed!] A dropped object accelerates at about 10 $\frac{m/s}{s}$ (10 $m/s^2 \approx 9.80\ m/s^2$) if air resistance is negligible. A Porsche can speed up from 0 to 60 mi/h in about 6 seconds. Which speeds up at a greater rate (i.e. which has the greater acceleration)? A dragster takes something like 4.5 seconds to get up to 60 mi/h. Could the dragster beat the freely falling object?

EXERCISE 2.7: (a) Move your hand from left to right at a speed of about 1 km/h. Is it possible? (b) Fill in the blanks: 30 m/s ≈ _____ mi/h ≈ _____ km/h . (c) About how many km/h equals 1 m/s?

There is a time for precision and a time for rough, quantitative thinking. Can you really visualize the difference between 5 and 6 mi/h? I bet you *can* visualize the difference between 10 and 30 mi/h.

Funny Business (careful distinctions)

- Distance ≠ magnitude of displacement (except when traveling in a straight line and not turning around)
- ave. speed ≠ magnitude of ave. velocity (except when dist. = mag. of displacement, as above)
- speed = magnitude of the instantaneous velocity (Always!)

These distinctions are important and will be discussed further in the next section.

Basic

Ch ... Ch ... Ch ... Changes

Now and forever, you must know the following notation: "Δ" means "change." It's the Greek capital delta. Quantitatively write

$$\begin{aligned}\Delta\,(\textit{whatever}) &= (\textit{final whatever}) - (\textit{initial whatever}) \\ &= (\textit{whatever})_{\text{final}} - (\textit{whatever})_{\text{initial}}\end{aligned}$$

It is surprisingly advantageous if you can *immediately* see that the negative change is

$$\begin{aligned}-\Delta\,(\textit{whatever}) &= (\textit{initial whatever}) - (\textit{final whatever}) \\ &= (\textit{whatever})_{\text{initial}} - (\textit{whatever})_{\text{final}}\end{aligned}$$

Furthermore, you calculus affectionatos know that "d," as in "dt," "dx," or "dv," stands for an infinitesimal (really, really, really small) amount. (Get it? Delta is the Greek d. They both stand for "de change!") Some folks say "dx is a differential element of x." That is, it's a really small bit of x.

Kinematics Definitions Expressed Mathematically:

Scalar (no direction in the definition)	**Vector** (includes direction)
	Position (+ or –): $\vec{\mathbf{r}}$; this choice of letter comes from "radial," since the position vector is radially outward from the origin
Distance (+): l ; for length of travel	**Displacement (+ or –):** $\vec{\mathbf{D}} \equiv \vec{\mathbf{r}}_2 - \vec{\mathbf{r}}_1$; (net change in position, $\Delta\vec{\mathbf{r}}$)
Ave. Speed (+): $S_{av} \equiv \dfrac{l}{\Delta t}$	**Ave. Velocity (+ or –):** $\vec{\mathbf{v}}_{av} \equiv \dfrac{\vec{\mathbf{D}}}{\Delta t} = \dfrac{\Delta\vec{\mathbf{r}}}{\Delta t}$
Instantaneous Speed(+): $\lvert\vec{\mathbf{v}}\rvert$ or v; the magnitude of the instantaneous velocity	**Instantaneous Velocity (+ or –):** $\vec{\mathbf{v}} \equiv \dfrac{d\vec{\mathbf{r}}}{dt}$
Magnitude of the Acceleration: as you know, we write this as $\lvert\vec{\mathbf{a}}\rvert$ or a.	**Ave. Accel.:** $\vec{\mathbf{a}}_{av} \equiv \dfrac{\Delta\vec{\mathbf{v}}}{\Delta t} = \dfrac{\vec{\mathbf{v}}_f - \vec{\mathbf{v}}_i}{\Delta t}$ **Inst. Accel.:** $\vec{\mathbf{a}} \equiv \dfrac{d\vec{\mathbf{v}}}{dt}$

Now you need to learn how to apply these definitions. Most texts provide a number of good lessons (see, for example, *Physics: The Nature of Things*, example 2.2 and exercise 2.1). Among the lessons is the need to push yourself to try to understand precisely what the equations mean and whether your results make physical sense. Blind manipulation of equations usually leads to frustration. Careful applications of the correct the equations are more likely to solve the problem at hand. Furthermore, **the equations can guide your thinking and your thinking can guide your use of the equations–it goes both ways, just as with any language.** Let's take a look at the common mistake pointed out in the aforementioned example and exercise and elucidate the correct thinking.

Ave Speed & Weighted Averages

Train yourself to recognize when it is best to use the basic definitions. Here is a good exercise that will help you gain some insight. It requires careful use of the definition of average speed. It's even possible to do this one in your head if you're very clever.

> EXERCISE 2.8: James Bond is trapped in an empty room where two walls are closing in on him. Ordinarily a simple predicament, this time he has a nervous breakdown and runs back and forth between the walls muttering obscenities. Your job, agent 3E8, is to calculate the total distance that Bond runs before he becomes an English tea biscuit. Initially, the walls are 10 m apart and Bond is at one of the walls. Each wall moves at a speed of 1 m/s. Bond runs frenetically from one wall to the other at an average speed of 4 m/s. How many miles does he put on his special agent shoes? Does his special agent heel cellular phone still work?

Let's look at another subtlety. Suppose that you are on a 6 kilometer driving trip. If you drive part of the trip at a speed of 10 m/s and the remaining part at 30 m/s, is your average speed 20 m/s? If you drive half the *distance* at 10 m/s and the other half at 30 m/s, is your average speed 20 m/s? Okay, so if you drive for half of the *time* at 10 m/s and the other half of the time at 30 m/s, is your average speed 20 m/s? These are very different situations. How do you account for each of them?

> EXERCISE 2.9: Suppose that you travel for one portion of your trip at 10 m/s and the remaining portion at 1000 m/s. (a) Can you imagine a scenario where your average speed is close to 10 m/s? (b) Can you imagine a scenario where your average speed is close to 1000 m/s?

I can think of 2 ways to solve this type of problem correctly and 2 ways to solve it incorrectly.

Correct Method #1: Careful use of the definition of S_{av}

(Used by Physics: The Nature of Things in Example 2.2.) First, write down what we know and what we want to find out. This could be any combination of partial distances, total distance, time for each portion of the trip, total time, ave. speed for each part, and ave. speed for the total trip. Sort it out.

Next, let's write down the applicable definition and make sure that we know what it means:

$$S_{av} \equiv \frac{l}{\Delta t} = \frac{\text{distance for part or all of trip}}{\text{time for part or all of trip}}$$

Now we can mathematically write out the relationships for *each* part and the *total* trip.

The last bit of information that you have is hidden (implicit) in this type of problem. You know that the following conditions surely must be true:

$$\Delta t_{total} = \Delta t_1 + \Delta t_2 \quad \text{and } l_{total} = l_1 + l_2$$

From this, we should be able to solve algebraically for the desired information. Ready for a solo flight? [Remember, you only learn by doing; so go for it!]

EXERCISE 2.10: Find the average speeds for the two cases cited in the opening paragraph of this section. (a) Travel at an ave. speed of 10 m/s for half of the *distance*, and then travel at an ave. speed of 30 m/s for the other half of the *distance*. What's your average speed? (b) Travel at an ave. speed of 10 m/s for half of the *time*, and then travel at an ave. speed of 30 m/s for the other half of the *time*. What's your average speed? Very interesting!

Disastrous Method #1: Blind averages

I hope that it is now clear that taking a simple average of the numbers will not get you the correct answer unless you happen to be lucky (i.e. each average speed is over the same amount of time). So now you know that

$$\text{*avoid*} \qquad S_{av} \neq \frac{S_{av,\,1} + S_{av,\,2} + S_{av,\,3}}{3} \qquad \text{*avoid*}$$

except when each is over the same amount of *time* (not not not distance).

Correct Method #2: Weighted averages

This approach turns out to have a number of applications in physics (see center of mass, for example). In fact, you've already derived it in Ex. 2.9. The average speed traveled for the longest period of time gets the most "weight." (In this context, "weight" has nothing to do with the force of gravity.)

$$S_{av} = S_{av,\,1} \cdot \left(\frac{\Delta t_1}{\Delta t_{total}}\right) + S_{av,\,2} \cdot \left(\frac{\Delta t_2}{\Delta t_{total}}\right) + S_{av,\,3} \cdot \left(\frac{\Delta t_3}{\Delta t_{total}}\right) + \ldots$$

Here the weighting fraction is the ratio of the time for that part divided by the total time. Notice that these fractions cannot have any units or we'd be back to "nonsense." Additionally, these fractions are positive, less than one, and add up to one. A common way to write this is to factor out the total time as follows:

$$S_{av} = \frac{S_{av,1} \cdot \Delta t_1 + S_{av,2} \cdot \Delta t_2 + S_{av,3} \cdot \Delta t_3 + \ldots}{\Delta t_{total}}$$

Do you see that $S_{av,1} \cdot \Delta t_1$ is just the distance for the first portion? Consequently the numerator is the total distance traveled and we're back to the definition for the total average speed! The greater the time interval for a particular ave. speed, the more "weight" it gets in calculating the total average. Pretty slick, huh?

EXERCISE 2.11: For a trip where one quarter of the time is spent with an average speed of $S_{av,1}$, one quarter with an average speed of $S_{av,2}$, one quarter at $S_{av,3}$, and one quarter at $S_{av,4}$, show that the total average speed can be obtained from the naive averaging of the average speeds for each portion. (That is, add the speeds and divide by 4–disastrous method #1.)

Disastrous Method #2: Weighted averages misused

An excellent question now is "Why can't we weight the speeds according to the distances traveled? After all, the units would still be those of speed!" As we have demonstrated, this simply would not make sense because this weighting is not based on the fundamental relation of ave. speed = total distance/total time. What physical meaning does the expression $S_{av,1} \cdot l_1$ have? So once again, for averaging speeds, we must weight according to the *time*, not not not distance.

avoid $$S_{av} \neq \frac{S_{av,1} \cdot l_1 + S_{av,2} \cdot l_2 + S_{av,3} \cdot l_3 + \ldots}{l_{total}}$$ *avoid*

Summarize the point that has just been made. Use your own words:

____

____

____

____

____

Another common problem arises with your old friend the minus sign. It bears attention.

"Oooh That *¤!*#@#*! Minus Sign!"

[It has been said that experimental physicists run around chasing nuts and bolts, and theoretical physicists run around chasing minus signs!] In your checkbook, a negative balance means that you're in debt (a well known state of affairs for most students). In kinematics, a minus sign means "the other way." You need to *define* one direction as the positive direction. Then the "other way" is the negative direction. Of course it really doesn't matter which way you say is +, just make any convenient choice and stick with it throughout the problem.

The Minus Sign for 1–D Motion

Position:

$x > 0 \Rightarrow$ on one side of the origin (x=0)

$x < 0 \Rightarrow$ on the other side of the origin (x=0)

Displacement:

$D = \Delta x > 0 \Rightarrow$ moved in the + direction (on either side of the origin)

$D = \Delta x < 0 \Rightarrow$ moved in the – direction (on either side of the origin)

Velocity:

$v > 0 \Rightarrow$ moving in the + direction (on either side of the origin)

$v < 0 \Rightarrow$ moving in the – direction (on either side of the origin)

Acceleration:

$a > 0 \Rightarrow$ speeding up while moving in the + direction
or slowing down while moving in the – direction (double –)

$a < 0 \Rightarrow$ slowing down while moving in the + direction
or speeding up while moving in the – direction

EXERCISE 2.12: I'll choose the origin, +/– directions, and units to be

Your task is to move from x = 3 m to x = 5 m and back to x = 1 m in a total of 2 seconds.

(a) Do you need to run or can you walk?
(b) What is your total distance traveled?
(c) What is your total displacement?
(d) What is your average speed?
(e) What is your average velocity?

EXERCISE 2.13: For the "point particle" (object whose size and shape are not taken into account; often drawn as a dot) shown in each part, fill in "+," "–," or "0," for each of the kinematics quantities in the columns on the right. (See the list of signs on the previous page.)

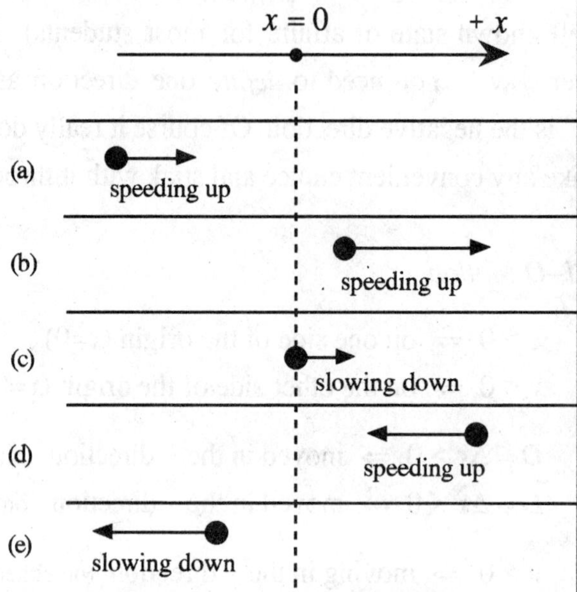

	+, –, or 0 ?			
	x	Δx	v	a
(a)				
(b)				
(c)				
(d)				
(e)				

EXERCISE 2.14: Calculate the average accelerations in each case. Be careful of the signs! State whether the object is increasing or decreasing its speed. Also state the initial and final directions (+ or –) and whether or not the object turns around during the time interval.

(a) $v_i = 4$ m/s, $v_f = 10$ m/s, $\Delta t = 2$ s
(b) $v_i = 10$ m/s, $v_f = 4$ m/s, $\Delta t = 2$ s
(c) $v_i = -4$ m/s, $v_f = 10$ m/s, $\Delta t = 2$ s
(d) $v_i = 4$ m/s, $v_f = -10$ m/s, $\Delta t = 2$ s
(e) $v_i = -10$ m/s, $v_f = 4$ m/s, $\Delta t = 2$ s
(f) $v_i = 10$ m/s, $v_f = -4$ m/s, $\Delta t = 2$ s
(g) $v_i = -4$ m/s, $v_f = -10$ m/s, $\Delta t = 2$ s
(h) $v_i = -10$ m/s, $v_f = -4$ m/s, $\Delta t = 2$ s

EXERCISE 2.15: Anthony walks up and down an aisle in a grocery store. When we start our stop watch, $t_o = 0$ s, he is halfway up the aisle. Let's call that position $x_o = 0$ m. When we look again he has moved up the aisle 4 meters in 4 seconds, i.e. $t_1 = 4$ s and $x_1 = 4$ m. He stops to pick out his favorite soy product. This takes him 3 s. He then turns around and rushes back down the aisle to get something he forgot. Let's label this event (his starting to walk again) as occurring at $t_2 = 7$ s and $x_2 = 4$ m. After 2 s, he passes his starting position, so $t_3 = 9$ s and $x_3 = 0$ m. He reaches the granola 1 s later ($t_4 = 10$ s and $x_4 = -2$ m), grabs it without stopping, slows down and exits the aisle at $t_5 = 13$ s and $x_5 = -6$ m.

(a) Draw an x–axis going up the page (*not* left to right) to represent the aisle. Mark the points $x = -6$ m to +4 m in 1 m increments. Draw dots on the x–axis to correspond with the snapshot events (positions and times) described above, and write the time by each dot.

For the following time intervals, find Anthony's distance traveled, displacement, ave. speed, and ave. velocity. Notice the difference in the definitions!

(b) Between $t_o = 0$ s and $t_1 = 4$ s.
(c) Between $t_o = 0$ s and $t_2 = 7$ s.
(d) Between $t_o = 0$ s and $t_4 = 10$ s.
(e) Between $t_o = 0$ s and $t_5 = 13$ s.
(f) Between $t_1 = 4$ s and $t_2 = 7$ s.
(g) Between $t_4 = 10$ s and $t_5 = 13$ s.

The Kinematics Graphs

There are three kinematic graphs for 1–D motion: x vs. t, v vs. t, and a vs. t. These graphical representations help to visualize the relationship between x, v, and a, as well as provide training in reading graphs. As noted in section B.3, you will need to pay attention to the **values, slopes, and concavity** of the graphs and the **area under the curves.** The meaning of these characteristics is different for each of the three types of graphs. All of this can be confusing at first. Keep studying and paying attention to the details and you will sort it out.

To begin with, keep in mind that x, v, and a may be any combination of 0, +, or –. For example, they could all be 0, they could all be +, or they could all be –. Then again, x could be –, v could be 0, and a could be +. *Any* combination is possible. Furthermore, if x is constant, v and a are 0. If v is constant, a is 0. However $a = 0$ implies nothing about v, just as $v = 0$ implies nothing about x. Think about the physical meanings of these statements and imagine some examples for each case. Make sure that you see this conceptually. Then see how the definitions of v and a support these statements.

x vs. t Graphs

- The **actual motion** is 1–D, *not* 2–D. Think of the motion as being up and down the x axis, *not* along the x vs. t curve! (See Anthony's motion in Ex. 2.15.)
- The **values** of the points tell us the time and corresponding **position**, (t, x). Hence $x = 0$ is simply seen where the graph crosses the t axis, since then $x = 0$. x is positive where the curve is above the t axis and negative where it is below.

Slope = 1st Derivative

- The **slope** of the x vs. t graph at any point gives the **velocity** at that time and position. A positive velocity is seen as a positive slope (up and to the left: /). A negative velocity is seen as a negative slope (down and to the left: \). The steeper the slope, whether it be + or –, the greater the *speed.* Instants when $v = 0$ can be found at maxima, minima, and points of inflection, since there the slope is 0. Finally, extended periods of no motion (rest) are seen as flat curve segments which may then head up or down, depending on whether the object moves in the positive or negative x directions after the pause.

Concavity = 2nd Derivative

- The **concavity** of the x vs. t graph (whether it *curves* up or down) reflects the rate of change of the slope–is the slope increasing or decreasing? And what concept does that reflect? Of course, the rate of change of velocity is the **acceleration**. An upward bowl-shaped segment corresponds to an increasing slope (velocity) and thus a positive acceleration. A downward bowl-shaped segment corresponds to a decreasing slope (velocity) and thus a negative acceleration.

Area under a curve has units & meaning equal to the product of the axis variables.

- The **area** under an x vs. t graph has the meaning of position (or displacement) *multiplied* by time with units of meter·second...what the heck is that? I don't know. It's not useful!!

EXERCISE 2.16: Can you "read" the information off of the following graph?

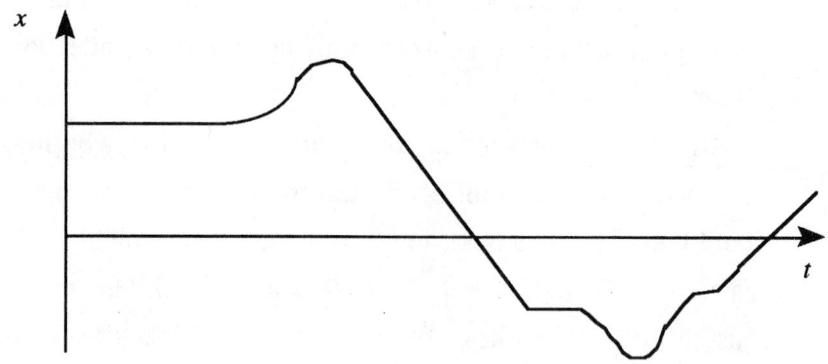

(a) Could you liken this motion to a roller coaster ride? (b) Show where the position is positive, where it is negative, and where it is zero according to our chosen coordinates. (c) Show where the velocity is positive, where it is negative, and where it is zero. (d) Show where the acceleration is positive, where it is negative, and where it is zero. (e) Is there any time period(s) where the object is taking a prolonged rest? (f) What does the area under this curve (i.e. between the curve and the t–axis) represent? ... Now you've got it!

AGAIN: take special note that the slope and the concavity may be +, –, or 0, above ($x > 0$) or below ($x < 0$) the t axis. Conceptually, you can be moving forward ($v > 0$), backward ($v < 0$), or not at all ($v = 0$) when you are at a positive or negative position (x). Similarly, you may be speeding up in the $+x$ direction ($a > 0$)while your position is + or –, speeding up in the $-x$ direction ($a < 0$)while your position is + or –, slowing down in $+x$ direction ($a < 0$)while your position is + or –, or slowing down in $-x$ direction ($a > 0$) while your position is + or –. Think about it!

v vs. t Graphs

- The **actual motion** is not *seen* here since there is no spatial axis.
- The **values** of the points tell us the time and corresponding **velocity**, (t, v). Hence $v = 0$ is simply seen where the graph crosses the t axis, since then $v = 0$. v is positive where the curve is above the t axis and negative where it is below.
- The **slope** of the v vs. t graph at any point gives the **acceleration** at that time and position. A positive acceleration is seen as a positive slope (up and to the left: /). A negative acceleration is seen as a negative slope (down and to the left: \). The steeper the slope, whether it be + or –, the greater the rate of change of the speed. Instants

when $a = 0$ can be found at maxima, minima, and points of inflection since the slope is 0 ("horizontal"). Finally, extended periods of constant velocity are seen as flat curve segments.

- (The **concavity** of the v vs. t graph (whether it *curves* up or down) reflects the rate of change of the slope. What concept does that reflect? On this graph, the rate of change of slope is the rate of change of acceleration. We won't deal with changing acceleration too much, but it is sometimes called the "**jerk**" since motion with changing acceleration can feel jerky. This is in contrast to the smooth speeding up or slowing down of constant acceleration.)
- The **area** under an v vs. t graph has the meaning of velocity multiplied by time with units of $^{m}/_{s} \cdot s = m$ ...what the heck is that? It's Δx, of course! Surely $v \cdot \Delta t$ rings a bell! See your text for a more complete discussion.

EXERCISE 2.17: Let's try it again. Using the same curve, different vertical axis variable, can you "read" the information off of the following graph?

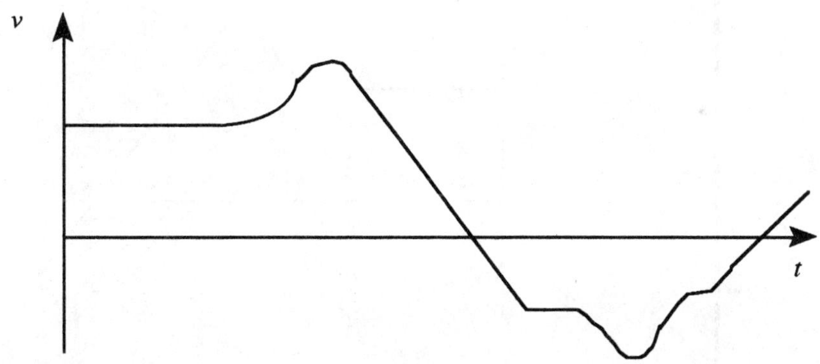

(a) Could you liken this motion to a roller coaster ride? (b) Show where the position is positive, where it is negative, and where it is zero according to our chosen coordinates. (c) Show where the velocity is positive, where it is negative, and where it is zero. (d) Show where the acceleration is positive, where it is negative, and where it is zero. (e) Is there any time period(s) where the object is taking a prolonged rest? (f) What does the area under (i.e. between the curve and the t–axis) this curve represent? (g) Does this curve represent the same motion as that represented in Ex. 2.16 (same curve with x vs. t axes)?

a vs. t Graphs

- Again, the **actual motion** is not *seen* here since there is no spatial axis.
- The **values** of the points tell us the time and corresponding **acceleration**, (t, a). Hence $a = 0$ is simply seen where the graph crosses the t axis, since then $a = 0$. a is positive where the curve is above the t axis and negative where it is below.
- (The **slope** of the a vs. t graph at any point gives the "**jerk** ." As I've noted before, this won't be a primary concept that we will use.)
- (The **concavity** of the a vs. t graph reflects the rate of change of the slope. On this graph, the rate of change of slope is the rate of change of jerk. Again, not something we'll be concerned with.)
- The **area** under an a vs. t graph has the meaning of acceleration multiplied by time with units of $\text{m/s}^2 \cdot \text{s} = \text{m/s}$ ... what the heck is that? It's Δv of course! Surely $a\,\Delta t$ rings a bell! See your text for a more complete discussion.

EXERCISE 2.18: And finally, using the same curve, different vertical axis variable, can you "read" the information off of the following graph?

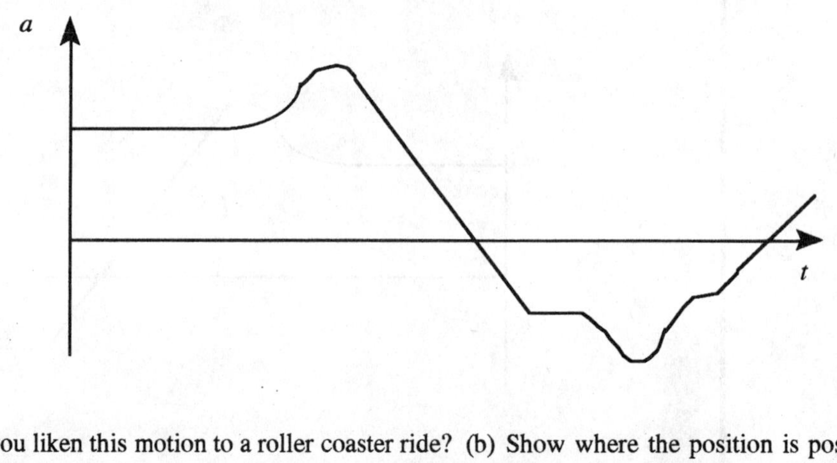

(a) Could you liken this motion to a roller coaster ride? (b) Show where the position is positive, where it is negative, and where it is zero according to our chosen coordinates. (c) Show where the velocity is positive, where it is negative, and where it is zero. (d) Show where the acceleration is positive, where it is negative, and where it is zero. (e) Is there any time period(s) where the object is taking a prolonged rest? (f) What does the area under (i.e. between the curve and the t–axis) this curve represent? (g) Does this curve represent the same motion as that represented in Ex. 2.17 (same curve with v vs. t axes)?

Now that you have a feel for the concepts, quantities, and mathematical definitions, let's practice using vectors to describe motion and take a look at the important special case of uniformly accelerated motion.

Kinematics Vectors

Always remember that any and all vector quantities:

1. are added/subtracted using the same methods; namely, by components (implementing trig. and the Pythag. th'm), by graphical means (using head to tail, protractors and rulers, or by geometry such as that contained in the law of cosines).

2. can ONLY be added/subtracted to other vectors with the **same physical meaning** and the **same units**. You can never add a displacement to a velocity! However, there may be instances where the *angles* of the displacement and velocity vectors are the same. (Ooooooh. Hidden information.)

Position & Displacement Vectors

The following discussion and diagram show you 3 important/helpful ways to think about and see the relation between $\vec{\mathbf{r}}$ and $\vec{\mathbf{D}}$.

$\vec{\mathbf{r}}$: The position vector. Drawn from the chosen origin to the position of interest. (Tail at origin, tip at position of interest.)

$\vec{\mathbf{D}} \equiv \Delta\vec{\mathbf{r}}$: The displacement vector. Drawn from the earlier position to the later position.

Consider the diagram shown below. We may look at this any of the following ways.

1. $\vec{\mathbf{r}}_1 + \Delta\vec{\mathbf{r}} = \vec{\mathbf{r}}_2$, that is, start at the tip of $\vec{\mathbf{r}}_1$, add the displacement ($\vec{\mathbf{D}} \equiv \Delta\vec{\mathbf{r}}$), and you end up at $\vec{\mathbf{r}}_2$.

2. $\Delta\vec{\mathbf{r}}\,(\text{aka }\vec{\mathbf{D}}) = \vec{\mathbf{r}}_2 - \vec{\mathbf{r}}_1$. This may be rearranged and written $\Delta\vec{\mathbf{r}} = (-\vec{\mathbf{r}}_1) + \vec{\mathbf{r}}_2$. So what? This now says "flip $\vec{\mathbf{r}}_1$, and add $\vec{\mathbf{r}}_2$. The resultant ($\Delta\vec{\mathbf{r}}$) is then from the tail (start) of $-\vec{\mathbf{r}}_1$ to the tip (end) of $\vec{\mathbf{r}}_2$. (Be sure that you see this.)

3. From (1 and 2) above, we see a short cut for drawing in $\Delta\vec{\mathbf{r}}$ if we draw $\vec{\mathbf{r}}_1$ and $\vec{\mathbf{r}}_2$ **tail to tail**, simply draw $\Delta\vec{\mathbf{r}}$ *from* the tip of $\vec{\mathbf{r}}_1$ *to* the tip of $\vec{\mathbf{r}}_2$.

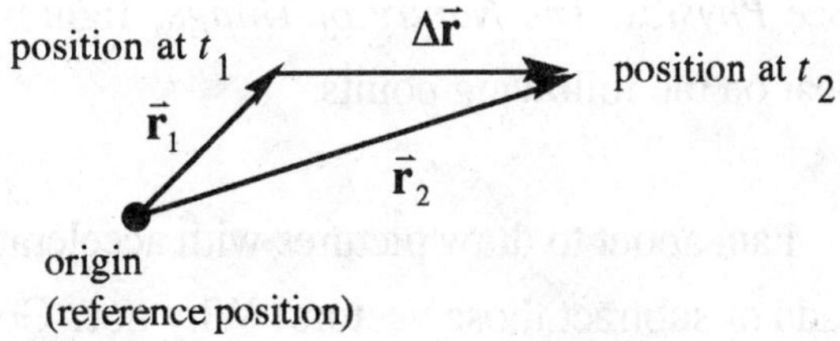

Got it? You'll need it!

Velocity Vectors

You will often see velocity vectors drawn 2 ways. This can be confusing at first unless you keep in mind that vectors can be moved around as you like to facilitate their addition (just as long as their magnitude and orientation are kept the same). The 2 ways that I am referring to are shown below (and on page 60 of *Physics: The Nature of Things* in figures 2.10 and 2.11, respectively).

1. Draw $\vec{v}$ at the location of the object and *tangent* to the path of the object. If $\vec{r}$ is drawn in, $\vec{v}$ is drawn at the end of $\vec{r}$ (the location the object).
2. Add velocity vectors graphically and forget about the location of the object. The diagram then only shows the orientation and scaled magnitude of the velocities.

The first diagram should absolutely NOT be interpreted as the addition of the position and velocity vectors, but rather the simultaneous display of an object's position and velocity vectors. The second way can seem odd if you are thinking along the lines of (1) above. In the first diagram, if the object is at different places at times t_1 and t_2, the velocity vectors will have their tails at different places. But as you know, we can move the velocity vectors around for a graphical analysis and place them tail to tail! The latter case is more abstract and tends to bother students when they first see it. Does it bother you now? I hope not, but get some help if it does. (Think of the first diagram as being drawn in physical space and the second diagram as being drawn in "velocity space"…if it helps.)

Acceleration Vectors

There are 3 ways that you will see $\vec{a}$ vectors drawn: along with $\vec{v}$ in either of the two ways noted above, or only with other $\vec{a}$ vectors and $\Delta\vec{a}$, like the second diagram for $\vec{v}$. (See *Physics: The Nature of Things*, figures 2.12–2.16.) It is absolutely essential to be clear on the following points.

I am about to draw pictures with acceleration and velocity vectors drawn together. Can I add or subtract those vectors? Why not? Good work.

What up?	Pictures	Object is ...
$\vec{a}$ parallel to $\vec{v}$	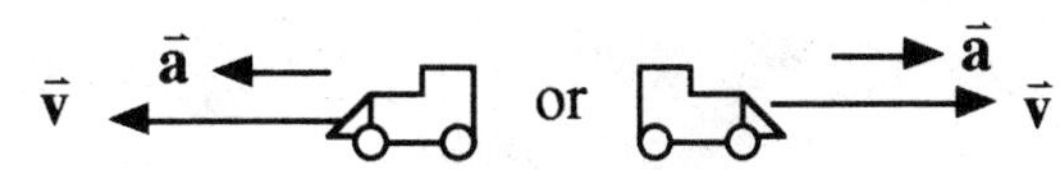	speeding up but not changing direction
$\vec{a}$ anti parallel to $\vec{v}$	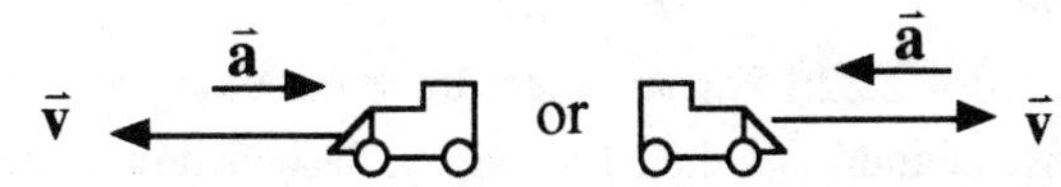	slowing down but not changing direction
$\vec{a}$ perpendicular to $\vec{v}$	Turning towards you; accel is out of page. Top view; turning left. 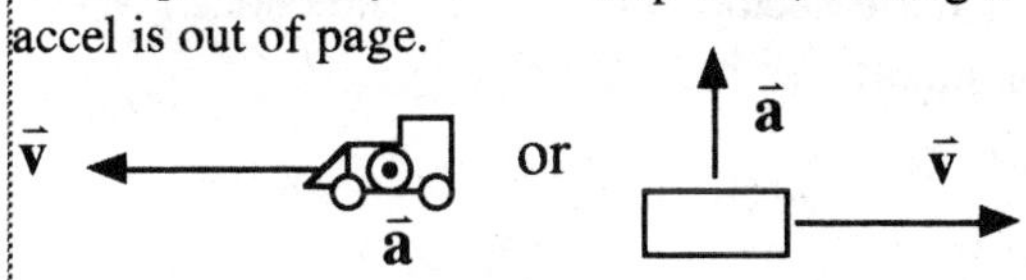	changing direction but not changing speed
$\vec{a}$ neither par. nor perp. to $\vec{v}$	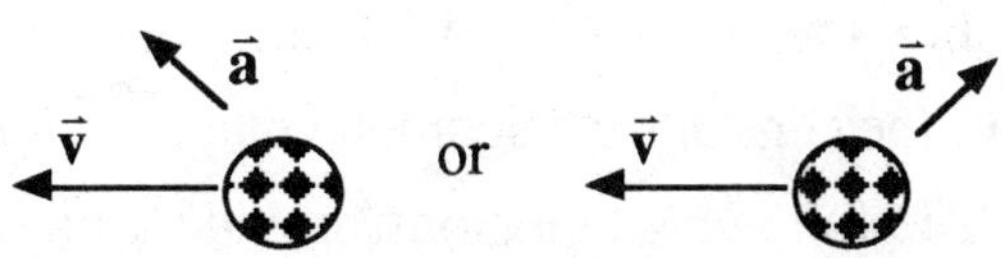	changing both speed & direction

Because of the physical significance of the direction of $\vec{a}$, it is common to break up acceleration vectors into components parallel to and perpendicular to $\vec{v}$.

EXERCISE 2.19: A crow flies from point A to point B along a straight line at some angle north of east. The crow ends up traveling 30 m east while it travels 40 m north. It flies at a constant velocity of 15 m/s. (a) Sketch a diagram and draw in the displacement and velocity vectors. (b) What is the magnitude of the total displacement (i.e. the distance "as the crow flies")? (c) What are the east and north components of the crow's velocity? (d) If the crow slows down with an acceleration of 3.0 m/s^2, what are the east and north components of the acceleration? Draw in the acceleration vector on a sketch like that in (a).

EXERCISE 2.20: The diagram shows a snapshot of an object whose velocity and acceleration are represented by the vectors shown. Based on this picture, (a) will the object be going faster, slower, or the same speed in the next moment? (b) will the object change directions? If not, why not? If so, in what direction will it start to move?

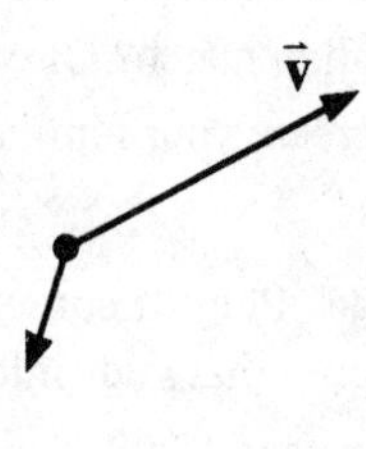

EXERCISE 2.21: For the path shown, can you say whether the object accelerates or not? If so, roughly what is the direction of the acceleration vector? If not, why?

EXERCISE 2.22: For an object traveling along a circular path at a constant speed, draw the object at any instant and show the circle with a dashed line. Now draw in the position vector using the center of the circle as the origin. Next draw in the velocity vector with the tail at the object. Finally, draw in the acceleration vector with the tail at the tail of the velocity vector. (Do NOT worry too much about the lengths of these vectors, just get the directions.) What is the resultant of adding these three vectors? [Trick question!]

1–D, Uniformly Accelerated Motion

The last big topic! The third of the four kinematic methods (graphing 1-D, using definitions, the constant acceleration equations, and the application of calculus to functions of time). The following equations provide a powerful template for setting up and solving many different kinematics problems. You will unequivocally want to have these equations committed to memory by the time you finish this chapter.

The Basic 1–D Kin. Eq'ns

(See table 2.4 of *Physics: The Nature of Things*.) Please excuse the fact that I may present these in an order different from your text. Your new friends:

0. $v_{av} = \frac{1}{2}(v_f + v_i)$
1. $\Delta y = v_i t + \frac{1}{2}at^2$
2. $v_f = v_i + at$
3. $2a\Delta y = v_f^2 - v_i^2$

The next few exercises are designed to help you get to know these equations.

EXERCISE 2.23: Check the units! This is always the first line of approach in getting to know a new equation. For each equation, check the units of each term (for example, $\frac{1}{2}\, a\, t^2$). Do the units of each term in the equation match? Of course you know that they should, but going through this exercise will immediately increase your familiarity.

EXERCISE 2.24: Plug in some numbers! This is the next line of approach in getting to know these equations. Suppose that your initial velocity (v_i) is 20.0 m/s and your acceleration (a) for the next 7.00 seconds is a constant 1.000 m/s^2. (a) How far do you travel? (b) What is your velocity at the end of those 7 seconds? (c) Verify that your numbers are consistent by using the equation numbered "3" in the list above. (d) What was your average velocity during that time interval?

EXERCISE 2.25: Construct some equations using the constant acceleration template equations! Now we're talking! You're on the road to kinematic success! You will need to modify these equations slightly to fit the particular situation that you are modeling. Be certain the acceleration used is either the constant acceleration for the time interval of interest or the average acceleration.

(a) Write the equations for the situation where $a = 0$ (constant velocity). How many "true" equations are you left with?
(b) Write the equations for the case where an object has been dropped, or thrown straight up or down into the air. Use $\vec{\mathbf{a}} = g$, downward = 9.8 $^{m}/_{s^2}$, downward and define up to be the positive direction. During what portions of this process can we use these equations?
(c) An object that is thrown up at an angle (a "projectile") can be described as composed of a horizontal motion (use x instead of y, $a_x = 0$) and a vertical motion (use y for position, a_y = g, down). Write the equations for projectile motion. Hey! You're doing chapter 3 already! Smokin'!

Intermediate

Kinematics Graphs

Getting One Graph From Another

For a particular motion, we can plot x vs. t, v vs. t, or a vs. t graphs. If we have the x vs. t graph, we can construct the other two. If we have the v vs. t graph, we can construct the a vs. t graph. (Going backwards, so to speak, from a vs. t to v vs. t to x vs. t, is incomplete because a vs. t tells nothing about the values of v and v vs. t tells nothing about the values of x.)

I will leave further investigation to you. This technique is nicely illustrated in most texts (see *Physics: The Nature of Things*. Example 2.11).

Kinematics Vectors

Differentiation of Vectors

As you know from the text, you differentiate a vector by differentiating the components separately (there are other ways, but I'll leave that for your later study of vector calculus). Let's just try one exercise here and a slightly tougher one in the advanced section.

EXERCISE 2.26: (a) For the position vector function $\vec{\mathbf{r}}(t) = [v_{ox}t]\hat{\mathbf{i}} + [v_{oy}t + \frac{1}{2}at^2]\hat{\mathbf{j}}$, where v_{ox} = 2.00 $^{m}/_{s}$, v_{oy} = 3.00 $^{m}/_{s}$, and a = −10.0 $^{m}/_{s^2}$, find the position, distance from the origin, x and y components of velocity, speed, x and y components of acceleration, and total acceleration when t = 2.00 s. (b) What is the displacement between t = 1.00 and t = 2.00 s?

NOTE: When writing a vector with unit vectors, it does NOT matter whether you put the unit vectors in front of or behind the components–they just tell you the direction:

$$\vec{\mathbf{r}}(t) = [x(t)]\hat{\mathbf{i}} + [y(t)]\hat{\mathbf{j}} = \hat{\mathbf{i}}[x(t)] + \hat{\mathbf{j}}[y(t)] = \hat{\mathbf{i}}[x(t)] + [y(t)]\hat{\mathbf{j}}, \text{ etc.}$$

1–D, Uniformly Accelerated Motion

Freeze Frame (Snapshot) Thinking

[Psst! Want to be an expert kinematicist? Here's the key to the game!] It's time to give these equations just a little more thought. What exactly are we doing when we use those 3 (or 4) "uniformly accelerated, linear motion" equations? I'll tell you after I define two important words. (You saw these words earlier.)

state – An object's *state* tells you what's going on with it at some *instant* – the *snapshot* data. The kinematic state is determined by the time, position, velocity, and acceleration.

stage – I'll refer to a *stage* in some process as some *time interval* of the process. (For example, the period when the acceleration is constant: Sitting in your car at a stop light would be one stage, speeding up would be another stage, and traveling at a constant velocity would be yet another stage.)

Now I can say:

> The 1–D constant accel. equations can be used to compare *any* two states within a given stage where the acceleration is constant. That is, take a mental snapshot at anytime that is useful or interesting and see what you know and want to know about that **state** (t_1, x_1, v_1). Then take another mental snapshot to check a later **state** (t_2, x_2, v_2). The 2 states whose data you are comparing must be within the same **stage**; that is, the acceleration must be constant between the states. The time variable, t, in the equations is actually the time *between* the two states (Δt), as is the acceleration variable, a.

Set up for kinematically relating an earlier state to a later state when the acceleration has been constant (or we use the average acceleration):

Any Earlier State (Subscripts may be i, 0, 1,...)	Stage Between States (Constant or ave. acceleration)	Any Later State (Subscripts may be f, 1, 2,...)
t_1	$\Delta t = t_2 - t_1$ ("t" for short)	t_2
x_1	$\Delta x = x_2 - x_1$	x_2
v_1		v_2
	a	

The equations then take the form:

1. $\Delta x = v_1 t + \frac{1}{2} a t^2$
2. $v_2 = v_1 + at$
3. $2a\Delta x = v_2{}^2 - v_1{}^2$ [Or write it as $v_2{}^2 = v_1{}^2 + 2a\Delta x$]

0. $v_{av} = \frac{1}{2}\cdot(v_1 + v_2)$, [I don't use this one much.]

common misconception – The "initial" state being compared does NOT have to be at the beginning of a stage or process. Likewise, the "final" state does NOT have to be at the physical end of the stage or process. They are simply *any* earlier and later instants within a stage that you need or want to correlate.

algebraic subtlety – As texts demonstrate, you can get the equation $2a\Delta y = v_f^2 - v_i^2$ by solving for t in $v_f = v_i + at$ and plugging that expression into $\Delta y = v_i t + \frac{1}{2}at^2$. Thus, this set of 3 equations is really only 2 *independent* equations (see B.4.c of this guide). Once you've used 2 of these, the other can provide no new information!! So don't waste your time fiddling with the remaining one!

EXERCISE 2.27: [Hot Tip!] Which equation would you use if you didn't have, need, or want time?

EXERCISE 2.28: Arinna enters a ski race. At the end of the first four seconds of the race, she is skiing at 20.0 m/s. For the next two seconds she slows down at a rate of 3.00 m/s² as she approaches a demonic mogul field. (a) What is her average acceleration during those first four seconds? (b) If her acceleration was constant for the first four seconds, what was her average speed? (c) How far along the slope does she travel while she slows down? (d) How fast is she going after she has slowed down?

There are three common types of 1–D kinematics problems that tend to pop up frequently. Let's take a look at them before leaving this topic.

- Vertical Free–Flight
- Multiple Stages
- Multiple Objects

Vertical Free–Flight

Some people call this motion "free *fall*," but I have trouble with that since the object could be traveling upward during part or all of it's time in the air. It is called "free" because we presume that there are no forces other than that of gravity acting on the object – air resistance must be negligible for this model to be any good, (Note that by "flight" I do not mean to say that the object is propelled by wings or jet engines; again, that's why it's "free.")

To do these problems, sort out the data as always (given, implicit, wanted), and work with the basic 1–D constant accel. equations to compare states *while in free–flight*.

- **Free Flight Stage**: occurs between the states just after release (launch) and just before landing or being caught. The equations tailored to fit this stage do NOT work for the launching or landing stages of the process!! We would need to construct different equations for those stages.

- **Physical Intuition**: objects in free–flight slow down by 9.8 m/s (≈ 10 m/s ≈ 22 mi/h) each second as they move upwards, and speed up by the same amount each second as they fall.

- **Data for a state**: t, y , v

- **Implicit data/information to watch for:**

"Dropped"	⇒	$v_{\text{start, y}} = 0$
"Thrown straight up"	⇒	$v_{\text{start, y}} > 0$, *if* up is designated as +
"Thrown straight down"	⇒	$v_{\text{start, y}} < 0$, *if* up is designated as +
"Top"	⇒	$v_{\text{top, y}} = 0$

Note that I've use "start" to subscript the velocities because that may or may not be the initial state that I am using in the equations, but it is the physical beginning of the flight.

- **1-D Constant Accel. Equations** ($a_y = -\,g$, taking up to be +):

0. $v_{av} = \frac{1}{2}(v_2 + v_1)$
1. $\Delta y = v_1 t - \frac{1}{2} g t^2$
2. $v_2 = v_1 - gt$
3. $-2g\Delta y = v_2^{\,2} - v_1^{\,2}$

Okay! That's all you've got in your kinematics tool box, so get to work applying the equations that we've constructed.

EXERCISE 2.29: Imagine that you drop an object from a tall building. Take a snapshot of the object every second as it falls. (Use $g \approx 10$ m/s²) (a) On a clean sheet of lined paper, sketch the object as it falls. Make five columns and label them at the top of the page as: "object," "t," "y," "v," and "a." Let one line width of your paper be equal to 5 m. Fill in the table by calculating the position, velocity, and acceleration, and drawing the object at t = 0, 1, 2, 3, 4 seconds. (b) About how many *stories* high must the building be to allow an object to fall for 4 s? [Neglecting air resistance.] (c) What's the final velocity after 4 s??

EXERCISE 2.30: At the edge of the top of a 78.4 m tall building, you throw an egg straight up with a velocity of 25.0 m/s. [Neglect your height and air resistance.] (a) At what speed does the egg hit the ground? (b) How long does it take to reach the ground? (c) How high above the ground is the egg when it reaches the top of its path? (d) At what speed does it pass you on the way down? (e) Answer (a) & (b) for the case where you throw the egg straight down at 25.0 m/s.

Multiple Stages

What if you're in a car and you speed up at a constant rate for a while and then need to apply your brakes and slow down (presumably at a nice, smooth, constant rate)? How many stages are there? Two: the speeding up stage and the slowing down stage. Can you use the 3 (well, 4) constant acceleration kinematics equations to compare your state when you started accelerating to the time when you're done decelerating? For example, could you use the first state described for the "initial" variables in the equations and the state after decelerating for the "final" variables in the equation to determine, say, the total distance and time traveled? Of course not! What would the a be? It is different for the 2 stages of the process. Consequently, you will **need to break the problem down into different stages.** Notice that the last state in the first stage is equivalent to the first state in the next stage. Don't forget that! A picture is worth a thousand words!

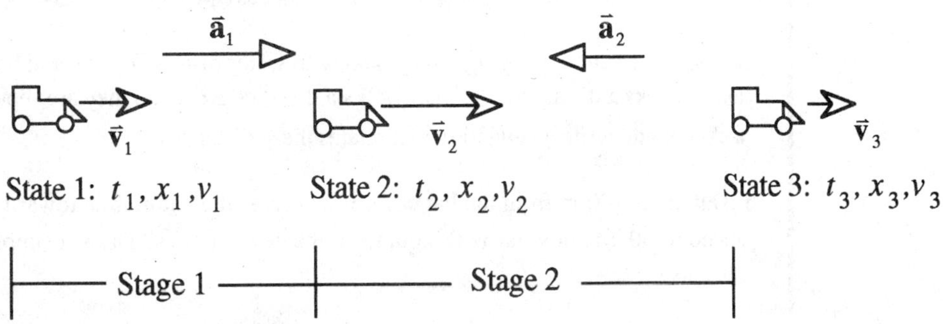

Possible kinematics equations:

Stage 1	Stage 2
$v_{av} = 1/2 \cdot (v_1 + v_2)$	$v_{av} = 1/2 \cdot (v_2 + v_3)$
$\Delta x = v_1 t + 1/2 a_1 t^2$	$\Delta x = v_2 t + 1/2 a_2 t^2$
$v_2 = v_1 + a_1 t$	$v_3 = v_2 + a_2 t$
$2a_1 \Delta x = v_2^2 - v_1^2$	$2a_2 \Delta x = v_3^2 - v_2^2$

We could use subscripts more extensively and write: $\Delta x_{1\to2}$, $\Delta t_{1\to2}$, $v_{av,\,1\to2}$, $a_{1\to2}$, etc.

We cannot relate states 1 and 3 directly…unless we use an average acceleration. Just remember the lesson about naive averages with velocities.

Notice that state 2 is the later state for stage 1, and the earlier state for stage 2! It is the link!

EXERCISE 2.31: [In *Physics: The Nature of Things*, problem 2.73 is a somewhat more difficult version of this problem.] Julia is relaxing on the beach in the hot sun when she decides to go for a swim. The kinematic details are as follows. Her initial position is 10 m directly to the water line, and she starts from rest. She walks along the beach at an angle of 30° from the straight line between her and the water. Due to the hot sand, she gradually picks up speed at a rate of 0.50 m/s^2. She hits the water running and begins swimming (assume her speed is the same as her final speed on the sand). She swims at an angle of 70° from the shoreline. If she swims with an average deceleration of 1.0 m/s^2, what is her speed after 3.0 seconds swimming?

Multiple Objects

If there are 2 (or more) objects involved in the process, write the kinematics equations for each object *separately*. Additionally, you will probably need to write an equation or two to describe certain *conditions* specified by the problem. For example, if the objects hit or cross paths, there is a time when they have the same position coordinates. (Note that this does NOT mean that they have had the same displacements in the time interval of the process! The displacements may, however, be related by some other known length–this a common ingredient in these types of problems.)

*****Make sure that you clearly sketch the process and show the essential states and stages.*****

EXERCISE 2.32: Relating displacements – an easy primer.

(a) Noah and Paul are having a play dual with water pistols. They stand back to back and begin walking. Noah walks a distance Δx_1, Paul walks a distance Δx_2, and they are separated by a distance of L. Make a sketch and write an equation that relates these distances.

(b) Dakota is 100 m from the bat cave. If he drives the bat mobile toward the cave in a straight line at a speed of 30 m/s, how far is Dakota from the cave in 1.2 s? Draw a simple sketch to visualize the distances.

(c) Sophie goes for a walk with her dog, Sam. They both start out in the same place and at rest. Sophie walks along a straight path at 1 m/s while Sam runs down the path at 3 m/s. After 10 s, how far is Sam from Sophie? Draw a simple sketch to train yourself for more difficult problems.

EXERCISE 2.33: Sage and Julia are playing double vertical catch. The way this works is that Sage goes to the third floor window of a building while Julia stays outside on the ground floor. Sage throws a ball down to Julia with an initial speed of 10 m/s. The ball starts out 9.0 m above Julia. At the same time that Sage throws her ball down, Julia throws a ball up to Sage at a speed of 10 m/s. The question is this: how long are the balls in the air before they pass?

Guidelines:

- Make a clear sketch showing the earlier states of the balls (at release) and the later states (when passing).
- Clearly show the distances that are involved.
- Gather and sort your data in a clear table.
- What type of problem is this?
- Write the pertinent equations for each ball. Be very very careful not to make minus sign errors!
- Write an equation that relates the distances involved. This relationship should be clear from your sketch.
- Solve
- Analyze

Advanced

Kinematics Vectors

Differentiation of Vectors

Here's your chance to flex a little calculus muscle.

EXERCISE 2.34: Starting with the definition for acceleration, use integration to derive equations for $\vec{\mathbf{v}}(t)$ and $\vec{\mathbf{r}}(t)$. Determine the meanings of the constants of integration. [HINT: Integrate each component separately.]

$$\vec{\mathbf{a}} \equiv \frac{d\vec{\mathbf{v}}}{dt} \xrightarrow{\text{integrate}} \vec{\mathbf{v}}(t) \xrightarrow{\text{integrate}} \vec{\mathbf{r}}(t)$$

EXERCISE 2.35: An object moves during the time interval from $t = 0$ and $t = 5$ s in a way that can be modeled by the position vector function $\vec{\mathbf{r}}(t) = \left[\frac{3}{2}t^2 - 2t + 8\right]\hat{\mathbf{i}} + \left[2t^2 - 17\right]\hat{\mathbf{j}}$. What is the displacement between $t = 0$ and the time when the object is moving at its minimum speed? At what time is the object traveling at its maximum speed?

Quiz

1. Select the situation that is *impossible*

(a) a positive position, negative displacement, and positive acceleration
(b) a positive velocity and a negative acceleration
(c) a positive & constant velocity and a positive acceleration
(d) a negative position, positive velocity, and positive acceleration
(e) zero velocity and a constant, non zero acceleration

2. It would be correct to say that an object that travels a negative *distance*

(a) has a negative speed
(b) has a negative velocity
(c) has a negative displacement
(d) ...it would *never* be correct to say!

3. Two objects, A and B, are thrown with the same speed out a window. Object A is thrown straight down. Object B is thrown straight up. Just before landing on the ground below, the velocity of A is: (Neglect air resistance)

(a) equal in magnitude but opposite that of B just before landing
(b) greater than that of B just before landing
(c) less than that of B just before landing
(d) equal in both magnitude and direction to that of B just before landing

4. $-\Delta v =$

(a) $v_f - v_i$ (b) 0 (c) $v_i - v_f$ (d) the acceleration

5. What condition must be satisfied for the average speed to equal the average velocity?

6. True or False? "The acceleration vector is always perpendicular to the velocity vector."

7. An object is thrown straight up into the air with an initial speed of 40 m/s. About how long will it be before the speed is 10 m/s? Is there a second answer to this question?

8. True or False? "It is possible for an object to be speeding up and have a negative acceleration."

9. True or False? "The constant acceleration kinematic equations may be applied to any motion."

10. True or False? "An object can never travel along a circular path and have no acceleration."

11. The area under a velocity vs time curve represents ____________.

12. True or False? "Zero velocity is seen on an x vs t curve only at the maxima and minima."

13. If we draw the position vector at t_1 and the position vector at t_2, the change in the position vector is draw from the tip of _____ to the tip of ______.

14. True or False? "Constant acceleration implies constant velocity."

15. True or False? "A stage occurs at an instant in time."

16. The component of the acceleration vector that is perpendicular to the velocity vector tells us how the ___________ is changing, while the component parallel to $\vec{v}$ tells us how the _____________ is changing.

17. True or False? "If the velocity is negative at some instant, the position is also negative."

18. True or False? "An object may have zero velocity and have non zero acceleration."

19. An object's position is described by the function: $x(t) = a - bt$. Which of the following are correct statements. ("a" and "b" are just some constants.)

(a) The acceleration is negative
(b) The velocity is constant.
(c) The velocity is zero.
(d) The acceleration is zero.
(e) The velocity is only zero at an instant, when the x vs t curve crosses the t axis.

20. An object is dropped from rest. The distance it falls during each successive second:

(a) is the same.
(b) increases.
(c) decreases.
(d) depends upon the alignment of the planets at the time.

Quiz Solutions

1. c ; constant velocity implies $a = 0$.

2. d ; distance is defined to be positive.

3. d ; when the object that is thrown up returns to its starting height it has the same speed (opposite direction) as when it started. But this is the same speed as the one thrown down!

4. c ; $\Delta v = v_f - v_i$, so $-\Delta v = v_i - v_f$

5. Average speed equals average velocity when the distance is equal to the displacement (no turning around!).

6. False. The acceleration vector may have any orientation relative to the velocity vector.

7. On the way up, it slows by about 10 m/s each second. Thus it will take 3 seconds. But there is a second answer. After it reaches the top, it turns around and speeds up 10 m/s each second. To get to this point, it takes 4 seconds to get to the top (where $v = 0$) and 1 more second to speed up to 10 m/s . So a total of 5 seconds for this alternate answer.

8. True. Speeding up while traveling in the negative direction is a negative acceleration.

9. False! So when we get to springs and pendula, which have changing accelerations, don't try and pull out these equations.

10. True. It's always changing direction.

11. The area under a velocity vs time curve represents displacement.

12. False. Zero velocity is also seen on an x vs t curve where it is flat (zero slope).

13. If we draw the position vector at t_1 and the position vector at t_2, the change in the position vector is draw from the tip of $\vec{\mathbf{r}}_1$ to the tip of $\vec{\mathbf{r}}_2$.

14. False! Precisely the opposite. Constant acceleration means that the velocity is changing.

15. False. A stage is over an interval of time. Thus it includes many states.

16. direction; speed

17. False. The position could be +, –, or 0; there is no correlation.

18. True. Otherwise, how would anything ever change its speed from 0 to non zero? The moment you begin to speed up, you are accelerating.

19. c, d. Take the derivative to see that the velocity is equal to – b. Differentiate again to see that $a = 0$.

20. b, increases. It is speeding up, after all.

Exercise Hints & Solution Plans

EX. 2.2: Think carefully about the definitions. Fuzzy thinking will get you in trouble. It is important to memorize these distinctions.

EX. 2.3: Use your imagination. Imagine being in a car and sticking your head out the window. Imagine being in a plane. Use your senses.

EX. 2.4: Again, use your imagination. Can you imagine a situation that satisfies the specified conditions? Think carefully about the definitions. Imagine looking at a speedometer. Ex. 2.5 should help.

EX. 2.6 & 2.7: Just use the rough equalities to make the comparison.

EX. 2.8: Think about altering the amount (time or distance?) that you travel at each speed. By the way, did you consider the magnitudes of these speeds? Would you likely be traveling in a car?

EX. 2.9: [Sol'n Plan]

Write down the information given to you ("givens"), and any implicit ("hidden") information.

Write down what you want to solve for ("unknowns").

Write down all pertinent relations between your knowns and unknowns.

Write down implicit conditions pertaining to the problem ($\Delta t_{total} = \Delta t_1 + \Delta t_2$, etc.).

Algebraically solve for the unknowns.

EX. 2.11: [Sol'n Plan]

Write down what's known and what's desired.

Write down the basic relation(s).

Write down any special conditions that are implicit.

Look at both objects (the walls and Bond).

Solve.

Exercise Solutions

EX. 2.1: [Quick Sol'n]

In describing motion, the 4 fundamental concepts are *time, space (position, distance, displacement), velocity (rate of change of position; roughly space/time)*, & *acceleration (rate of change of velocity)*.

EX. 2.2: [Quick Sol'n]

(a) distance = 1 mile; displacement = 0 mile.

(b) NO! Since ave. speed is defined as the distance divided by the time, it is non zero here. The ave. velocity is the displacement divided by the time, which is zero in this case.

EX. 2.3: [Quick Sol'n]

(a) Velocity can be detected 1) visually by looking at other objects that you pass, 2) visually by looking at a speedometer, 3) tactilely by feeling the air pass your skin, and 4) auditorally by listening to sounds that you associate with movement. Did you come up with others? Notice that our determination of movement is *always in reference to other objects*. But who is to say they aren't moving and you aren't at rest? Can you tell that you are flying through space, along with the Earth, at about mi/h as you sit there and read this?

(b) Acceleration is another story. When you accelerate, your body (mass) resists the acceleration and you feel it. You feel jerked around, thrown forwards, or thrown backwards. If you are riding in a vehicle over a bumpy road, those bumps provide little changes to your velocity–little accelerations. (This can also give you a clue that you are moving ... over the bumpy road.) Hence, unlike velocity, acceleration is easily and unequivocally detected.

EX. 2.4: [Quick Sol'n]

(a) YES! Imagine sitting in your car at a stop light. The speedometer reads 0 mi/h. You step on the gas. For an instant, the speedometer still reads 0 mi/h but you're speeding up because the next instant your speedometer doesn't read 0 mi/h. Throw an object straight up into the air. At the top of its path, the velocity is 0 m/s. But in the previous instant and the next instant the velocity is not 0. The velocity is continuously changing–it's accelerating.

(b) Sure. Any time you travel in a straight line at a constant speed, your velocity is not zero but your acceleration is.

EX. 2.5: [Quick Sol'n]

(a) NO! In a snapshot you can't tell what will happen the next instant. The odometer could be in motion and changing, which would indicate a non zero speed even at the instant the odometer reads 000000.0 miles. With a video image we could see if the odometer is moving and how fast it is moving, if it is.

(b) No again. We need to see the *change* in the speed or direction to see the acceleration. If the speedometer moves up or down in the next instant, as seen in a video, we can say that the car is accelerating. Notice that the speedometer could not change and the car could still be accelerating. How? Right! By changing direction. So, you could have a zero velocity and an acceleration of 0 $^{m}/_{s^2}$, +10 $^{m}/_{s^2}$, or – 500 $^{m}/_{s^2}$. Similarly, you could have a speed of 100 $^{m}/_{s}$ and an acceleration of 0 $^{m}/_{s^2}$, +10 $^{m}/_{s^2}$, or – 500 $^{m}/_{s^2}$.

(c) The odometer would be moving. This shows a change in position.

(d) The speedometer would be moving. This shows a change in speed. You would need to check the compass for a change in direction.

EX. 2.6: The Porsche has an acceleration of (60 – 0) $^{mi}/_{h}$/6 s = 10 $^{mi}/_{h}$/s . Since 1 $^{m}/_{s}$ ≈ 2 $^{mi}/_{h}$, this is roughly 5 $^{m}/_{s}$/s = 5 $^{m}/_{s^2}$. A dropped object with negligible air resistance wins easily. In fact, it takes the dropped object about 3 seconds to get to 30 $^{m}/_{s}$ ≈ 60 $^{mi}/_{h}$! It'll even beat that dragster!

EX. 2.7: [Quick Sol'n]

(a) Sure! Move your hand about 1 foot in the time it takes to say "one thousand one."

(b) 30 $^{m}/_{s}$ ≈ 60 $^{mi}/_{h}$ ≈ 120 $^{km}/_{h}$ ... freeway speeds. Remember that a km is less than a mile.

(c) Since 1 $^{km}/_{h}$ ≈ 1/2 $^{mi}/_{h}$ and 1 $^{m}/_{s}$ ≈ 2 $^{mi}/_{h}$, 4 $^{km}/_{h}$ ≈ 2 $^{mi}/_{h}$ ≈ 1 $^{m}/_{s}$. So 4 $^{km}/_{h}$ ≈ 1 $^{m}/_{s}$.

EX. 2.8: [Quick Sol'n]

(You should draw a picture!)

Bond is toast when the walls come together.

The walls meet halfway and so travel 5 m at a rate of 1 $^{m}/_{s}$. Thus the walls meet in 5 s.

Bond runs at an average of 4 $^{m}/_{s}$ for 5 s, thus he runs for a total of **20 m**.

Converting to miles using 1000 m = 1 km = 0.62 mi to form the conversion factor 6.2×10^{-4} $^{mi}/_{m}$ gives $20 \cdot 6.2 \times 10^{-4} = \mathbf{1 \times 10^{-2}}$ **miles** on those shoes (to one sig. fig.).

EX. 2.9: If you were to travel a greater portion of the trip at 10 $^{m}/_{s}$, you might expect that your average would be closer to 10 $^{m}/_{s}$. If you traveled all but an insignificant amount of the trip at 10 $^{m}/_{s}$, you would rightly expect the average to be close to 10 $^{m}/_{s}$. Similarly, if you traveled all but an insignificant amount of the trip at 1000 $^{m}/_{s}$, you would rightly expect the average to be close to 1000 $^{m}/_{s}$. Now the question remains: what exactly do we mean by a greater portion? Do we mean a greater distance or a greater amount of time? Or does the distinction matter? Let's see!

EX. 2.10: [Full Sol'n]

(a) [MODEL] A simple kinematics problem in which we carefully use the definition of average speed

[SETUP]

Data:	Given	Hidden (implicit)	Want(unknown)
	$S_{av,\,1} = 10\ ^{m}\!/\!_{s}$		$S_{av,\,total} = ?$
	$l_1 = 3{,}000$ m		
	$S_{av,\,2} = 30\ ^{m}\!/\!_{s}$		
	$l_2 = 3{,}000$ m		
	$l_{total} = 6{,}000$ m		

Relations/Equations:

$$S_{av} \equiv \frac{l}{\Delta t} \quad \text{(definition of average speed)}$$

$$\Delta t_{total} = \Delta t_1 + \Delta t_2 \quad \text{and } l_{total} = l_1 + l_2 \quad \text{(implicit conditions)}$$

[SOLVE] Use the definition for each part and for the total trip separately.

$$S_{av,\,1} \equiv \frac{l_1}{\Delta t_1} \xrightarrow{\text{solve for unknown}} \Delta t_1 = \frac{l_1}{S_{av,\,1}} = \frac{3000\text{ m}}{10\ ^{m}\!/\!_{s}} = 300\text{ s}$$

$$S_{av,\,2} \equiv \frac{l_2}{\Delta t_2} \xrightarrow{\text{solve for unknown}} \Delta t_2 = \frac{l_2}{S_{av,\,2}} = \frac{3000\text{ m}}{30\ ^{m}\!/\!_{s}} = 100\text{ s}$$

$$\Delta t_{total} = \Delta t_1 + \Delta t_2 = 400\text{ s}$$

$$S_{av,\,total} \equiv \frac{l_{total}}{\Delta t_{total}} = \frac{6000\text{ m}}{400\text{ s}} = 15\frac{\text{m}}{\text{s}}$$

[ANALYZE] Did you catch my error? It was really a choice. I chose to write 15 $^{m}\!/\!_{s}$ rather than 20 $^{m}\!/\!_{s}$ even though 20 $^{m}\!/\!_{s}$ uses 1 sig. fig. in accordance with the data given. I suppose I should start writing my data with 3 sig. fig.'s, if you don't mind. In this exercise, I'll take my numbers to be exact. Then we notice that the average speed is *not* 20 $^{m}\!/\!_{s}$, as you might have guessed upon blind use of averages.

Oh yes, and about how fast are we talking here? 10 $^{m}\!/\!_{s} \approx$ 22 $^{mi}\!/\!_{h}$, 15 $^{m}\!/\!_{s} \approx$ 33 $^{mi}\!/\!_{h}$, and 30 $^{m}\!/\!_{s} \approx$ 66 $^{mi}\!/\!_{h}$—pretty standard driving speeds.

(b) [MODEL] A simple kinematics problem in which we carefully use the definition of average speed

[SETUP]

Data:	Given	Hidden (implicit)	Want(unknown)
	$S_{av,\,1} = 10\ \text{m/s}$	$\Delta t_1 = \Delta t_2 = 1/2\ \Delta t_{total}$	$S_{av,\,total} = ?$
	$S_{av,\,2} = 30\ \text{m/s}$		
	$l_{total} = 6{,}000\ \text{m}$		

Relations/Equations:

$$S_{av} \equiv \frac{l}{\Delta t} \quad \text{(definition of average speed)}$$

$$\Delta t_{total} = \Delta t_1 + \Delta t_2 \quad \text{and } l_{total} = l_1 + l_2 \quad \text{(implicit conditions)}$$

[SOLVE] Use the definition for each part and for the total trip separately.

$$S_{av,\,1} \equiv \frac{l_1}{\Delta t_1} \xrightarrow{\text{solve for unknown}} l_1 = S_{av,\,1} \cdot \Delta t_1 = 10\ \text{m/s} \cdot \frac{1}{2} \Delta t_{total}$$

$$S_{av,\,2} \equiv \frac{l_2}{\Delta t_2} \xrightarrow{\text{solve for unknown}} l_2 = S_{av,\,2} \cdot \Delta t_2 = 30\ \text{m/s} \cdot \frac{1}{2} \Delta t_{total}$$

$$l_{total} = l_1 + l_2 = 20\ \text{m/s} \cdot \Delta t_{total}$$

$$S_{av,\,total} \equiv \frac{l_{total}}{\Delta t_{total}} = \frac{20\ \text{m/s} \cdot \Delta t_{total}}{\Delta t_{total}} = 20 \frac{\text{m}}{\text{s}}$$

[ANALYZE] Very, very interesting! Here the blind averages *would* have worked. That's only because we spent the same amount of *time* traveling at each of the speeds (on average). Further notice that the result here is indeed different than that obtained in part (a). If you just apply the relations carefully, you will obtain the correct solution. You needn't memorize these solutions in detail, but they are worth studying.

EX. 2.11: [Quick Sol'n]

Simply apply the weighted average formula:

$$S_{av} = \frac{S_{av,\,1} \cdot \Delta t_1 + S_{av,\,2} \cdot \Delta t_2 + S_{av,\,3} \cdot \Delta t_3 + S_{av,\,4} \cdot \Delta t_4}{\Delta t_{total}}$$

$$= \frac{S_{av,\,1} \cdot \frac{1}{4} \Delta t_{total} + S_{av,\,2} \cdot \frac{1}{4} \Delta t_{total} + S_{av,\,3} \cdot \frac{1}{4} \Delta t_{total} + S_{av,\,4} \cdot \frac{1}{4} \Delta t_{total}}{\Delta t_{total}}$$

$$= \frac{\left(S_{av,\,1} + S_{av,\,2} + S_{av,\,3} + S_{av,\,4}\right) \cdot \frac{1}{4} \Delta t_{total}}{\Delta t_{total}}$$

$$= \frac{\left(S_{av,\,1} + S_{av,\,2} + S_{av,\,3} + S_{av,\,4}\right)}{4}$$

EX. 2.12:

(a) You can either wait to see what you calculate the average speed (not velocity) to be, or think about the distance and time required. Most people walk about 2 steps (≈ 1 m) in about 1 s. So a "walk" would get you about 2 m in 2 s. We'll need to go faster than that! Run! Let's see just how fast.

(b) Total distance = 2 m + 4 m = 6 m (Note that distance is always positive.)

(c) Total displacement is between the end points. You start at 3 m and end up at 1 m, so you went back 2 m. Thus tot. displ. = –2 m. In equation form: $\Delta x = x_f - x_i = 1\text{ m} - 3\text{ m} = -2\text{ m}$.

(d) Ave. speed = tot. dist./ tot. time = 6 m/2 s = 3 m/s (3 times walking speed, a doable run).

(e) Ave. vel. = tot. displ./ tot. time = –2 m/2s = –1 m/s.

EX. 2.13:

Notice that the signs correspond to our choice of positive direction. In particular, notice that a positive acceleration does not necessarily mean that an object is speeding up! If an object starts out with a *speed* of 3 m/s in the negative direction (v = – 3 m/s) and ends up speeds up to 7 m/s in the negative direction (v = —7 m/s), the change in velocity (not speed) is $\Delta v = v_f - v_i = -7\text{ m/s} - (-3\text{ m/s}) = -4\text{ m/s}$. Hence the acceleration would also be negative (since Δt is positive). Make sure that this point is clear.

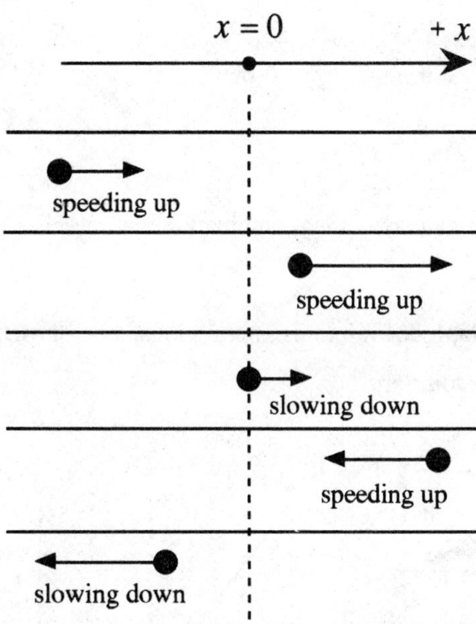

	+, –, or 0 ?			
	x	Δ*x*	*v*	*a*
(a)	–	+	+	+
(b)	+	+	+	+
(c)	0	+	+	–
(d)	+	–	–	–
(e)	–	–	–	+

EX. 2.14: Let's tabulate the information requested.

	directions (sign of v)		Δ direction		
	initial	final	Turns around?	Δ speed $= (v_f - v_f)$	ave. accel. $= \Delta v/\Delta t$
(a)	+	+	no	$(10 - 4) = 6$ m/s	$+3$ m/s²
(b)	+	+	no	$(4 - 10) = -6$ m/s	-3 m/s²
(c)	–	+	yes	$(10 - (-4)) = 14$ m/s	$+7$ m/s²
(d)	+	–	yes	$((-10) - 4) = -14$ m/s	-7 m/s²
(e)	–	+	yes	$(4 - (-10)) = 14$ m/s	$+7$ m/s²
(f)	+	–	yes	$((-4) - 10) = -14$ m/s	-7 m/s²
(g)	–	–	no	$((-10) - (-4)) = -6$ m/s	-3 m/s²
(h)	–	–	no	$((-4) - (-10)) = 6$ m/s	$+3$ m/s²

EX. 2.15: Part (a) asks for a vertical, not horizontal x - axis.

(a)

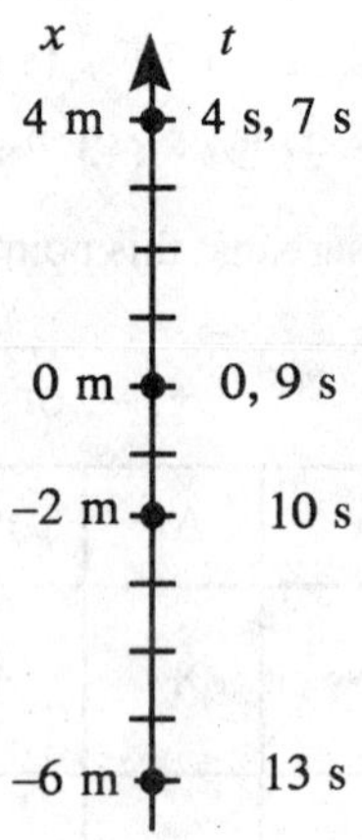

	dist. walked	displacement ($\Delta x = x_f - x_i$)	ave. speed (dist./Δt)	ave. velocity ($\Delta x/\Delta t$)
(b)	4 m	4 m	1 m/s	1 m/s
(c)	4 m	4 m	4/7 m/s	4/7 m/s
(d)	10 m	–2 m	1 m/s	–1/5 m/s
(e)	14 m	–6 m	14/13 m/s	–6/13 m/s
(f)	0 m	0 m	0 m/s	0 m/s
(g)	4 m	–4 m	4/3 m/s	–4/3 m/s

Investigate any other combination that interests you.

EX. 2.16: I'll break the graph down into distinct regions, label those regions with Roman numerals, and then refer to them in the solution.

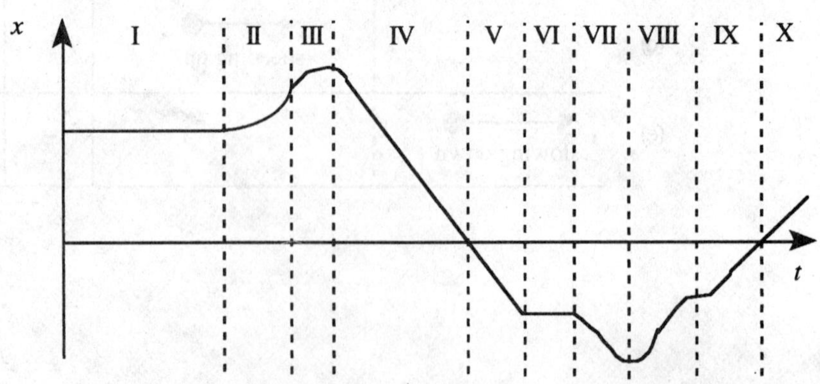

(a) This is *not* like a roller coaster ride. Remember that the motion is not along this curve, it is along the x axis, like walking up and down a shopping aisle.

	Positive Regions	Zeros	Negative Regions
(b) Position:	I – IV; X	between IV & V and IX & X	V – IX
(c) Velocity:	II; III; VII – X	I; between III & IV; VI; between VII & VIII between VIII & IX	IV; V; VII
(d) Accel.:	II; VII; VIII; beginning of IX	I; IV; V; VI; IX; X	III; end of VIII

(e) Prolonged rest when $v = 0$ for an extended period (flat curve, not max or min): I, IV

(f) Area means nothing here!

EX. 2.17: Same curve, very different meaning!

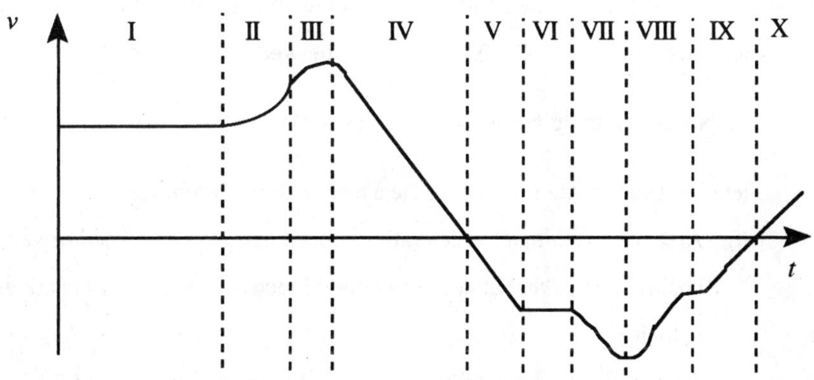

(a) It's hard to say what the motion looks like. If this is 2–D motion, then all we are plotting is the speed, not the direction. But then what's the meaning of a negative speed? This is again 1–D motion, not a roller coaster ride.

	Positive Regions	Zeros	Negative Regions
(b) Position:	I know nothing about the position from this graph!!		
(c) Velocity:	I – IV; X	between IV & V and IX & X	V – IX
(d) Accel.:	II; III; VII – X	I; between III & IV; VI; between VII & VIII between VIII & IX	IV; V; VII

Notice anything familiar about the solution? The velocity regions here are the same as the position regions for the previous exercise; the accel. regions are the same as the previous exercises velocity regions. Keep in mind that this graph represents a very different motion than in the previous exercise. So why the connection? (Unsure? Ask someone!)

(e) Prolonged rest when $v = 0$ for an extended period. Here that would mean that the curve was along the t axis for an extended length. Nowhere is that the case, so no extended rest here!

(f) Aha. Area has meaning of velocity times time, which is distance or displacement. The area above the t axis is positive, while that under is negative. Since distance is never negative, the area must represent the displacement.

(g) As I've said, this motion is quite different from that in exercise 2.16. Just look at the velocity patterns!

EX. 2.18: Again, same curve, very different meaning! The regions are the same, only the axes have been changed to vex, I mean protect, the innocent. So look to the graph in the previous solution, but make sure that you change the axes to acceleration vs time. (Continued, next page.)

(a) This is again 1–D motion, not a roller coaster ride.

	Positive Regions	Zeros	Negative Regions
(b) Position:	I know nothing about the position from this graph!!		
(c) Velocity:	I know nothing about the velocity from this graph!!		
(d) Accel.:	I – IV; X	between IV & V and IX & X	V – IX

Notice anything familiar about the solution? Why *is* that?!

(e) That information is not on the accel. vs time graph.

(f) Area has meaning of acceleration times time, which is speed or velocity. The area above the t axis is positive, while that under is negative. Since speed is never negative, the area must represent the average velocity.

(g) This motion is quite different from that in exercises 2.16 and 2.17

EX. 2.19: While I am going to draw in displacement, velocity, and acceleration vectors on the same diagram, you know that they can't be added, except with vectors that have the same units.

(a)

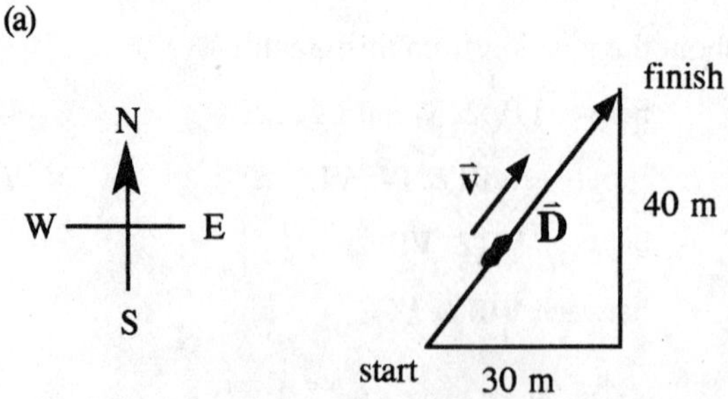

(b)

Total displacement brought to you by Pythag.

... or the 3–4–5 rt. triangle → 50 m.

(For deciding on the significant figures, I'll treat these numbers as exact.)

(c) Sketch the velocity vector with components:

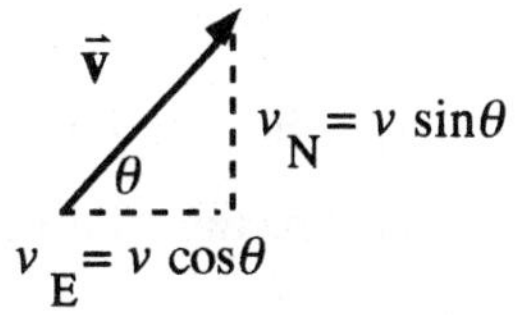

We know $v = 15$ m/s, but what's that angle? In this particular case, the angle of the velocity vector is the same as that of the displacement vector. (This is by no means always the case.) This allows us to get the angle from the displacement triangle: $\theta = \tan^{-1}\left(\frac{40\text{ m}}{30\text{ m}}\right) = 53^\circ$

Thus, $v_N = v\sin\theta = (15\text{ m/s})\sin 53^\circ = \mathbf{12}$ **m/s**

$v_E = v\cos\theta = (15\text{ m/s})\cos 53^\circ = \mathbf{9.0}$ **m/s**

(d) A sketch here is critical! (It doesn't have to be fancy.) I'm going to first use my original diagram:

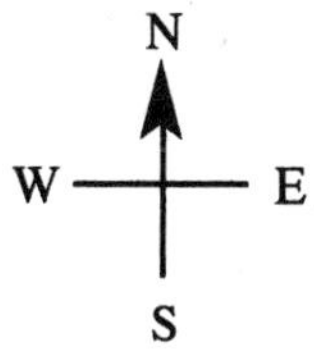

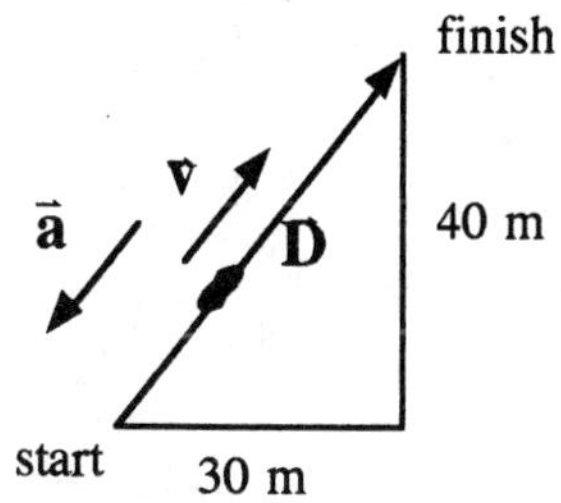

So why is the accel. vector opposite the velocity vector? It's important to notice that right when you read a problem – "slowing down without a change in direction" needs to scream at you.

Now a quick sketch of the accel. vector alone:

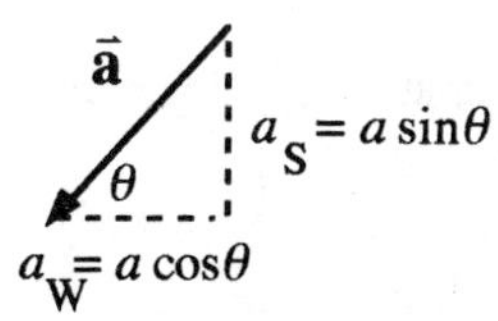

Again, the very nice person that designed this problem has set it up so that we can still use the angle found from the displacement triangle. Certainly not always true. Also notice your choices with the negative signs of the components

Here, $a_S = a\sin\theta = (3.0\text{ m/s})\sin 53^\circ = \mathbf{2.4}$ $\mathbf{m/s^2}$

$a_W = a\cos\theta = (3.0\text{ m/s})\cos 53^\circ = \mathbf{1.8}$ $\mathbf{m/s^2}$

If I gave these as north and east components, I would need to put in a minus sign.

EX. 2.20: (a) The component of $\vec{a}$ parallel to $\vec{v}$ tells us what's happening to the speed. Since the parallel component of $\vec{a}$ is opposite $\vec{v}$, it will be going slower in the next moment. (b) The perpendicular component of $\vec{a}$ is down and to the right of $\vec{v}$, so the object will arc downward to the right.

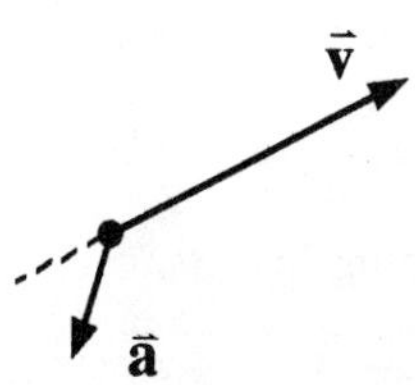

EX. 2.21: It definitely accelerates on that path. Why? Because it is changing direction. We can't tell anything about the component of accel. parallel to the velocity (i.e. if it's changing speed), but we know that there is a component of the acceleration that is perpendicular to the path. In fact, we know that it is down and to the right (into the arc), because that is the direction of the change in direction.

EX. 2.22: Here's mine:

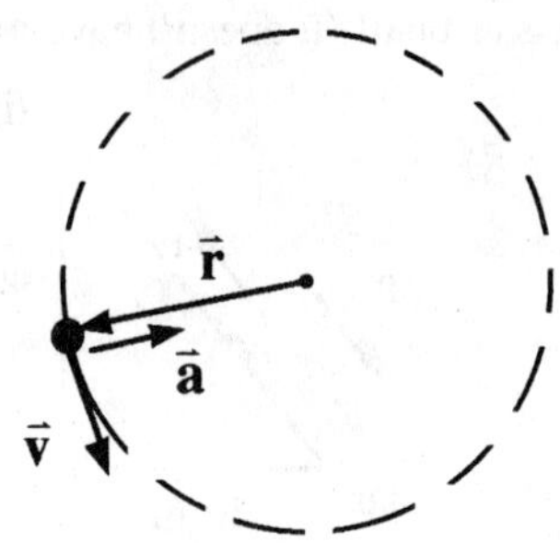

I hope I didn't fool you. The *sum* of these vectors is "NONSENSE!" But we can still draw them on the same diagram if we like. In fact, it is often informative to do so.

EX. 2.23: If you're having trouble here, see chapter 1. If that doesn't help ask your instructor, TA, or classmate to walk you through this. It *is* important.

EX. 2.24: [Full Sol'n]

[MODEL] Constant acceleration motion always allows us to use the very handy set of kinematics equations given. We assume constant acceleration here.

[SETUP]

Data:	Given	Hidden (implicit)	Want(unknown)
	$v_i = 20.0$ m/s		$\Delta x\ (t = 7.00\text{ s}) = ?$
	$t = \Delta t = 7.00$ s		$v\ (t = 7.00\text{ s}) = ?$
	$a = 1.000$ m/s^2		$v_{av} = ?$

Relations/Equations:

0. $v_{av} = \frac{1}{2}(v_f + v_i)$
1. $\Delta x = v_i t + \frac{1}{2}at^2$
2. $v = v_i + at$
3. $2a\Delta x = v_f^2 - v_i^2$

[SOLVE] Simple plug ins.

(a) $\Delta x = (20.0\text{ m/s})(7.00\text{ s}) + \frac{1}{2}(1.000\text{ m/s}^2)(7.00\text{ s})^2 = 165\text{ m}$

(b) $v\ (t = 7.00\text{ s}) = (20.0\text{ m/s}) + (1.000\text{ m/s}^2)(7.00\text{ s}) = 27.0\text{ m/s}$

(c) I'll check by using the velocity just obtained and see if we get the same displacement.

$\Delta x = (v_f^2 - v_i^2)/2a = [(27.0\ ^m\!/\!_s)^2 - (20.0\ ^m\!/\!_s)^2]/2(1.000\ ^m\!/\!_{s^2}) = 165$ m, Check!!

(d) Using the "naive" average works in the special case of constant acceleration. $v_{av} = 23.5\ ^m\!/\!_s$

[ANALYZE] First, if we traveled for 7 s at 20.0 $^m\!/\!_s$, we would travel 140 m. So it makes sense that we go a bit further when we accelerate. Second, an acceleration of 1 $^m\!/\!_{s^2}$ means that each second, the velocity changes by 1 $^m\!/\!_s$. Thus in 7 seconds, it makes sense that we would go from 20.0 $^m\!/\!_s$ to 27.0 $^m\!/\!_s$. Third, we see that the equations are consistent and that, once we use 2 of equations 1, 2, and 3, we have used the equations up … unless we want to continue and use one or two different states in the equations.

EX. 2.25: You need to be able to write these quickly.

(a) $\Delta y = v_i t$ is really the only equation. The others whittle down to $v_f = v_i$, which is implicit in the statement that $a = 0$.

(b) Since you are asked to choose up as + and the acceleration is down, we insert $a = -g = -9.8\ ^m\!/\!_{s^2}$ into the templates: (1) $\Delta y = v_i t - {}^1\!/\!_2 g t^2$, (2) $v = v_i - gt$, (3) $-2g\Delta y = (v_f^2 - v_i^2)$.

(c) It's just those 4 equations above, but use subscripts to distinguish between the vertical (y, v_y, $a_y = \pm g$) and horizontal (x, v_x, $a_x = 0$) directions. So 1 horizontal equation and 3 vertical equations. Now see how quickly you can write them.

EX. 2.26: This approach is very general.

So I don't have to keep reading the problem, let me summarize what we know:

[MODEL] Really an exercise in the mathematics of vectors, although the position vector does just happen to describe a projectile.

[SETUP]

Data:	Given	Want(unknown)
	$v_{ox} = 2.00\ ^m\!/\!_s$	position and distance from the origin at t = 2.00 s
	$v_{oy} = 3.00\ ^m\!/\!_s$	x and y components of velocity and speed at t = 2.00 s
	$a = -10.0\ ^m\!/\!_{s^2}$	x and y components of accel. and total accel. at t = 2.00 s

Relations/Equations:

$$\vec{\mathbf{r}}(t) = [v_{ox}t]\hat{\mathbf{i}} + [v_{oy}t + \tfrac{1}{2}at^2]\hat{\mathbf{j}}$$

[SOLVE] Key: you must differentiate *before* putting in the particular time.

Position at t = 2.00 s: we can just put in the numbers here.

$$\vec{\mathbf{r}}(t) = [v_{ox}t]\hat{\mathbf{i}} + [v_{oy}t + \tfrac{1}{2}at^2]\hat{\mathbf{j}} \Rightarrow$$

$$\vec{\mathbf{r}}(t = 2.00\text{ s}) = [(2.00\ \tfrac{\text{m}}{\text{s}})(2.00\text{ s})]\hat{\mathbf{i}} + [(3.00\ \tfrac{\text{m}}{\text{s}})(2.00\text{ s}) + \tfrac{1}{2}(-10.0\ \tfrac{\text{m}}{\text{s}^2})(2.00\text{ s})^2]\hat{\mathbf{j}}$$

$$= (4.00\text{ m})\hat{\mathbf{i}} + (-14.0\text{ m})\hat{\mathbf{j}}$$

Distance from the origin at t = 2.00 s: this is the length of the position vector at that time.

$$r(t=2.00\text{ s}) = |\vec{\mathbf{r}}| = \sqrt{(4.00\text{ m})^2 + (-14.0\text{ m})^2} = 14.6\text{ m}$$

Now we are concerned with the velocity. Before we plug in the time, we had better find the velocity function by differentiating (with respect to time) each component separately .

$$\begin{aligned}\vec{\mathbf{v}}(t) &= \frac{d}{dt}\vec{\mathbf{r}}(t) = \left[\frac{d}{dt}(v_{ox}t)\right]\hat{\mathbf{i}} + \left[\frac{d}{dt}\left(v_{oy}t + \tfrac{1}{2}at^2\right)\right]\hat{\mathbf{j}} \\ &= [v_{ox}]\hat{\mathbf{i}} + [v_{oy} + at]\hat{\mathbf{j}}\end{aligned}$$

x and y components of velocity at $t = 2.00$ s are shown with the unit vector notation:

$$\vec{\mathbf{v}}(t) = \left[2.00\tfrac{\text{m}}{\text{s}}\right]\hat{\mathbf{i}} + \left[3.00\tfrac{\text{m}}{\text{s}} + \left(-10.0\tfrac{\text{m}}{\text{s}^2}\right)t\right]\hat{\mathbf{j}} \quad\Rightarrow\quad \vec{\mathbf{v}}(t=2.00\text{ s}) = \left[2.00\tfrac{\text{m}}{\text{s}}\right]\hat{\mathbf{i}} + \left[-17.0\tfrac{\text{m}}{\text{s}}\right]\hat{\mathbf{j}}$$

Speed = magnitude of the velocity vector at that time:

$$v(t=2.00\text{ s}) = |\vec{\mathbf{v}}| = \sqrt{\left(2.00\tfrac{\text{m}}{\text{s}}\right)^2 + \left(-17.0\tfrac{\text{m}}{\text{s}}\right)^2} = 17.1\ \tfrac{\text{m}}{\text{s}}$$

Finally, we arrive at the acceleration by differentiating the velocity function (*not* the velocity at that time< since we would then be differentiating a number!)

$$\begin{aligned}\vec{\mathbf{a}}(t) &= \frac{d}{dt}\vec{\mathbf{v}}(t) = \left[\frac{d}{dt}(v_{ox})\right]\hat{\mathbf{i}} + \left[\frac{d}{dt}(v_{oy} + at)\right]\hat{\mathbf{j}} \\ &= [0]\hat{\mathbf{i}} + [0 + a]\hat{\mathbf{j}} \\ &= \left[-10.0\tfrac{\text{m}}{\text{s}^2}\right]\hat{\mathbf{j}}\end{aligned}$$

The acceleration is a constant, so at any time it has a magnitude of 10.0 m/s^2 and is in the $-y$ direction.

(b) What is the displacement between t = 1.00 and t = 2.00 s? Here we need to recall that displacement is the change in the position vector. Hence,

$$\begin{aligned}\vec{\mathbf{D}} &= \Delta\vec{\mathbf{r}} = \vec{\mathbf{r}}(t=2.00\text{ s}) - \vec{\mathbf{r}}(t=1.00\text{ s}) \\ &= \left[(4.00\text{ m})\hat{\mathbf{i}} + (-14.0\text{ m})\hat{\mathbf{j}}\right] - \left[(2.00\text{ m})\hat{\mathbf{i}} + (-2.00\text{ m})\hat{\mathbf{j}}\right] \\ &= \left[(4.00\text{ m} - 2.00\text{ m})\hat{\mathbf{i}} + (-14.0\text{ m} - (-2.00\text{ m}))\hat{\mathbf{j}}\right] \\ &= \left[(2.00\text{ m})\hat{\mathbf{i}} + (-12.0\text{ m})\hat{\mathbf{j}}\right]\end{aligned}$$

We can find the magnitude and angle of this vector if we want, but I've had enough. How about you?

EX. 2.27: The one without the t ! $2a\,\Delta y = \ldots$ Keep this in mind so that you can save time on the exam!

No need using two equations when one will do.

EX. 2.28: [Full Sol'n] You may want to read ahead to the section on "multiple stages."

[MODEL] There are two distinct stages in this process. She starts from rest, speeds up, then slows down. Clearly we are dealing with a kinematics analysis where one stage we use an average acceleration and the next stage we have a constant acceleration. Model this as simple 1-D motion even though she is probably moving over a very uneven surface.

[SETUP]

Visualize & Sketch

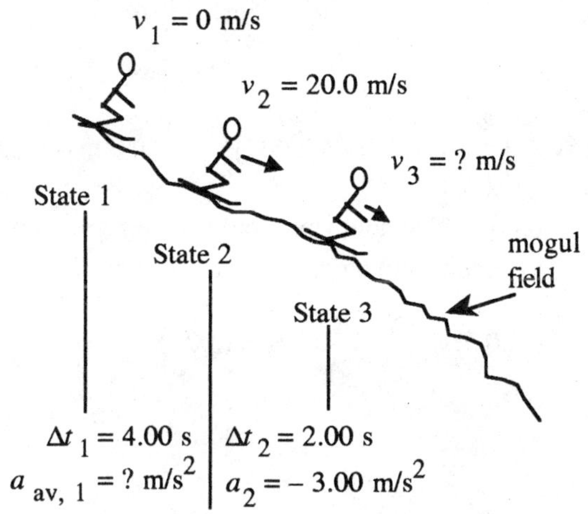

Gather & Sort Data

I've gathered and sorted most of the data on the sketch. The Hidden info. is that she starts at rest–as in most races. Just to keep me on track and save me time, I'll list what we want:

(a) $a_{\text{av, stage 1}} = ?$

(b) $v_{\text{av, stage 1}} = ?$

(c) $\Delta x_{\text{stage 2}} = ?$

The distance here is taken simply as the distance along the slope.

Method/Tool Box:

I will use the definitions for the average kinematic quantities and the constant acceleration equations.

Relations/Equations:

Definitions of Ave. Kin. Quantities

$$S_{av} \equiv \frac{l}{\Delta t} \quad \text{or} \quad \vec{\mathbf{v}}_{av} \equiv \frac{\vec{\mathbf{D}}}{\Delta t} = \frac{\Delta \vec{\mathbf{r}}}{\Delta t}$$

$$\vec{\mathbf{a}}_{av} \equiv \frac{\Delta \vec{\mathbf{v}}}{\Delta t} = \frac{\vec{\mathbf{v}}_f - \vec{\mathbf{v}}_i}{\Delta t}$$

Constant Acceleration Equations

0. $v_{av} = \frac{1}{2}(v_f + v_i)$
1. $\Delta y = v_i t + \frac{1}{2}at^2$
2. $v_f = v_i + at$
3. $2a\Delta y = v_f^2 - v_i^2$

[SOLVE]

(a) $a_{\text{av, stage 1}} = ?$ A word of caution: we are asked for the average acceleration for the first stage. Strictly, we need to address the direction of the acceleration and we need to include the directions of the velocity vectors. But since we don't know anything about the slope, we are forced to deal with this in the simplest manner possible, namely, look at the difference in speed divided by time.

$$a_{\text{av, stage 1}} = \Delta v / \Delta t = (v_2 - v_1)/\Delta t = (20.0 \text{ m/s} - 0 \text{ m/s})/4.00\text{s}$$

$$= 5.00 \text{ m/s}^2$$

(b) If her acceleration for the first stage was constant, we may easily find her average speed for that stage:

$$v_{\text{av, stage 1}} = \tfrac{1}{2}\,(20.0\ \text{m/s} + 0\ \text{m/s}) = 10.0\ \text{m/s}$$

(c) Using "Δy" to represent her displacement along the slope, we use the constant acceleration equations. In this case, we can get away with using one simple equation.

$$\Delta y = v_i\,t + \tfrac{1}{2}\,a\,t^2 \xrightarrow{\text{becomes}} \Delta y_{\text{stage 2}} = v_2\,t + \tfrac{1}{2}\,a\,t^2\ ,$$

where t is $\Delta t_{\text{stage 2}} = 2.00$ s

and a is $a_{\text{stage 2}} = -\,3.00$ m/s² (it's negative!)

$$\Delta y_{\text{stage 2}} = (20.0\ \text{m/s})(2.00\ \text{s}) + \tfrac{1}{2}\,(-\,3.00\ \text{m/s}^2)(2.00\ \text{s})^2 = 40.0\ \text{m} - 6.00\ \text{m} = 34.0\ \text{m}$$

(d) You can do this one! Which equation fits? (Caution: you should not expect all problems to be solved as simply as this by finding a single equation and plugging in the right numbers!)

$$v_3 = v_2 + a\,t = (20.0\ \text{m/s}) + (-\,3.00\ \text{m/s}^2)(2.00\ \text{s}) = 14.0\ \text{m/s}$$

EX. 2.29: To see the simple pattern, use $g \approx 10$ as instructed. I'll take *down* to be positive. Note that "drop" means that $v_i = 0$. You might also want to put in the numbers that you get when you use 9.8 .m/s²

object	t	$y = \frac{1}{2} g\,t^2$	$v = g\,t$	$a = g$
	0 s	0 m	0 m/s	10 m/s²
	1 s	5 m	10 m/s	10 m/s²
	2 s	20 m	20 m/s	10 m/s²
	3 s	45 m	30 m/s	10 m/s²

You finish it.

Note that $t = 0$ is the instant that the object is released, NOT before. Then, even though the speed is 0, it is changing and hence there is an acceleration.

EX. 2.30: [Full Sol'n]

[MODEL] Simple 1-D kinematics problem, neglecting air resistance and your height. This problem only has one stage: the free flight stage with acceleration due to gravity. We are not asked anything that requires looking at the throwing or landing processes. There are four states of interest that should be shown on an initial sketch.

Key Point Of This Problem:

Be *very* careful with the ± signs of displacement, velocity, and acceleration!! Very Careful.

[SETUP]

<u>Visualize & Sketch</u>

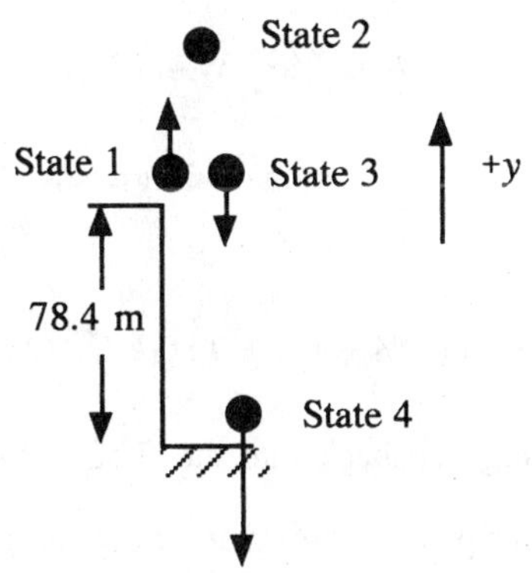

<u>Gather & Sort Data</u>

Given	Hidden	Want
$v_1 = +25.0$ m/s	$a = -g = -9.80$ m/s²	(a) $v_4 = ?$
building ht = 78.4 m	$v_{top} = 0$ m/s	(b) $\Delta t_{1\to4} = ?$
	$\lvert\Delta y_{1\to3}\rvert = 0$	(c) $\lvert\Delta y_{2\to4}\rvert = ?$
		(d) $v_3 = ?$
<u>for part (e) use:</u>		
$v_1 = -25.0$ m/s		$v_{ground} = ?$
		$t_{ground} = ?$

Notice that the diagram for part (e) would have the first state at state 3 and the final state at state 4. You could relabel accordingly, just be clear when and where the state of interest is.

Method/Tool Box:

The **constant acceleration kinematics equations** should do the trick! (By drawing a clear picture that shows the states, selecting and showing a positive direction, and laying out the givens, hiddens, and wants–paying careful attention to the ± signs–we have avoided many of the common errors that waste so many students' time.)

Relations/Equations:

Constant Acceleration Equations

Tailored to Fit 1-D Vertical Free Flight ($a_y = g$, down)

0. $v_{av} = \frac{1}{2}(v_f + v_i)$
1. $\Delta y = v_i t - \frac{1}{2}gt^2$
2. $v_f = v_i - gt$
3. $-2g\Delta y = v_f^2 - v_i^2$

Remember that these equations can relate the kinematic variables of *any* two states in a constant acceleration stage. Do note that the acceleration is the same *everywhere* (going up, top, going down).

[SOLVE]

(Now entering Mathematic Land!! Sometimes we get one equation and one unknown, while other times we end up with a system of multiple equations and multiple unknowns. Make sure that you have enough equations to solve for your unknowns.)

There is usually more than one way to apply the constant acceleration equations. With experience, you learn to choose the easiest and fewest equations to accomplish the task. I'll point out some alternate methods, but I won't work them all out.

(a) We are asked to find out something about state 4. We already know the speeds at states 1 and 2. So we can relate state 1 to state 4 or state 2 to state 4. Here are some different approaches.

METHOD 1: Using states 1 & 4.

$v_4 = v_1 - g\,t$ (KNOWN: $v_1 = +\,25.0$ m/s , $g = +\,9.80$ m/s^2 , WANT: v_4 , NEED: t)

We have one equation and two unknowns. We need an equation that will give us t between states 1 & 4. You don't have many choices!!!

$\Delta y_{1\to4} = v_1 t - \frac{1}{2}\,g\,t^2$ (KNOWN: $v_1 = +\,25.0$ m/s , $g = +\,9.80$ m/s^2 , $\Delta y_{1\to4} = -\,78.4$ m)

We can solve for t in this second equation and plug that back into the first. NOTICE: First, $\Delta y_{1\to4}$ must be negative to be consistent with the choice of positive being up. Second, g is entered as positive because we have already put the negative sign in. It is usually best to keep $g = +\,9.80$ m/s^2 and put the negative sign in the equations if we choose up to be positive. Third, solving for $t_{1\to4}$ answers part (b), so we are doing two things at once–a good thing. But what if we were not asked for the time?

* METHOD 2: Using states 1 & 4, but avoiding the need to find $t_{1\to4}$!

Hey, how about that equation that doesn't have time in it? After all, it came algebraically from the two equations used in Method 1 and so is not really a new equation. Let's see if we can luck out with only one eq'n and 1 unk'n:

$v_4^2 = v_1^2 - 2g\,\Delta y_{1\to4}$ (Hey! We know everything but the one variable that we seek, v_4. Don't forget to square v_1 and to put $\Delta y_{1\to4}$ in as **negative** 78.4 m. Also, remember to take the square root at the end.).

I get: v_4 = **46.5 m/s** (check that I didn't make a mistake!)

METHOD 3: Using states 2 & 4.

We could use the fact that $v_2 = 0$, but what's the problem? We don't yet know $\Delta y_{2\to4}$. Of course we could find it, but this seems pretty twisted.

METHOD 4: Using states 3 & 4.

To do this, you need to pick up on the hidden (implicit) information that $v_3 = -v_1 = -25.0$ m/s . Try this and compare the steps to method 2 to gain some insight.

***** OK! Do you see that there are many ways to go, but that there is often one that is more straightforward? I'll leave it to you to investigate alternate solutions for the next parts. Here are some quick solutions that get us to the heart of the matter.**

(b) I'll relate states 1 and 4, since we want the time interval between them. If I can, I'll choose 1 eq'n with the 1 unk'n that we seek.

$\Delta y_{1\to4} = v_1 t - \frac{1}{2} g t^2$ (KNOWN: $v_1 = +25.0$ m/s , $g = +9.80$ m/s^2 , $\Delta y_{1\to4} = -78.4$ m)

We have a quadratic equation to solve! We can rewrite this a number of ways. I'll look at it as:

$\frac{1}{2} g t^2 - v_1 t + \Delta y_{1\to4} = 0$ (where the quadratic coefficients are a = 4.90 m/s^2, b = – 25.0 m/s, and c = – 78.4 m)

From the quadratic equation I find: $t = +7.30$ s and -2.19 s. Clearly the negative solution is meaningless, so $\Delta t_{1\to4} = \mathbf{+7.30}$ s

Note: we could avoid the quadratic by finding v_4, but I don't think it's worth it.

(c) Have you got the game now? What states are good to relate here?

- Relate states 2 & 4 via the "timeless" equation? $v_4^2 = v_2^2 - 2g\,\Delta y_{2\to4}$ ... we would need v_4.
- Relate states 2 & 4 via $\Delta y_{2\to4} =$ ....., but then we need to find $\Delta t_{2\to4}$.
- Relate states 1 & 2 via $\Delta y_{1\to2} = \ldots$? But do we'll need $\Delta t_{1\to2}$.
- Relate states 1 & 2 via $v_2^2 = v_1^2 - 2g\,\Delta y_{1\to2}$ (since we know $v_2 = 0$), and then account for the building height? Yes! Do you see from the diagram that $|\Delta y_{2\to4}| = |\Delta y_{1\to2}| + 78.4$ m?

I find: $\Delta y_{1\to2} = 31.9$ m, so $|\Delta y_{2\to4}| = \mathbf{110\ m}$

(d) Simple answer: speed = 25.0 m/s, direction is down. So $v_3 = \mathbf{-25.0}$ **m/s**. Want to prove it? Relate states 1 & 3 via the "timeless" eq'n and notice the hidden information ... $\Delta y_{1\to3} = 0$!!! Of course there are other ways to obtain the same result.

(e) I'll leave this to you to both do and check your work with others. Just make sure that you account for the *physically different situation* by making the **initial velocity negative** (since I chose up to be positive).

EX. 2.31:

[MODEL] Because of the way the question is asked, this is still essentially a 1-D kinematics problem. There are two stages that need to be linked together in order to solve the problem.

Key Point Of This Problem:

You cannot relate two states with the constant accel. eq'ns unless you have a constant accel. throughout the interval between the two states (or you know, or can find out, the average accel.).

[SETUP]

Visualize & Sketch

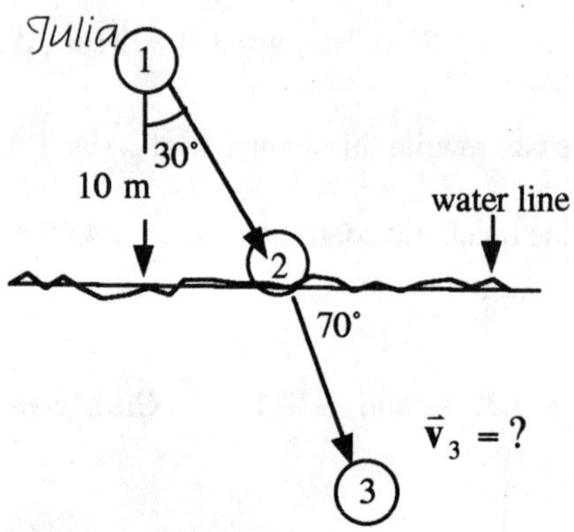

Gather & Sort Data

Given	Hidden	Want
$v_1 = 0$ m/s		$v_3 = ?$
dist. to water = 10 m		
angles shown		
$a_{\text{stage 1}} = 0.50$ m/s²		
$a_{\text{stage 2}} = -1.0$ m/s²		
$\Delta t_{2\to3} = 3.0$ s		

Method/Tool Box:

The **constant acceleration kinematics equations** carefully applied to two different stages. Note that I will choose the positive direction to be in the direction of her motion for each stage. The positive direction is chosen to be different for each stage so that we can just use the constant accel. 1-D eq'ns. This will work as long as we are careful to refer to the diagram and not forget what is physically going on.

Relations/Equations:

Constant Acceleration Equations

(+x is taken to be along her path at each stage)

0. $v_{\text{av}} = \frac{1}{2}(v_{\text{f}} + v_{\text{i}})$
1. $\Delta x = v_{\text{i}}t + \frac{1}{2}at^2$
2. $v_{\text{f}} = v_{\text{i}} + at$
3. $2a\Delta x = v_{\text{f}}^2 - v_{\text{i}}^2$

Now we apply these to each stage by carefully adjusting the subscripts and referring to the diagram.

[SOLVE]

(*Now* it's time to enter Mathematic Land!! Notice that I absolutely do NOT start here. Many beginning physics students want to jump to the equations right away, but problems usually aren't that easy. The set up is where we sort out the physics and choose a method, or methods, of approach. Only then are we ready to tailor equations to the problem and finally move to solving for the unknowns.)

I'll just get to it. If you haven't done this yourself, why don't you take it from here? The math is pretty easy for this problem.

We want v_3, so I'll work backwards to find out what we need. You can go about this any way that is valid and works. (NOTE: I will assume that "10 m," "30°" and "70°" are accurate to 2 sig. fig's to be consistent with the other data.)

STAGE 2: Relating states 2 & 3.

Glancing up at the data table, I see that: $a_{\text{stage } 2} = -1.0\ \text{m/s}^2$ and $\Delta t_{2\to3} = 3.0$ s. The easiest equation for v_3 is

$$v_3 = v_2 + a_2\,\Delta t_{2\to3}.$$

Applying the other equations to stage 2 won't help, so I'll move to stage 1 to find v_2.

STAGE 1: Relating states 1 & 2.

We need v_2 and have v_1 and a_1. We can either find $\Delta t_{1\to2}$ or $\Delta x_{1\to2}$. Look at the equations. To find $\Delta t_{1\to2}$, I'll need $\Delta x_{1\to2}$ anyway. Applying basic trig, I find

$$\Delta x_{1\to2} = (10\ \text{m})/\cos 30^\circ = 12\ \text{m}$$

I've given $\Delta x_{1\to2}$ to 2 sig. fig's, but you could keep 11.547... in the calculator to use in the next equation. I wonder what the difference would be... Let's find out!!

Avoiding t and going straight for v_2:

$$v_2^2 = v_1^2 + 2a_1\,\Delta x_{1\to2}$$

Since $v_1 = 0$,

$$v_2 = \sqrt{2\,a_1\,\Delta x} = \sqrt{2\cdot 0.50\cdot 11.547...} = 3.398....\text{m/s}$$

Or using 12 m, we find

$$v_2 = \sqrt{2\,a_1\,\Delta x} = \sqrt{2\cdot 0.50\cdot 12} = 3.464....\text{m/s}$$

Rounding to 2 sig. fig's would give 3.4 and 3.5, but "2 sig. fig's" means that there is uncertainty in the last digit. Since we are not asked to give a number for v_2, we can proceed to find v_3 and see how this error propagates.

$$v_3 = v_2 + a_2\,\Delta t_{2\to3} = v_2 + (-1.0\ \text{m/s}^2)\,(3.0\ \text{s}) = v_2 - 3.0\ \text{m/s}.$$

The error does show up! Using $v_2 = 3.398$ gives $\boldsymbol{v_3} = \mathbf{0.40}$ **m/s**, while $v_2 = 3.464$ gives $\boldsymbol{v_3} =$ **0.46 m/s**. Which do you trust more? (Notice that I DO use 2 sig. fig's for the final result.)

[ANALYZE]

Did we notice that you didn't use the 70° angle given? We didn't need it. That happens sometimes. People call that extra information a "red herring," since we throw it back like an unwanted fish.

The speed seems reasonable since Julia is slowing down from drag and or fatigue. She is moving at less than half walking speed. In fact, Olympic swimmers swim at rates much slower than running speed. This brings out a major error in this model: it is very unlikely that Julia would start swimming at the same speed that she was running before entering the water. When she dives in the water she will have a substantial deceleration.

The investigation of significant figures was an interesting side note. Had I rounded to $v_2 = 12$ m/s, I might give the answer as $v_3 = (0.46 \pm 0.05)$ m/s. But even this does not quite capture 0.40 m/s in the window of expected precision. Such is the tenuous nature of sig. fig. rules! It's a good thing to be aware of.

EX. 2.32:

(a)

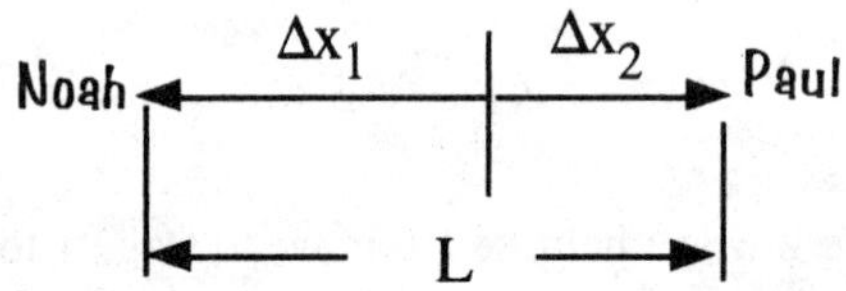

We could write this a few ways:

$|\Delta x_1| + |\Delta x_2| = L$

$\Delta x_2 - \Delta x_1 = L$, where $\Delta x_2 > 0$ and $\Delta x_1 < 0$

$\Delta x_1 - \Delta x_2 = L$, where $\Delta x_1 > 0$ and $\Delta x_2 < 0$

(b) With $a = 0$, a kinematic equation will give us $\Delta x = v\,\Delta t = (30$ m/s$)\,(1.2$ s$) = 36$ m, but that is NOT the distance that we are asked for. While you may be able to do this one in your head, you may miss this simple step in more difficult problems unless you make a sketch.

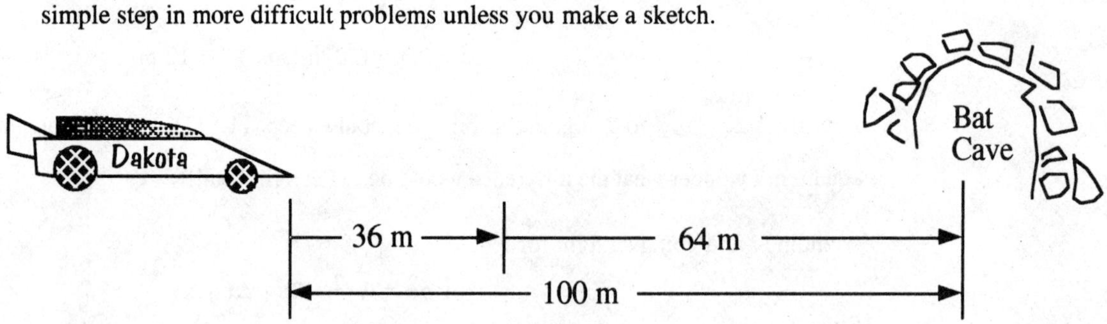

(c) Again, you may be able to do this one without a sketch, but train yourself here for harder problems.

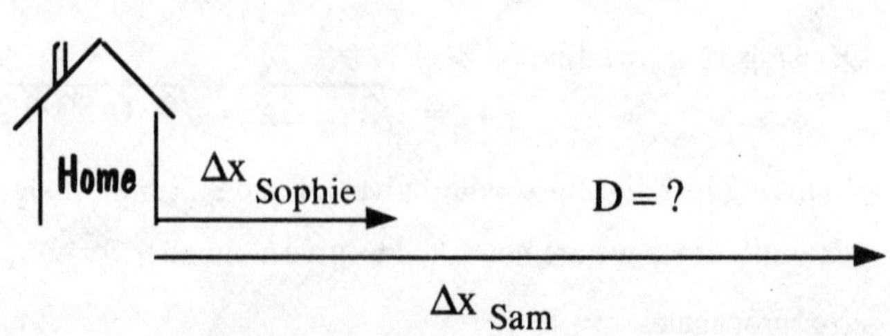

$\Delta x_{\text{Sam}} - \Delta x_{\text{Sophie}} =$ Distance Sam is from Sophie

The solution is simple: Sophie travels $\Delta x_{\text{Sophie}} = v\,\Delta t = (1$ m/s$)\,(10$ s$) = 10$ m, while Sam travels $\Delta x_{\text{Sam}} = v\,\Delta t = (3$ m/s$)\,(10$ s$) = 30$ m. Sam is 20 m away from Sophie.

EX. 2.33:

[MODEL] 1-D kinematics problem with a = constant = g, down. There are 2 objects which eventually pass the same placed at the same time.

Key Point Of This Problem:

Again, be *very* careful with the ± signs of displacement, velocity, and acceleration!! Especially be careful to see where the displacements are measured from. Finally, the phrase "they pass" carries an implicit ("hidden") equation/relation. Ready? Let's set it up!

[SETUP]

Visualize & Sketch

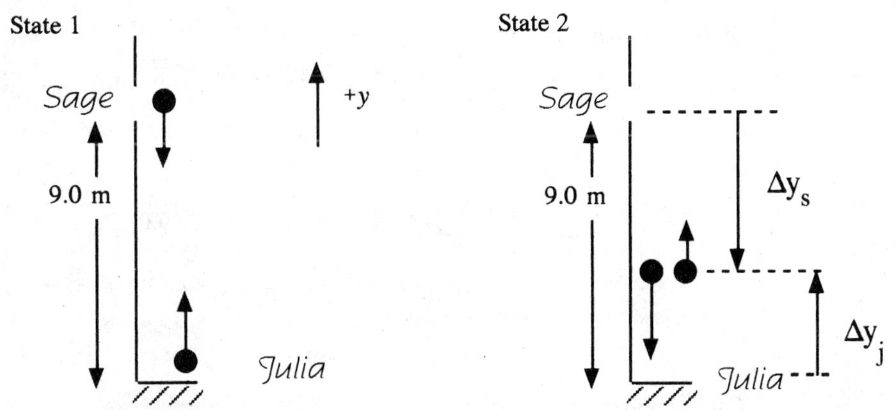

NOTE! Keep the variables straight gets to be difficult when you have multiple objects. I will use "s" and "j" to refer to the balls thrown by Sage and Julia, respectively. I will use "1" and "2" to refer to the state of the process. Thus $v_{j,1}$ refers to the initial velocity of the ball that Julia threw.

Gather & Sort Data

Given	Hidden	Want
$v_s = -10.0$ m/s	$a = -g = -9.80$ m/s²	$\Delta t_{1\to2} = ?$
$v_j = +10.0$ m/s	they pass at state 2	
3rd floor ht = 9.0 m		

Method/Tool Box:

Write the **constant acceleration kinematics equations** for 1–D vertical free flight for each object. Then mathematically express the condition that they pass (same place at state 2).

Condition For Passing At State 2:

Notice that Δy_s is negative (down) and Δy_j is positive. This show be clear from the diagram. Once we notice this, we can get the relation correct:

$$\Delta y_j - \Delta y_s = 9.0 \text{ m}$$

There are plenty of other ways of writing this.

Relations/Equations:

Constant Acceleration Equations

Tailored to Fit 1-D Vertical Free Flight ($a_y = g$, down)

Sage's Ball

1. $\Delta y_s = v_{s,1}t - \frac{1}{2}gt^2$
2. $v_{s,2} = v_{s,1} - gt$
3. $-2g\Delta y_s = v_{s,2}^{\ 2} - v_{s,1}^{\ 2}$

Julia's Ball

1. $\Delta y_j = v_{j,1}t - \frac{1}{2}gt^2$
2. $v_{j,2} = v_{j,1} - gt$
3. $-2g\Delta y_j = v_{j,2}^{\ 2} - v_{j,1}^{\ 2}$

[SOLVE]

Since both balls are in the air for the same length of time, the "t"'s in the equations are the same. We don't know the velocities at state 2, but we could easily find them. If we take equation number 1 for each ball and put the right side into the condition for pass at state 2, we'll be in business!

$$\Delta y_j - \Delta y_s = 9.0 \text{ m}$$

$$(v_{j,1}\, t - \tfrac{1}{2}\, g\, t^2) - (v_{s,1}\, t - \tfrac{1}{2}\, g\, t^2) = 9.0 \text{ m}$$

$$v_{j,1}\, t - v_{s,1}\, t = 9.0 \text{ m} \Rightarrow (v_{j,1} - v_{s,1})t = 9.0 \text{ m}$$

Careful with those signs!!!!

$$[(+10 \text{ m/s}) - (-10 \text{ m/s})]\, t = 9.0 \text{ m} \Rightarrow t = (9.0 \text{ m})/(20 \text{ m/s}) = \mathbf{0.45} \text{ s}$$

Pretty slick solution, huh? You can get to it other ways as well. Try it out if you want to flex your kinematics muscle.

[ANALYZE]

Let me just check to see if this makes sense. Let's see how far each ball travels in 0.45 s:

- $\Delta y_s = v_{s,1}\, t - \tfrac{1}{2}\, g\, t^2 = (-10 \text{ m/s})(0.45 \text{ s}) - 4.9 \text{ m/s}^2 (0.45 \text{ s})^2 = -5.5 \text{ m}.$
- $\Delta y_j = v_{j,1}\, t - \tfrac{1}{2}\, g\, t^2 = (+10 \text{ m/s})(0.45 \text{ s}) - 4.9 \text{ m/s}^2 (0.45 \text{ s})^2 = +3.5 \text{ m}.$

HEY, IT WORKS!! Just check the condition for passing!

Finally, Did you notice that I drew the diagram with a bit of insight? How did I know that Sage's ball would travel further than Julia's? ... Because Sage's ball is speeding up while Julia's is slowing down!

EX. 2.34:

Bring on the calculus!

$$\vec{\mathbf{a}} = \frac{d\vec{\mathbf{v}}}{dt} \Rightarrow \int \vec{\mathbf{a}}\, dt = \int \frac{d\vec{\mathbf{v}}}{dt}\, dt \Rightarrow \int \vec{\mathbf{a}}\, dt = \int d\vec{\mathbf{v}}$$

We can just skip that second step and pretend that we are multiplying by dt. Since we are looking for constant acceleration problems, we can pull the acceleration out of the integral:

$$\vec{\mathbf{a}}\int dt = \int d\vec{\mathbf{v}} \Rightarrow \vec{\mathbf{a}}\,\Delta t = \Delta\vec{\mathbf{v}} \Rightarrow \vec{\mathbf{a}}\,\Delta t = \vec{\mathbf{v}}_{\mathrm{f}} - \vec{\mathbf{v}}_{\mathrm{i}}$$

$$\vec{\mathbf{v}}_{\mathrm{f}} = \vec{\mathbf{v}}_{\mathrm{i}} + \vec{\mathbf{a}}\,\Delta t$$

The constant of integration here is the initial velocity. We could also do this component by component as follows:

$$\vec{\mathbf{a}} = \frac{d\vec{\mathbf{v}}}{dt} \Rightarrow a_{\mathrm{x}}\hat{\mathbf{i}} + a_y\hat{\mathbf{j}} + a_z\hat{\mathbf{k}} = \frac{dv_x}{dt}\hat{\mathbf{i}} + \frac{dv_y}{dt}\hat{\mathbf{j}} + \frac{dv_z}{dt}\hat{\mathbf{k}}$$

We proceed one by one:

$$\int a_{\mathrm{x}}\,dt = \int \frac{dv_x}{dt}\,dt \Rightarrow \int a_{\mathrm{x}}\,dt = \int dv_x$$

Which gives

$$v_{x,\mathrm{f}} = v_{x,\mathrm{i}} + a_{\mathrm{x}}\,t$$

Continuing on:

$$\vec{\mathbf{v}} = \frac{d\vec{\mathbf{r}}}{dt} \Rightarrow v_{\mathrm{x}}\hat{\mathbf{i}} + v_y\hat{\mathbf{j}} + v_z\hat{\mathbf{k}} = \frac{dx}{dt}\hat{\mathbf{i}} + \frac{dy}{dt}\hat{\mathbf{j}} + \frac{dz}{dt}\hat{\mathbf{k}}$$

Let's just do the x-component, since the others are identical.

$$v_{\mathrm{x}} = \frac{dx}{dt} \Rightarrow \int dx = \int v_{\mathrm{x}}\,dt \Rightarrow \Delta x = \int\left(v_{\mathrm{x,i}} + a_{\mathrm{x}}t\right)dt$$

Easy!

$$\Delta x = v_{\mathrm{x,i}}t + \frac{1}{2}a_{\mathrm{x}}t^2$$

Look familiar?!

The constant of integration that pops up is the initial position. So the two constants of integration are the initial conditions (position and velocity). Do you see how this all requires that the acceleration be constant? Otherwise, we couldn't take that first step. So when the acceleration is not constant, we can't use these equations. However, if we have the acceleration as a function of time, we can integrate that function to get the velocity function, and then integrate the velocity function to get the position function.

EX. 2.35:

You need to be clear on the definitions and apply the powerful tools of calculus. Recall how to maximize and minimize a function.

$$\text{Displacement } (\vec{\mathbf{D}}) = \Delta\vec{\mathbf{r}} = \vec{\mathbf{r}}\,(\text{at } t_{\mathrm{f}}) - \vec{\mathbf{r}}\,(\text{at } t_{\mathrm{i}})$$

Since the earlier time of interest is at $t = 0$, it is easy to find the position–just plug in for t :

$$\vec{\mathbf{r}}\,(\text{at } t_{\mathrm{i}} = 0) = (8\text{ m})\,\hat{\mathbf{i}} - (17\text{ m})\,\hat{\mathbf{j}}$$

We now need the times when the speed is minimum (to find the displacement requested) and when it is maximum (a separate question that we are to answer).

To find the extremes of the velocity, we need to differentiate the velocity function and set it equal to zero. First we need the velocity as a function of time!

$$\vec{\mathbf{v}}(t) = \frac{d\vec{\mathbf{r}}}{dt} = \left(\frac{dx}{dt}\right)\hat{\mathbf{i}} - \left(\frac{dy}{dt}\right)\hat{\mathbf{j}} = \left(\frac{d\left(\frac{3}{2}t^2 - 2t + 8\right)}{dt}\right)\hat{\mathbf{i}} - \left(\frac{d\left(2t^2 - 17\right)}{dt}\right)\hat{\mathbf{j}}$$

So,

$$\vec{\mathbf{v}}(t) = (3t - 2)\hat{\mathbf{i}} - (4t)\hat{\mathbf{j}}$$

To find the max and min of this function, we take the derivative and set it equal to zero (recall that the function has a zero slope at max and min on a graph).

$$\frac{\mathrm{d}}{\mathrm{dt}}\vec{\mathbf{v}}(t) = (3)\hat{\mathbf{i}} - (4)\hat{\mathbf{j}}$$

Hey, that function can't be zero! Yup, I tricked you. What does that function represent? **The acceleration.** So what do we now know? **The acceleration is constant, so the min velocity is at the start and the max is at t = infinity!**

There are many possible problems based on this technique. I could have given you a function where there was a max and min velocity. Then you would set the derivative of the velocity function (the acceleration) equal to zero, solve for t, and then determine if the time(s) obtained give max or min velocities. We could then plug the t for min velocity into the position function and then find the displacement requested (change in position vector).

BRAVO!!!

Ch. 3: Advanced Kinematic Models

Pop quiz-study technique check! What are the 3 most important study techniques that will save you time, greatly increase your comprehension, and make lectures easier to follow? If you don't know, and even if you do, you might want to skim over Part A again to see if there is anything in there that can make life easier for you. [OK, my top 3 are: 1) **Preview** the material before lecture, 2) Write down your **questions** as you read/work in a study journal. Make sure that you get all of those little, nagging, "stupid" questions answered one way or another, and 3) Every, I say *every*, day do some work in each of your classes. For physics, an off-day will consist of a minimum of **15 minutes** of working problems. Consistency pays much greater dividends than you can imagine until you have done it for a while.]

STUDY TIP: *"Active learning" means that you are taking responsibility for your development. To participate, try the following:*

1. *Preview the text sections and read the conceptual and basic sections of the study guide before the topic is covered in class. It's going to save you lots of time!*
2. *Work through the exercises in the study guide and keep your work in a study journal. If you get stuck on anything – and you probably will if you are human – write your question down and get it answered soon. Remember the wise Sophocles:*

**"One must learn by doing the thing;
for though you think you know it,
you have no certainty until you try."**

Overview

3 important applications of kinematics coming at you! Your task now is to learn how to describe (kinematics) **projectile** motion, **circular** motion, and **relative** motion. This will both reinforce what you learned in chapters 1 and 2, and give you tools that you will use frequently in analyzing more complex systems. I also include some important distinctions in equation types. Make sure you read about those distinctions. They will be referred to throughout the guide. That's about it. Let's get to it!

Conceptual

Projectile Motion

Projectile Kinematics – General Approach

A projectile is anything tossed into the air that doesn't fly (like a bird or plane) or float (like a hang glider, paper airplane, or feather).

> EXERCISE 3.1: Can you come up with the 6 general directions in which you can "project" (throw, kick, shoot, etc.) an object? For each case, sketch the initial velocity vector of the object just as it takes to the air and note whether the motion will be 1-D or 2-D. [Hints: Don't distinguish between left and right, and don't forget about the simplest case!]

One way to analyze the motion of things tossed into the air is to simply describe the motion – i.e., do a kinematics analysis.

Method: Compare any two states during the free flight stage using the constant acceleration equations. Using what you know about the states, you can find out what you don't know. (The states are labeled either as "initial" and "final," or by number, as in the next exercise.)

Stage: The "stage" of free fall (I'll use "fall" like everyone else if you remember that this includes both rising and falling objects, deal?) … Now where was I? Oh yes. The "stage" of free fall includes all the states between the time *just* after the object is released to *just* before it hits something (like the ground). It does *not* include what's happening while it is being projected nor after it hits something, since the free fall equations would not pertain to those situations.

State: At any instant while the object is in the air you might know or want to know the time (t), position (x, y), and velocity ($\vec{v} = v_x\hat{i} + v_y\hat{j}$) of the object. You will always know the acceleration of the object (unless you have to calculate for a projectile on another planet).

EXERCISE 3.2: Which states shown and numbered below cannot be compared with the others by using only the projectile equations? That is, which states are not in the main free fall stage?

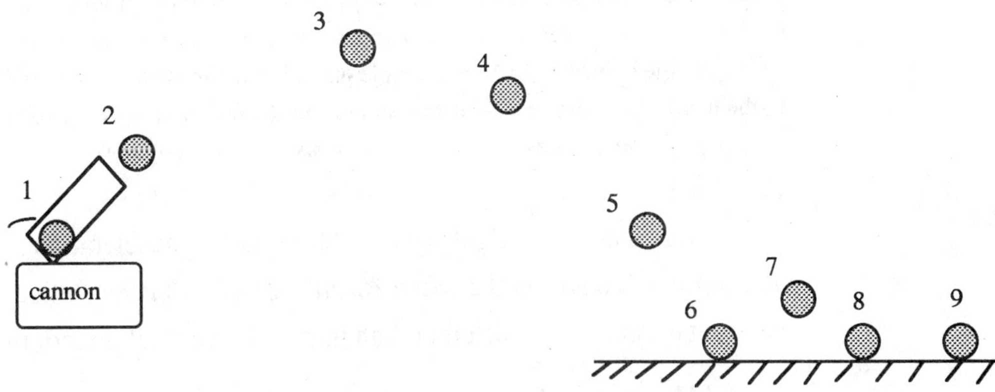

Projectile Kinematics – General Details

Huge conceptual leap: one can analyze a projectile's vertical motion and the horizontal motion separately! Think of the one motion of the projectile as two combined motions. While this may seem like no big deal, it was one of the enormous – and not at all obvious – leaps that those Renaissance guys needed to make in order to break free of the Aristotelian worldview.

So when you ask "Hey, what's going on with that there projectile?", what you really need to know is its kinematic state as given by the following information:

<u>Projectile Kinematic State (at any instant, t)</u>

	Vertical Motion (y-direction)	Horizontal Motion (x-direction)
position	y	x
speed	v_y	v_x

Note that the total speed (v_{total}) and angle of travel (θ) are directly related to the component speeds (v_x and v_y), but we'll get to that soon enough. "Are we going to make any special, simplifying assumptions?" You bet! And that'll tell us something about the constant acceleration of the free fall stage.

<u>Special Conditions Assumed</u>

1. No propulsion or floating
2. Not describing any rotations in the air…yet
3. Negligible air resistance (so $a_x = 0$)
4. Constant vertical accel. ($a_y = \pm g$)

EXERCISE 3.3: When *is* air resistance negligible? It's pretty easy to get a rough, qualitative answer to this question. Take out 2 identical sheets of paper. You're going to drop them. (a) Hold one sheet with one hand so that it makes a horizontal plane (parallel to the ground). With your other hand, hold the other sheet to form a vertical plane. Drop them at the same time. In which case is air resistance more "negligible"? Why? (b) Now crumple one up into a loose ball, and the other into a very tight ball. Drop them from the same place at the same time. In which case is air resistance more "negligible"? Why? (c) How good is the assumption that air resistance is negligible?

Notice that free flight (AKA "free fall") acceleration of a projectile, given the assumptions above, is the same for all objects. Further notice that the acceleration in the horizontal direction is different than that in the vertical direction! In particular,

- Horiz. Accel. (a_x) = 0 m/s² ⇒ Coasting! Constant horiz. speed (v_x)
- Vert. Accel. (a_y) = g, downward = 9.8 m/s², downward
  ⇒ Vert. speed (v_y) slows down as object rises
  ⇒ Vert. speed (v_y) speeds up as object drops

What does this mean? What does this say about the motion of all projectiles? It means that the side-to-side motion is like an object floating in space, while the vertical motion is like that of an object either dropped, thrown straight down, or thrown straight up. The next exercise provides a visualization aid.

EXERCISE 3.4: Try to visualize projectile motion as composed of separate horizontal and vertical motions. (a) Imagine that you are in a room that has very bright lights over head. You toss a ball up and across the room, and it makes a shadow on the floor as it flies through the air. Make a sketch of the position of the shadow at equal increments of time. [Hint: Does the spacing stay the same, get bigger, or get smaller as the object move across the room?] (b) Now the bright lights are behind you so as to make a shadow on the opposite wall. Sketch the position of the vertical shadow as the object flies through the air.

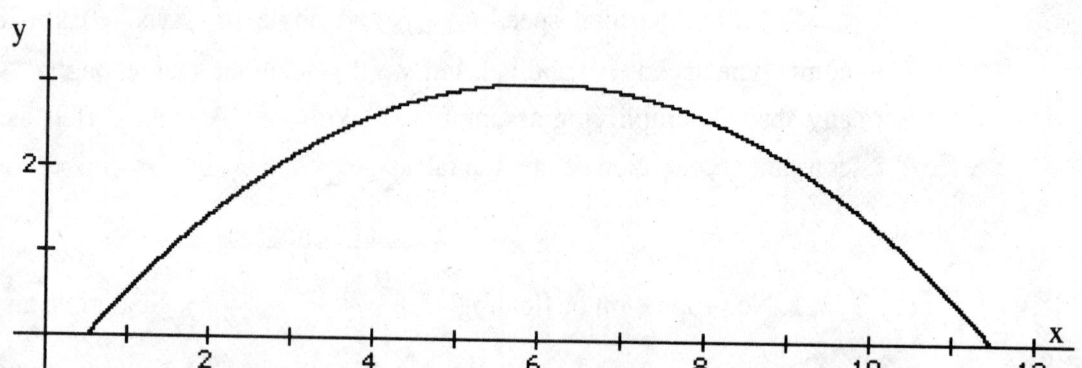

So in analyzing a projectile's motion, we will almost always split the motion into horiz. and vert. motions. The next exercise gives you a chance to do this without requiring equations.

EXERCISE 3.5: Kym wants to measure how fast she can throw a baseball, but she doesn't have any fancy radar device to use. In fact, all she has is a baseball and a metric tape measure. What can she do? Aha! She goes to the beach and climbs up 4.9 m (say 5 m) on a lifeguard's tower. From there, she throws the ball so that it starts out perfectly horizontal. She then gets down and measures the horizontal distance to be 20 m. Now she knows how fast the ball was traveling as she released it! Do you? What was the speed at release? [See the Hint section if you need some help.]

An Instructive Special Case

Sometimes you throw something up and it lands above you. Sometimes it lands below you. Here, assume that the object lands at the exact same height that it was released. OK? (See the figure for Exercise 3.4.) Further assume that, while the object may be projected at different angles, it leaves with the same initial speed each time (for example, 20 m/s). The next exercise will help you get a feel for projectile motion in general and the particular results that you will derive mathematically some time this week.

EXERCISE 3.6: Project an object at angles of 30°, 45°, and 60° above the horizontal. Assume that the object lands at the same height that it was released and starts with the same initial speed each time.

(a) For which incident angle is the time in the air ("hang time") the greatest?
(b) For which incident angle is the initial vertical speed the greatest?
(c) For which incident angle is the initial horizontal speed the greatest?
(d) What is the "range" (horizontal distance traveled before getting back to the same height) for 0° and 90° launches?
(e) Given that increased time in the air helps increase the range and increased horizontal velocity also helps the object move sideways more (faster), what angle would you guess gives the maximum range for this special case?

Circular Motion

We now turn our attention to describing *circular* motion (object whirled around at the end of a string, a car traveling around a circular corner, roughly the moon around the earth, etc.) and *rotational* motion (a spinning ice skater, a spinning record or CD, a merry-go-round, etc.). To do this, we use concepts similar to those used for the motions that we've already described. You need to learn how to describe the amount that you've gone around a circular path, the rate of travel, and the accelerations for changing direction and/or speed.

Distance/ Amount Around

Notice in the diagram that there are *three ways to describe the amount around*: the distance along the circle (ℓ), known as the "arc length"; the number of revolutions completed (partial or full); and the amount of angle rotated around ($\Delta\theta$). Also note that the radial distance away from the center pivot point of rotation (R) plays a ... pivotal role :-). (That radial distance turns out to make a big difference in all of the concepts regarding rotational motion! You often get to play "Where, oh where, do I put that *%@$#*!* R?")

- arc length = (depends on R) m
- # rev's = 1/4 rev
- angle of rotation = $\pi/2$ rad (90°)

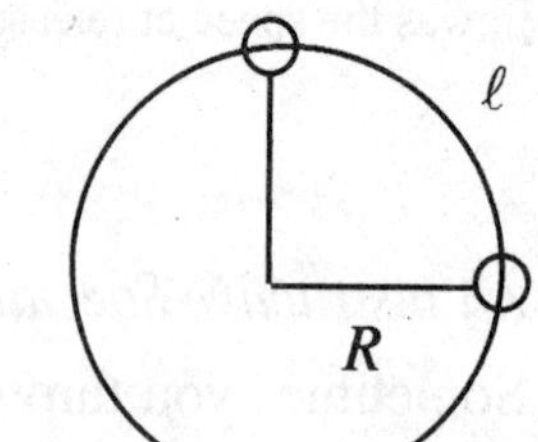

> EXERCISE 3.7: Riddle. Is it possible for two people to stand still on a merry-go-round and both rotate around by 30°, but travel different distances? Explain and use a diagram to show the situation.

Rate of Rotation

Now notice that there are three corresponding ways to answer the question "How fast is it rotating?" Basically, it's **amount around/time to get around**. These are all closely related, but they are not identical.

- *Linear* Speed (S_{av}) = $\dfrac{\text{distance (along the circular arc)}}{\text{time}}$

  Measured in m/s, miles/h, etc.

- *Rotational* Speed (*Frequency*, f) = $\dfrac{\text{\# of revolutions (i.e. \# times around)}}{\text{time}}$

  Measured in revs/s (AKA cycles/s ≡ Hz), revs/min (also written "RPM"), etc.

- *Angular* Speed (ω) = $\dfrac{\text{angular displacement}}{\text{time}}$, ω is the Greek lower case "omega"

  Measured in radians/s, NOT NOT NOT degrees/s!!

> EXERCISE 3.8: Now, is it possible for two people to stand still on that same merry-go-round, complete one revolution in the same amount of time, yet be traveling at different speeds? Explain and use a diagram to show the situation.

Frequency (f) & Period (T), What's the Difference?

These are two ways to describe the rate of rotation. Not a big difference here. That's why people don't care much if you use one or the other. Still they are a little different. Here is a table that compares and contrasts the ideas. No big whoop, just make sure you keep it straight.

Definition :	freq. $(f) \equiv \frac{\text{\# rev.}}{\text{time}}$	$\xleftrightarrow{\text{invert}}$	period $(T) \equiv \frac{\text{time}}{\text{\# rev.}}$
Units :	$\frac{\text{rev.}}{\text{s}}$ (aka "Hz")	$\xleftrightarrow{\text{invert}}$	$\frac{\text{s}}{\text{rev}}$ (aka "Hz^{-1}")
Fast Rotations :	big f	$\xleftrightarrow{\text{invert}}$	small T
Slow Rotations :	small f	$\xleftrightarrow{\text{invert}}$	big T
Relationship :	$f = \frac{1}{T}$	$\xleftrightarrow{\text{invert}}$	$T = \frac{1}{f}$

So the period is the period of *time* to complete a cycle: bigger period $\Rightarrow$ more time $\Rightarrow$ a slower rotation. The frequency is a measure of how many revolutions are completed in one second (or one minute): bigger frequency $\Rightarrow$ more revolutions $\Rightarrow$ a faster rotation. They are simply the inverse of one another.

Acceleration For Circular Motion

As you well know, there are two ways to accelerate, by changing speed and by changing direction. When it comes to changing speed, we can look at the rate of change of linear speed ($\Delta v/\Delta t$) and that of angular speed ($\Delta\omega/\Delta t$). When it comes to changing direction, the sharpness of the turn and the speed with which the object takes the turn are what matter.

EXERCISE 3.9: REVIEW - The velocity vector shows the direction of travel at an instant. What is the direction of the acceleration vector relative to the velocity vector in the three cases: (a) only increasing speed, (b) only decreasing speed, (c) only changing direction?

Relative Motion

Simple Case

If a person is driving down the highway at 60 mph and sticks their hand out the window, will it hurt you if you slap their hand? It depends, doesn't it? If you are standing on the road, it's going to really hurt! If, however, you are in the passenger seat of a car traveling at 61 mph, it won't hurt a bit. What's important here is the *relative* velocity of the hands. Thus, in some situations it is important to specify the frame of reference. The next example should help clarify what is meant by this.

EXERCISE 3.10: Suppose that you can row a boat at a speed of 4 m/s through still water. If you row in a river that flows with a speed of 3 m/s, what would your velocity relative to someone on the river bank be in the following situations? [Hint: Assume that you would float with the river if you did not row. Then, since you move with the river, you would be still relative to the river. Thus when you row, you will always row at 4 m/s in the frame of the river. Your velocity relative to the bank is the sum of your velocity (4 m/s, whatever direction) and that of the river.]

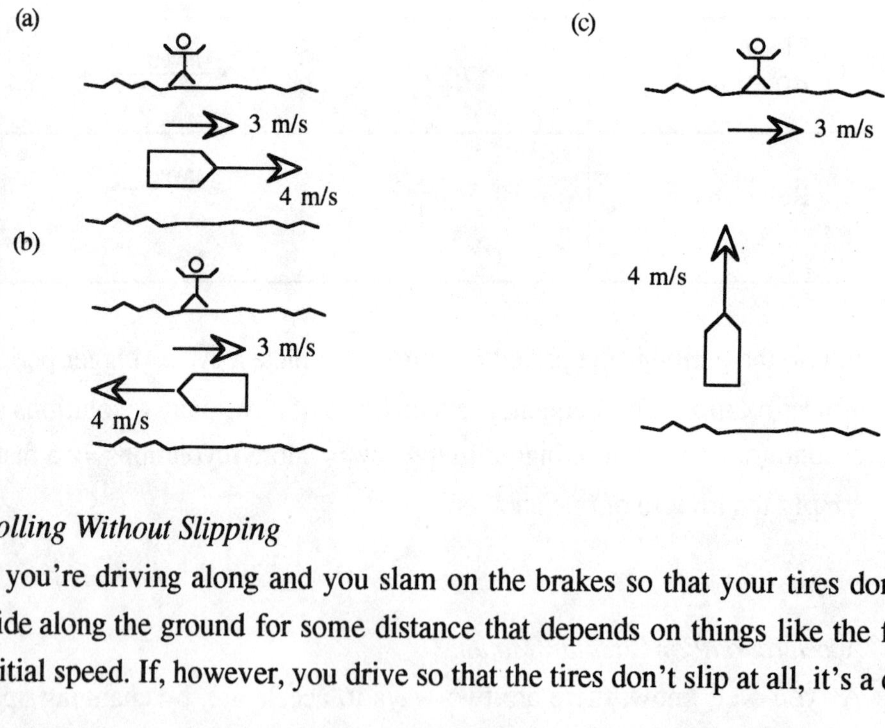

Rolling Without Slipping

If you're driving along and you slam on the brakes so that your tires don't rotate, you'll slide along the ground for some distance that depends on things like the friction and your initial speed. If, however, you drive so that the tires don't slip at all, it's a different story.

EXERCISE 3.11: A tire has a circumference of 2.5 m, and rolls along the ground without slipping. (a) If it rolls and completes one full revolution, how far does it travel? (b) What if it make half a rev.? [Hint: Imagine that the tire has wet paint on the tread. How long will the paint strip on the road be?]

So, in the special case of rolling without slipping, the rotation is directly related to the linear movement: 1 revolution (2π rad) $\Rightarrow$ 1 circumference of linear distance traveled. Thus a tire with a larger radius will travel further in one revolution that one with a smaller radius. It also means that if the tire rotates at say 2 rev's/s, it will travel a linear distance equal to twice the circumference every second.

Basic

Equation Types

Not all equations carry the same importance or are the same type. In **Physics: The Nature of Things**, *the authors have thoughtfully color coded the equations according to which must be memorized, which you should be very familiar with, and so on. Below, I make distinctions between types of equations. These distinctions are helpful to be aware of, but there is no need to memorize these or formally classify all the equations that you run into.*

1. **Defining Relations** - *For example, average speed is defined as distance traveled divided by the time taken* ($S_{av} \equiv \frac{\text{distance}}{\Delta t}$)*, average velocity is defined as the displacement divided by the time* ($\bar{v}_{av} \equiv \frac{\text{displacement}}{\Delta t} = \frac{\bar{r}_f - \bar{r}_i}{\Delta t}$)*, and rotational frequency is defined as the number of revolutions completed divided by the time taken* ($f \equiv \frac{\#\text{ rotations}}{\Delta t}$).

2. **Variable Relations** - *For example, the relationship between the circumference (c), radius (r), and total angle (2π) of a circle is* $c = 2\pi r$. *The relation between the frequency of revolution (f) and period of revolution (T) is* $T = 1/f$.

3. *** Equation Templates *** - *Examples of this type of equation include the constant acceleration kinematic equations, "F = ma", conservation of energy, and so on. This type of equation provides the starting point for many problems. They are "templates" because you use them to* construct *equations that fit the specifications of your particular problem. Generally, you can't just plug into these equations. Again, you have to tailor them to your needs.*

4. **Conditions & Constraints** - *These are often implied, but not explicitly stated. You have to notice what's going on, decide what that means in terms of the physics going on, and sometimes construct an equation that reflects the condition or constraint. For example, if two object are connected by a taut rope, the magnitudes of their velocities and accelerations must be the same (the directions may be different):* $|v_1| = |v_2|$, $|a_1| = |a_2|$. *If an object is constrained to slide along an inclined surface, it doesn't accelerate in the direction perpendicular to the surface. And exercises 2.32 and 2.33 showed how to relate various displacements and lengths. You'll meet some more as we go along.*

5. **Derived Expressions** - *These are obtained after performing mathematical operations (algebra, calculus, etc.) on equations of the types above.*

Projectile Motion

Projectile Equations

EXERCISE 3.12: Go back and do exercise 2.25. You need to be able to write these equations quickly and without hesitation. They are simply the constant acceleration kinematics equations for the horizontal ($a_x = 0$) and vertical ($a_y = g$) motion of a projectile. They will be the equation templates (starting points) for most of the projectile equations that you do – and that's good to know!! Write the equations below.

Projectile Equation Templates (from ex. 3.--- above)

	Vertical Motion ($a_y = g$, down)	Horizontal Motion ($a_x = 0$)
position:	$\Delta y =$	$\Delta x =$
speed:	$v_y =$	($v_x =$)
sans time:	$v_y^2 =$	

EXERCISE 3.13 Suppose that you know the velocity components (speeds in the vertical and horizontal directions, v_y and v_x.) at some instant during the free fall stage. Find general expressions for the total speed (magnitude of the velocity vector) and the direction of travel at that instant. A general sketch of a velocity vector will help. Write the solution below.

total speed = $v_{total} =$

angle of travel at an instant = angle of velocity vector

=

EXERCISE 3.14: Finally, suppose that you know both the magnitude and direction of the velocity at an instant. Write down general expressions for the vertical and horizontal components of the velocity. Write the solution below. [Just simple trig.!]

$v_y =$ $v_x =$

Hidden Information

Often students get stumped by projectile problems. The first step is to know (Did I say without hesitation?) the relations that you just wrote down. They are your projectile tool box. But what if you still can't get it to work out? Then you are probably neglecting some of the information given in the problem. Look for special conditions that are stated in words and translate that info. into symbols.

EXERCISE 3.15: Fill in the implicit data table below. You must be able to translate these key words and phrases into symbols.

IMPLICIT DATA

"Dropped" ⇒ $v_{\text{start, y}}$ =__ , $v_{\text{start, x}}$ =__

"Thrown straight up" ⇒ $v_{\text{start, y}}$ __ , $v_{\text{start, x}}$ =__ , (if up is chosen as +)

"Shot straight down" ⇒ $v_{\text{start, y}}$ __ , $v_{\text{start, x}}$ =__ , (if up is chosen as +)

"Shot horizontally" ⇒ $v_{\text{start, y}}$ =__ , $v_{\text{start, x}}$ = ________ , θ_{start} = __

"Top of path" ⇒ $v_{\text{top, y}}$ =__ , $v_{\text{start, x}}$ = (________), θ_{top} = __

Comparing 2 states ⇒ Δy =__ ,

at the same height $v_{\text{moving down, y}}$ = ________ , v_x = (________),

$\theta_{\text{moving down}}$ = ________

"2 objects hit" (or pass) ⇒ ...

Circular Motion

Distance/ Amount Around

You already know the relationship between the arc length, the angle, and the radius, because you know the equation for the circumference of a circle. The circumference (c) is related to the radius of the circle (R) and the angle of one complete circle ($\Delta\theta = 2\pi$ rad) by:

$$c = (\Delta\theta_{\text{one revolution}}) \cdot R = 2\pi \cdot R$$

BULLETIN!! Keep this relation at the tip of your gray matter! It will help you sort out an otherwise confusing set of relations. It even makes them easy to remember. Here's one. Since the arc length (ℓ) is a portion of the circumference, we simply use a portion of the total (2π) angle:

$$\ell = \Delta\theta \cdot R$$

I mentioned this relation so many times before because it is very important to memorize it. (Refer to Part B - Mathematic Land, to remind yourself about where the number π and the units of radians come from.) Notice that a radian multiplied by (*not* divided by) a meter is a meter. This is important because you do NOT NOT want to write $\ell = \Delta\theta / R$. That would be absolute nonsense! Hard to remember? NO!! That the circumference is 2π times R, not divided by R, immediately reminds you that you multiply radians by R to get the arc length.

Rate of Rotation

Take a quick look at those three ways of describing the rate of rotation. Notice that the expressions for the linear and angular speeds are average values. (They are also the instantaneous values *when* they are constant. For instantaneous speeds, the Δ is written as a d and we obtain $\omega = d\theta/dt$, etc.)

- *Linear* Speed (S_{av}) = $\dfrac{\text{distance (along the circular arc)}}{\text{time}} \Rightarrow S_{av} = \dfrac{\ell}{\Delta t}$

  Measured in meters/sec, miles/h, etc.

- *Frequency* (f) = $\dfrac{\text{\# of revolutions (i.e. \# times around)}}{\text{time}} \Rightarrow f = \dfrac{\text{\# rev's}}{\Delta t}$

  Measured in revs/s (AKA cycles/s ≡ Hz), revs/min (also written "RPM"), etc.

  Alternatively,

  Period (T) = $\dfrac{1}{f} = \dfrac{\text{time}}{\text{\# of revolutions}}$

- *Angular* Speed (ω) = $\dfrac{\text{angular displacement}}{\text{time}} \Rightarrow \omega_{av} = \dfrac{\Delta\theta}{\Delta t}$, ω is the lower case "omega"

  Measured in radians/s, not not not degrees/s!!

Acceleration By Changing Speed

How can we describe the situation where an object goes round and round faster and faster or slower and slower? You really have two, somewhat interchangeable, choices. You can discuss the rate of change of the speed (v), or the rate of change of the angular speed (ω), since they both must change. (We generally don't use the rate of change of frequency.)

- Acceleration: $a_{av} = \dfrac{\Delta v}{\Delta t} \Leftrightarrow a = \dfrac{dv}{dt}$; units : $\dfrac{m/s}{s} = \dfrac{m}{s^2}$

Preview Concept!

- Angular Acceleration: $\alpha_{av} = \dfrac{\Delta\omega}{\Delta t} \Leftrightarrow \alpha = \dfrac{d\omega}{dt}$; units : $\dfrac{rad/s}{s} = \dfrac{rad}{s^2}$

You won't see α in the text until later. It fits in nicely though, doesn't it? Also, remember that this acceleration (a) is the component of the acceleration vector that is *parallel* to the velocity vector. It only accounts for the change in rate, not the change in direction!

Acceleration By Changing Direction

"If an object is moving on a circular path, it must be changing direction. Right?" Of course. So let's talk about the component of the acceleration (perpendicular to the velocity vector) that describes the change in direction. For now, let's say that the speed doesn't change. The result will be valid either way. ... Here, you figure it out.

EXERCISE 3.16: You drive a car around a circular turn as in the four cases shown below. Without any calculations, order them according to which case has the greatest acceleration and which has the least. Assume that the speeds are constant as the car rounds the corner. What 2 factors determine the acceleration due to changing direction? [Remember, no equations yet! The question is: in which case is the *rate of change of direction* the most, and which the least? Can you distinguish the in between cases without an equation?] Do you think that the acceleration is twice as much when the speed is doubled?

CASE 1: $v = 20$ mph (≈ 10 m/s)

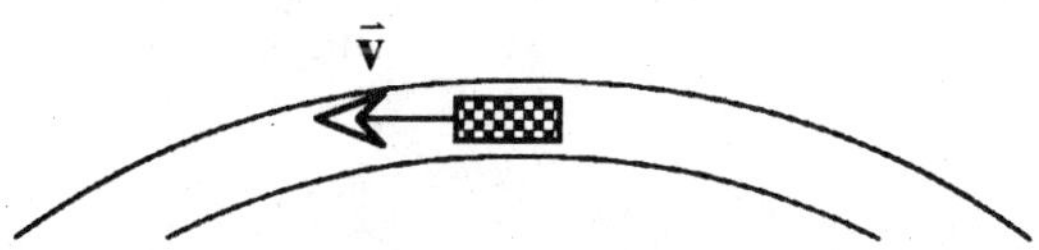

CASE 2: $v = 40$ mph (≈ 20 m/s)

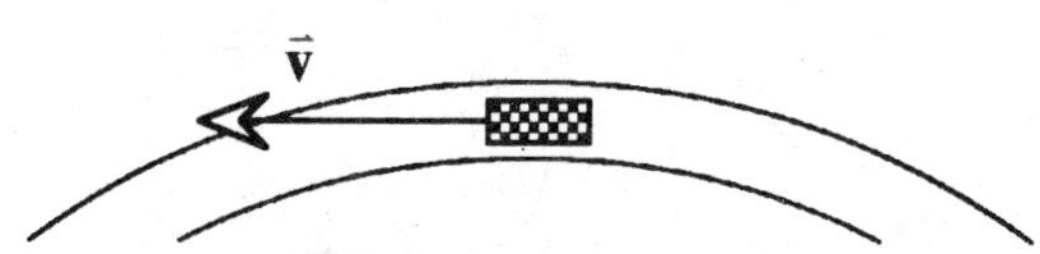

CASE 3: $v = 40$ mph (≈ 20 m/s)

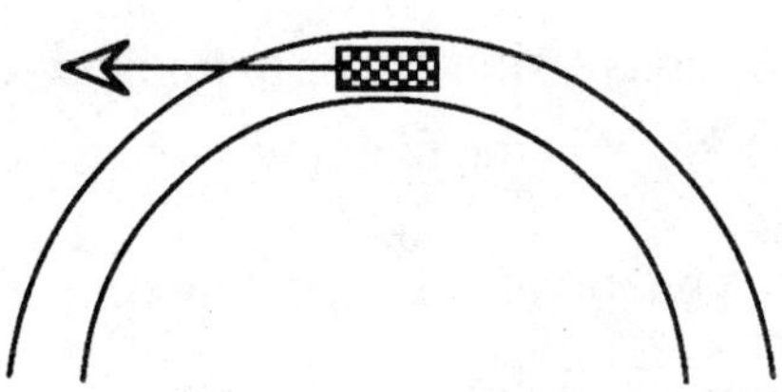

CASE 4: $v = 20$ mph (≈ 10 m/s)

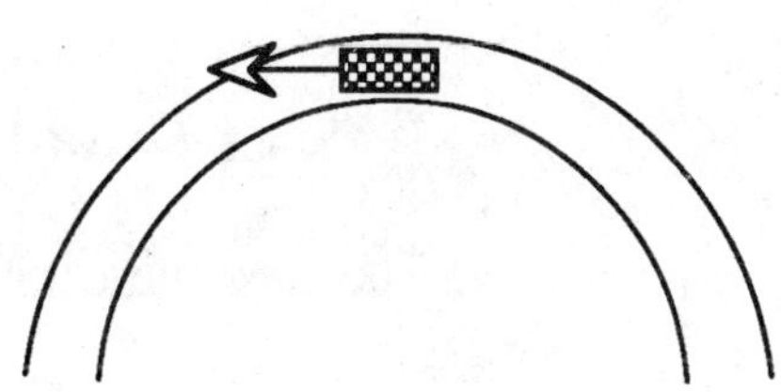

EXERCISE 3.17: Guess the equation! From the previous exercise, you know the two factors that determine the magnitude of the acceleration due to changing direction, namely v and R. And you know the units of those two factors. So what combination of v and R give the units of acceleration? [Hint: First get the seconds right.] ... PS. - this actually turns out to give the right equation!

The component of acceleration that corresponds with the rate of change of direction

1. has direction that is perpendicular to the direction of travel, i.e. $\perp$ to $\vec{v}$.
2. has magnitude $a_{\perp}$ = _____ (fill in the blank)

Relations Between the Concepts

Get a clue!!
R and 2π
are the links.

"All these concepts for just displacement, velocity, and acceleration! How can I ever keep it straight?" Let me show you. A factor of R here, a factor of 2π there, and you're set. Just get the logic and you will always be able to figure out where the R or 2π are supposed to go!

> EXERCISE 3.18: There is a fast way to remember how to go between these related concepts. If you don't catch on to this, you'll constantly be tripped up. So catch on!
>
> (a) How do you "convert" from radians to meters? That is, if you have the angle ($\Delta\theta$), how do you get the arc length (ℓ)?
>
> (b) How do you get the angle ($\Delta\theta$) if you know the arc length(ℓ)? [BIG HINT: check the units and notice that m·m = m^2, not radians.]
>
> (c) Use what you learned from parts a and b to find the relation between v in m/s, and ω in rad/s. [Hint: Again, where should the R go to make the units right?]
>
> (d) How do you convert the angle to the # of rev's (i.e. from radians to revolutions) ? [Hint: How many rad's in a rev?]
>
> (e) Use part d to see how you convert f in Hz to ω in rad/s?
>
> (f) So, how would you convert from f to v?

Now you know. No?

- Since 2π rad = 1 rev, 2π is the conversion factor to go between rev's and radians.
- Since $\ell = \Delta\theta \cdot R$ (from $c = 2\pi R$), R is the factor to go between radians and meters.

The next little table is to help you organize your thinking about the relationships between these variables. Get the simplicity of it.

Displacements:	# rev's	rev's × 2π → ← $\Delta\theta / 2\pi$	$\Delta\theta$	$\Delta\theta \times R$ → ← ℓ / R	ℓ
rates of rot.:	$f \equiv \dfrac{\text{\# rev's}}{\Delta t}$	$f \times 2\pi$ → ← $\omega / 2\pi$	$\omega \equiv \dfrac{\Delta\theta}{\Delta t}$	$\omega \times R$ → ← v / R	$v \equiv \dfrac{\ell}{\Delta t}$

It's a great advantage to be able to see that $\ell = \Delta\theta\, R = (2\pi \cdot \text{\# rev's})\, R$, and by identical reasoning, $v = \omega R = 2\pi f\, R$. Now is it clear that $\omega = 2\pi/T$? (Just look at the units of the numerator and denominator! You've got radians/s staring right at you!) If not, study the relations until it is clear. This will pay big big dividends now and in later chapters.

As for the accelerations, you need to keep clear whether you are talking about an acceleration ($\vec{\mathbf{a}}_{\parallel}$) corresponding to a change in speed, which is in the same or opposite direction as $\vec{\mathbf{v}}$, or an acceleration ($\vec{\mathbf{a}}_{\perp}$) corresponding to a change in direction, which is perpendicular to $\vec{\mathbf{v}}$ (i.e. perpendicular to the instantaneous direction of travel).

Relative Motion

Notational Wonders

There's a nice trick to getting this stuff straight. As you know, when you want to say how fast something is moving you need to specify the velocity relative to some other object. (A car traveling at 65 mph relative to the ground may creep by you at 5 mph if you're going 60 mph relative to the ground.) Luckily, we don't always have to be so careful. But when we do, here's the notation of choice:

$$(\text{displacement, velocity, or accel.})_{\text{of (object A) relative to (object B)}}$$

In shorthand:

$$(\text{displacement, velocity, or accel.})_{\text{(object A), (object B)}}$$

Now watch this. If we know the velocity of a bird relative to a plane, we denote this as $\vec{\mathbf{v}}_{\text{bird, plane}}$. If we know the velocity of the plane relative to the ground, we denote this $\vec{\mathbf{v}}_{\text{plane, ground}}$. If we now want the velocity of the bird relative to the ground, we write the relationship between the velocities as:

$$\vec{\mathbf{v}}_{\text{bird, ground}} = \vec{\mathbf{v}}_{\text{bird, plane}} + \vec{\mathbf{v}}_{\text{plane, ground}}$$

Want to see a quick way to always get this right? We wanted "bird to ground," so we added "bird to plane" and "plane to ground." It's kind of like the plane cancels out in the middle. Similarly, a displacement from me to you followed by a displacement from you to the door would result in a displacement from me to the door: me to you + you to door = me to door. But displacements of me to you + door to you would not have such a clear meaning. Looking at subscripts: p to m + m to k = p to k! This always works. Try it!

EXERCISE 3.19: Just by looking at the notation, determine which of these equalities are valid and which are invalid.

(a) $\vec{\mathbf{D}}_{\text{A, C}} = \vec{\mathbf{D}}_{\text{A, B}} + \vec{\mathbf{D}}_{\text{B, C}}$

(b) $\vec{\mathbf{a}}_{1,\,3} = \vec{\mathbf{a}}_{2,\,3} + \vec{\mathbf{a}}_{1,\,2}$

(c) $\vec{\mathbf{v}}_{\text{dog, cat}} - \vec{\mathbf{v}}_{\text{cat, rat}} = \vec{\mathbf{v}}_{\text{rat, dog}}$

(d) $\vec{\mathbf{D}}_{\text{you, me}} = \vec{\mathbf{D}}_{\text{me, N.Y.}} + \vec{\mathbf{D}}_{\text{me, S.F.}}$

(e) $\vec{\mathbf{D}}_{\text{you, me}} = -\vec{\mathbf{D}}_{\text{me, you}}$

(f) $\vec{\mathbf{v}}_{\text{A, H}} = -\vec{\mathbf{v}}_{\text{H, A}}$

(g) $\vec{\mathbf{a}}_{5,\,3} = \vec{\mathbf{a}}_{5,\,4} - \vec{\mathbf{a}}_{3,\,4}$

(h) $\vec{\mathbf{v}}_{\text{d},\,2} = \vec{\mathbf{v}}_{\text{d},\,1} - \vec{\mathbf{v}}_{1,\,2}$

Of course, you want to be able to set up and understand the relationships using physical reasoning and diagrams. But this sure helps get the ball rolling... speaking of which ...

Rolling Without Slipping

Keep in mind that this is another special case that is particularly easy to write equations for. You know that if you accelerate too quickly, you'll lay some rubber on the road because the tire "slipped." Similarly, if you apply the brakes too abruptly, you'll lay some rubber due to the tire sliding along the road. None of that is described here. In our case, when the tire rotates once, it travels a distance equal to one circumference of the tire. So, if the tire rotates 5 times a second, it will travel 5 times the circumference a second. This provides a simple relationship between the rotational motion of the tire and the linear motion of the tire (and car, or whatever is attached to the tire).

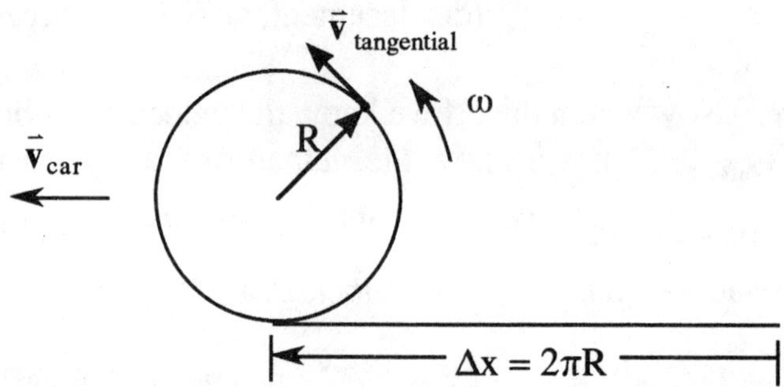

amount rotated ($\Delta\theta$)	$\ell = \Delta\theta \cdot R$ → ← $\Delta\theta = \ell / R$	arc length distance moved by a point on the outer edge of the tire (ℓ)	$\ell = \Delta x$ → ←	distance traveled along the road (Δx)
angular rate of rotation (ω)	$v_{tang.} = \omega \cdot R$ → ← $\omega = v_{tang.} / R$	tangential speed of a point on the outer edge of the tire ($v_{tangential}$)	$v_{tang.} = v_{car}$ → ←	speed of car (v_{car})

EXERCISE 3.20: I push baby Julia in a carriage that has tires with a radius of 12 cm. The tires rotate around at 2.0 rev's per second and do not slip as they rotate. How far does the carriage travel in one minute if this rate is constant? Are these numbers that I've made up reasonable or ridiculous? [See Hint section for sol'n plan.]

Intermediate

Projectile Motion

Problem Solving Layout

Now you're almost ready to start solving some hefty problems. First let me show you a nice layout for applying the problem solving routine to projectiles. This is just a nice bookkeeping method that I think streamlines the work. See what you think. Use what works for you, but do be methodical. [This will work for every projectile problem!]

[MODEL] Projectile motion with negligible air resistance and no floating or self-propulsion.

[SETUP]

Diagram: Show the path and states of interest in the problem and label the states as 1, 2, 3, ...

Data: Instead of breaking this down in terms of:

Given | Hidden (implicit) | Want(unknown)

Organize the data according to the state. You may have or want any of the following data:

State 1		State 2		State 3		State 4		...
$t_1 =$		$t_2 =$		$t_3 =$		$t_4 =$		
$x_1 =$	$y_1 =$	$x_2 =$	$y_2 =$	$x_3 =$	$y_3 =$	$x_4 =$	$y_4 =$	
$v_{1,x} =$	$v_{1,y} =$	$v_{2,x} =$	$v_{2,y} =$	$v_{3,x} =$	$v_{3,y} =$	$v_{4,x} =$	$v_{4,y} =$	
$v_1 =$		$v_2 =$		$v_3 =$		$v_4 =$		
$\theta_1 =$		$\theta_2 =$		$\theta_3 =$		$\theta_4 =$		

Look out for hidden, encoded messages from the physics-problem gods!! ...What's true at the TOP of the path?... This includes ...* Special conditions: For example, if the ball just makes it over a 5 m fence, then $y = 5$ m at the instant it goes over. Write this info. down!

Relations/Equations: Always start by quickly writing down the basic projectile equations that come from the constant acceleration templates. You might also include the relations between the velocity vector components, magnitude, and direction. [Write them below. If you can't do it from memory, then write them down once every hour for a day, then four times a day for a week. Soon it should take less than 1 minute – seriously. Believe me, this will buy you time!! ... See the Basic section to check.]

- Are there any special conditions that you can express as equations? If so, write them down.
- Choose either up or down to be positive so that $a_y = -g$ or $+g$, respectively.

Vertical Motion Equations

1.

2.

3.

Horizontal Motion Equation

4.

Velocity Vector Relations

[SOLVE] What do you want to find? You can either play with the equations algebraically before plugging in the numbers or plug in the numbers right away and then do the algebra. The former is preferred by most physicists and is a must if you are deriving a general expression. While I confess that there are occasions where the latter is faster, the former is often faster, can be more readily checked for algebraic and reasoning errors, and can minimize round-off errors.

[ANALYZE] Does your result make physical sense? If you derived a general expression, does it make sense in the extreme physical limits? What else can you say about the process?

I know that you've got plenty of homework problems to try, but make sure that you do enough projectile problems so that you streamline the process. It gets knee-jerk fast with just a bit of effort. (Of course, if your math skills aren't up to speed, it's all going to be a rough ride.) **Suggestion**: just set up a bunch of problems from the back of the chapter. Notice what knowns and unknowns you have but don't solve them all. This will help you see the similarity between projectile problems.

EXERCISE 3.21: (Use the problem solving template given above to do this one!) If an object lands at the same height that it is launched, how far away does it land? This is called the "range" and is sometimes labeled R. You don't need to know anything else to derive a general expression. Assume that it is launched with a speed of v, and at an angle of θ. For a set launch speed, what angle will give the greatest range? [See hint/sol'n plan section if you need help.]

Detective or all knower?

If you're hesitating, you may be one of those that has trouble writing anything down before you know how it's all going to work out. That's a big problem. It's not a workable approach. Drop it or the 3 ghosts of physics past, present, and future will visit you tonight! You rarely know how it will exactly work out until the end. (Remember the detective's approach!) You know all of the basic equations and the given data, and you know to look for hidden info. That's a lot! Don't try to solve it before laying out the problem!! Once you get to the "Solve" stage, go into Mathematic land and crank the wheel. Handle this now or you *will* be haunted throughout your physics (and engineering, etc.) courses!

EXERCISE 3.22: If an object is shot upwards at an angle of θ, how long does it take to get to the top of its path and how high does it go? We may as well leave out the numbers for now. [See hint/sol'n plan section if you need help.]

Advanced

Combined Motions

What types of motions have you learned to describe (kinematics)? Write a list before you read on. Got it? Here's mine: motion along a line ("linear translation") with constant speed (AKA constant velocity motion); motion along a straight line with constant acceleration (i.e. changing speed, not direction); projectile motion, which combines a constant velocity horizontal motion with a constant acceleration vertical motion; objects moving along circular paths, with the speed both constant and changing; rolling motion, which combines circular and linear motion. Did I leave anything out? It might be useful for you to rewrite this list in bulleted fashion so that you can visualize just what you've accomplished. How many categories did I come up with?

METHOD: to analyze combined motion, you use the old adage, "Divide and conquer!" You break the problem up into the different motions, for example, the constant velocity and constant acceleration parts, as with projectiles, or linear and circular parts, as with rolling motion, and so on.

EXERCISE 3.23: The Olympic Diver. [I'm going to make up some numbers for this problem. Part of your job is to see whether my numbers are realistic or not.] Alissa makes a gold medal dive off of a 10.0 m platform. Her take-off velocity is measured to be 3.89 m/s at an angle of 80.0°. When she reaches the top of her jump, she tucks into a ball and does 4.50 rev's before entering the water.

(a) How high does she (actually her center) jump above the platform?
(b) How far away, horizontally, does she land from beneath the platform?
(c) What is her average angular speed while she is spinning?
(d) Approximate the tangential velocity of her head as she spins. [Hint: What is her approx. radius when curled up?]
(e) Using the approximation made in part (d), what is the acceleration of her head as she spins? [Ooh, you'll need to be careful to receive the gold medal on this part.]

Quiz

1. True or False? "For an object thrown up and over into the air, the velocity at the *top* of the path is zero."

2. When someone says that an object is "thrown horizontally," you know that this means ... what?

3. How many basic equations are in your projectile tool box?

4. I threw an apple in to the air, 4 seconds later it landed I know not where... :-(But I do know that I gave it an initial horizontal speed of 20 m/s. What was its horizontal speed 2.345 s after I threw it?

5. True or False? "Any two points on a rotating tire have the same speed measured in m/s."

6. True or False? "The acceleration for a projectile at the top of its path is zero."

7. Object 1 is dropped from rest. Object 2 is projected horizontally from the same height and at the same time. What happens is that ...

   (a) both objects land at the same time, although the projected one travels a greater horizontal distance.
   (b) the dropped object hits first, since the projected one goes further.
   (c) the projected one hits first, since it has a greater initial speed.
   (d) most students in the room get a puzzled look on their face.

8. True or False? "An object projected at an angle of 45° will always travel a greater horizontal distance than if projected at any other angle."

9. Which equations (plural) are correct? [Hint: you could, oh, I don't know ... check the units!]

(a) $\omega = 2\pi \cdot T$	(e) $\Delta\theta = 2\pi \cdot R$	(i) $\Delta\theta = \pi \cdot R^2$	(m) $f = 2\pi/T$
(b) $\omega = R \cdot v$	(f) $T = 2\pi/\omega$	(j) $\omega = \pi \cdot f$	(n) $\Delta\theta = \omega/v$
(c) $\omega = v/R$	(g) $v = \omega \cdot R$	(k) $v = 2\pi R/T$	(o) $\ell = \Delta\theta \cdot R$
(d) $v = 2\pi f R$	(h) $\omega = 2\pi \cdot v$	(l) $f = 1/T$	(p) $\Delta\theta = \ell/R$

10. True or False? "When test tubes filled with a liquid are spun in a centrifuge, the liquid accelerates outward."

11. True or False? "Air resistance would actually tend to decrease the speeds and distances of a projectile. This effect depends on the size, shape, and mass of the projectile."

12. True or False? "An expert driver can drive a car around a smooth, winding race track without accelerating." If true, explain how this would be achieved.

13. If you drive around a circular track at a constant speed, does your velocity change? Does you acceleration change? (Be careful!)

14. True or False? "The direction of motion at any instant is equivalent to the direction of the velocity vector at that instant."

15. Two people run a foot race around a circular track. Person A must run on the inside lane while B runs on the outside lane. If they both complete one revolution in the same amount of time, does A run slower, faster, or the same speed as B? Explain.

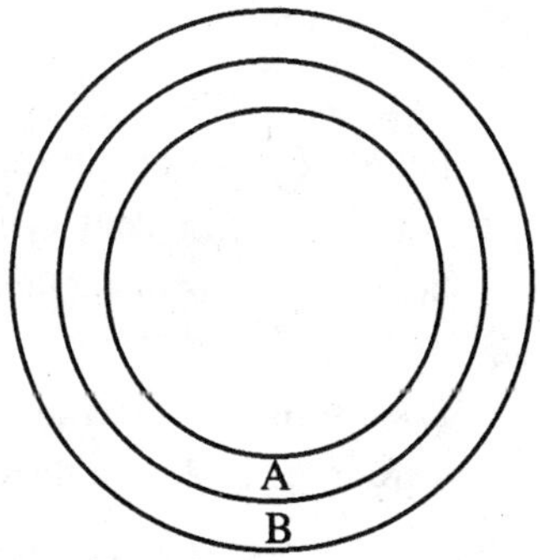

16. The foot race in #15 above is run again, but this time the runners are known to run at the exact same speed. Who finishes first? Why?

17. Select the statement which is completely true.

(a) The acceleration from changing speed has a magnitude of dv/dt and a direction perpendicular to the velocity vector.
(b) The acceleration from changing speed has a magnitude of dv/dt and a direction perpendicular to the direction of travel.
(c) The acceleration from changing direction has a magnitude of dv/dt and a direction perpendicular to the velocity vector.
(d) The acceleration from changing direction has a magnitude of dv/dt and a direction parallel to the velocity vector.
(e) The acceleration from changing speed has a magnitude of v^2/R and a direction parallel to the velocity vector.
(f) The acceleration from changing speed has a magnitude of v^2/R and a direction perpendicular to the direction of travel.
(g) The acceleration from changing direction has a magnitude of v^2/R and a direction perpendicular to the direction of travel.
(h) The acceleration from changing direction has a magnitude of v^2/R and a direction parallel to the velocity vector.

18. True or False? "An object traveling in circular motion may have a tangential component to its acceleration."

19. If you aim your slingshot directly at an apple on a tree, and the apple drops just as you shoot,

(a) the rock will hit the apple.
(b) the rock will go above the apple.
(c) the rock will go below the apple.
(d) the result depends on the initial speed of the rock.

20. You shoot some rocks so that they launch and land at the same height. They all get the same initial speed. Answer the following questions.

(a) Which is in the air longer, the rock shot at 30°, 45°, or 60°?
(b) Which has a greater horizontal speed, the rock shot at 30°, 45°, or 60°?
(c) Which lands the furthest distance away, the rock shot at 30°, 45°, or 60°?

Quiz Solutions

1. False! If you threw it straight up, the velocity would be zero at the top. If you throw it up and over, only the vertical component of the velocity is zero at the top. At that instant, the object still has a horizontal component to the velocity (which is the same as at any other time while in flight).

2. It's just like the situation at the top of the path in the previous question, only this time it is at the start of the projectile stage.

   "Shot horizontally" $\Rightarrow$ $v_{start,\ y} = 0$, $v_{start,\ x} = v_{start,\ total}$, $\theta_{start} = 0°$

3. I would say 4: 3 for the vertical motion, 1 for the horizontal motion. But you could include the velocity vector relationships between the magnitude, angle, and components. You might also contest that the 3 vertical equations are really just 2 independent equations, since the $v_y{}^2 = \ldots$ equation can easily be gotten from the other two. Still, I'd personally say 4.

4. 20 m/s. If you know the horizontal speed at *any* time, you know it at *every* time! ($a_x = 0$)

5. False. A point further from the center of rotation (i.e. with a larger radius) travels a greater distance to complete a revolution and thus has a greater tangential v.

6. False! While in the air, the acceleration is always g, downward. This means that the vertical speed is always changing. An instant before reaching the top an object has some vertical speed, at the top it has no vertical speed, and an instant later it has vertical speed again – it slows, stops & turns around, and speeds up. That's acceleration! It surely doesn't get to the top and hang there at rest for an extended period of time!

7. a. The only difference is in the horizontal motion. The change in height is the same, the vertical acceleration is the same, and the initial vertical speed is the same … so the time to fall (vertically) is the same! It doesn't even matter if the object is shot out of a gun! (Neglecting air resistance, of course.)

8. False! This is only true when the object is launched and lands at the same height. Imagine throwing a shot-put. It lands below the launch height. Something like 42° turns out to maximize the range in this case. Try working it out! Get help from your instructor during office hours. (You are making use of their office hours, aren't you?)

9. Since radians ($\Delta\theta$) are defined as the arc length divided by the radius, they are "dimensionless" and often not written with the units. It can, however, help to keep them around until you're certain that you've got the relations written down correctly. Use the following units to verify the results:

   - ℓ m
   - v m/s
   - f rev/s
   - R · m/rad
   - $\Delta\theta$ rad
   - ω rad/s
   - T s/rev
   - 2π rad/rev

   Correct relations: a, c, d, f, g, k, l, o, p.

   Notes: (j) has the correct relation in terms of units, but as you know, a unit check won't tell you if you are off by a constant factor or not. What is true is that $\omega = 2\pi{\cdot}f$.

(k) is a funny one. First change $1/T$ to f. Then rewrite $2\pi f$ as ω. Now does it make sense? Alternatively, think of this as the $\text{m/rev} \div \text{s/rev} = \text{m/rev} \cdot \text{rev/s} = \text{m/s}$.

10. False! More on this in the next two chapters, but the liquid is moving in a circle so the acceleration is directed into the center of the circle. Think about it.

11. True.

12. False. If you're changing directions, you're accelerating. End of story.

13. The velocity vector is tangent to the path. The **velocity does change** ... in direction, not magnitude. Now, the acceleration vector is directed in towards the center of the circle. (This is because you're only changing direction, not speed, and so you have no component of acceleration parallel to the velocity vector.) The **acceleration also changes** ... in direction, not in magnitude (v^2/R).

14. True. Asked for the direction of travel? Find the direction of the velocity vector.

15. A runs slower! B has to run a greater distance to complete one revolution.

16. A finishes first, since she has a shorter distance to run.

17. g.

18. True. If the speed is increasing, there is a component of acceleration in the direction of $\vec{\mathbf{v}}$. If the speed is decreasing, there is an acceleration component in the direction opposite $\vec{\mathbf{v}}$.

19. a. The rock will hit the apple every time. Give the rock a greater initial speed and it will hit the apple sooner (before the apple falls much). Give the rock less initial speed and it will hit the apple later (after the apple falls a while). Now it is possible that you shoot the rock so slowly that it doesn't make it over to the apple. If there were no ground to stop the rock and apple, the rock would eventually hit the apple well below the ground level.

20. Remember this exercise?

(a) The 60° shot stays in the air longer, since it has more initial vertical speed and goes higher.

(b) The 30° shot is directed horizontally more than the others, so its side to side speed is greater.

(c) The 45° shot has the greatest range under these circumstances. It's the balance between more hang time (60°) and more horizontal speed (30°).

Exercise Hints & Solution Plans

EX. 3.5: [Hint] First think of the horizontal motion. What do you know that is always true about the horizontal speed? What is the average horizontal speed? ... Well what do you need to know to find it out? How are you going to find that out? Look to the vertical motion! How far did the ball fall? What is the initial *vertical* speed (right at the release)? So how long does it take to fall vertically as far as it falls? ... OK, if you've answered those questions, you can solve the puzzle.

EX. 3.20: [Sol'n Plan] There are two ways to go here. METHOD 1. First convert the given frequency to the angular rate, ω. Then convert the angular rate to the linear speed (also the tangential speed of the outer edge of the tire). And finally, get the distance. METHOD 2: Find out how many rev's are completed in 1 minute. Then convert that to the number of radians spun around. From that, get the distance.

EX. 3.21: [Hint] You know the solution plan comes out of the problem solving template given. Now you just need to think about what specifically is going on in this problem? What is the implicit information here? There are 2 ways to get the same result!!

Method 1: What do you want? ... OK, you want R, i.e. Δx. You assume you know v, and θ, so you need ...... Δt !! Get an expression for it by using Δy ... what? (Notice that you can factor out a Δt!]

Method 2: You still want Δx, so you still need Δt. You don't have many equations besides Δy to use ... to be specific, you've only got one other choice! (You need to notice this sort of thing! There are only 4 main projectile equations! Know this now and forever!) What do you about v_y at the states just after launch and just before landing? They are the same magnitude and opposite directions! Write it out! ... So $v_{1,y} = v \sin\theta$, and $v_{2,y} = -v \sin\theta$. Solve for Δt and plug it into the Δx equation.

As for finding the angle for maximum range (given a set launch speed), you now have an equation for the range in terms of the angle. How do you maximize an equation? Take the derivative with respect to the think that is variable (here, the angle), set the expression equal to zero, and find the value for the variable that gives this max (or min). Plug this value into the original equation to get the max (or min) value for the thing you are maximizing (or minimizing). This is one of the most basic procedures in first semester calculus.

EX. 3.22: [Hint] What does "top" imply? It is really really really easy to find the time if you choose the right one of the 4 basic projectile equations to look at. Use that implicit meaning of "top" to help you choose. Also, which is easiest and applicable? Go ahead and try them all quickly to help you see why one is the best. Remember, you assume that you know v and θ. ... Once you have t, you could go to the $\Delta y = ...$ equation, but is there another that would be easier, given what "top" means? These should be very easy, mathematically.

Exercise Solutions

EX. 3.1: These are the sorts of distinctions that you need to start making.

1. Just drop it! 2. Straight down. 3. Down and over. 4. Horizontal launch. 5. Up and over. 6. Straight up.

In cases 1, 2, and 6, the motion will be one dimensional; hence, we only need to describe the vertical motion. In cases 3, 4, and 5, the motion will be two dimensional; hence we will need a set of equations that describes the vertical motion and a set that describes the horizontal motion.

EX. 3.2: We want to compare any earlier ("initial") state to a later ("final") state of the same free fall stage. Let's look at some of the subtleties and possible misconceptions.

Clearly free fall states (the only force is gravity): 2, 3, 4, 5

State 1: Cannot! Either the cannon hasn't exploded yet and the ball is at rest, or the cannon has just been ignited and the ball is starting up the tube by the expansion of the exploding gun powder. Either way, it is certainly not in free fall and we must describe this part of the process with different physics.

State 6: It depends! If the instant sketched is supposed to depict the ball *just before* it hits the ground, then it *is* the very last state of free fall and *is* part of the free fall stage. If the sketch depicts the instant just *after* hitting, state 6 is not a free fall state. (It is common to look at the state just *before* hitting.)

State 7: Cannot! Even though this is a free fall state, it is occurring during a different free fall stage than states 2-5. So, for instance, you could not call state 3 the initial state and state 7 the final state in the same projectile equations.

States 8 & 9: Cannot! I take it this is clear.

EX. 3.3: You can really see that the less air the object catches, the more negligible air resistance becomes. Furthermore, the more mass the object has, the less effect the air resistance has on the object. To say quantitatively when air resistance is negligible, we need to know the accuracy required for the experiment. For this class, you can assume that it is negligible unless told otherwise or you're obviously dealing with the effects of air, such as with a falling leaf.

EX. 3.4: The horizontal shadow moves along the ground with constant speed (no accel.), so in equal amounts of time, it will travel equal distances. The vertical shadow on the wall will look just like an object that is thrown straight up – it will slow down on the way up and speed up on the way down. This motion was studied in Ch. 2. In equal increments of time, the spacing gets closer together on the way up and further apart on the way down.

[See diagram on the next page → .]

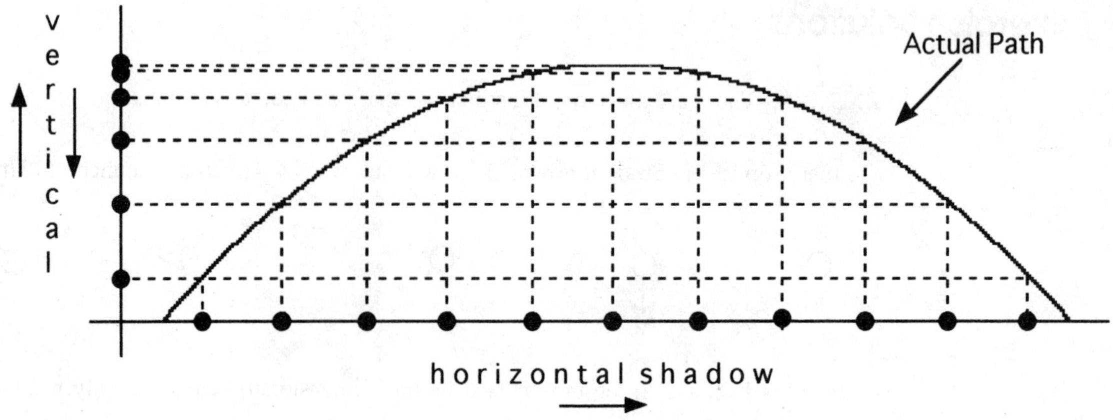

EX. 3.5: [Quick Sol'n]

- You want the horizontal speed as it leaves Kym's hand. But the horizontal speed never changes, so if we find the average horiz. speed, that will be the same as the initial horiz. speed.
- The average horiz. speed is the distance (20 m) divided by the time (unknown). You need the time while in flight! $v_{x,\ av} = {}^{\Delta x}/_{t} = {}^{20\ m}/_{t} = ?$
- Now look at the vertical motion. What do you know? It starts with no vertical speed, since it is thrown horizontally, and it falls a distance of 4.9 m (≈5 m). How long does it take for a dropped object to fall 4.9 m? 1 second! (Remember that? If not, $\Delta y = {}^{1}/_{2}gt^2 = {}^{1}/_{2}\,(4.9\ {}^{m}/_{s^2})t^2$ with $t = 1$ s will give it to you.) So, from the vertical motion you know that the ball was in the air for 1 s.
- Back to the horizontal motion, the ball went 20 m (horizontally) in 1 s, so the average speed is 20 m/s . And since the horiz. speed is constant, she threw the ball at **20 m/s**!

Note that the calculation can all be done in your head. If she threw it (horizontally) from 5 m above the ground and it landed 36.5 m away, her speed would've been 36.5 m/s. Make sure that you can think this through quickly.

EX. 3.6: First imagine it, then go throw a ball into the air to try it out.

(a & b) The 60° shot stays in the air longer, since it has more initial vertical speed and goes higher. It should be apparent that a greater angle of projection will give a greater height and vertical speed.

(c) The 30° shot is directed horizontally more than the others, so its side to side speed is greater. It should be apparent that a smaller angle of projection will increase the horizontal component of the velocity.

(d) A 0° shot either won't go anywhere in the air, or will immediately drop below the level of release. Either way, $R = 0$. A 90° shot will not move horizontally, so again $R = 0$.

(e) The 45° shot has the greatest range under these circumstances. It's the balance between more hang time (60°) and more horizontal speed (30°).

EX. 3.7: You bet it's possible! If one stand close to the center of the merry-go-round and the other stands near the edge. While their angular displacements are equal, $\Delta\theta = 30^\circ$, the distance moved along the circle is greater for the one further out. That's why it's more fun to ride on the outside!

Again, same angular displacement, different distances traveled. So, although the two complete one revolution together, the person on the outer edge goes around a bigger circle and thus travels farther.

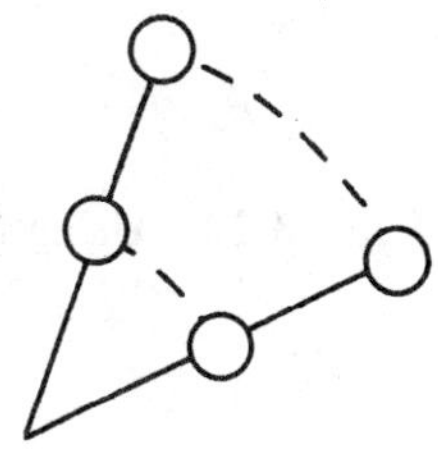

EX. 3.8: Two people on the same merry-go-round will have the same frequency (rev's/time) and the same angular speed ($\Delta\theta$/time), but may have different linear speeds. If one person is twice as far from the center as the other (i.e. twice the radius), then they will travel along a circle with twice the circumference. Since they will be traveling along a bigger circle, they will travel a greater distance. Thus a greater radius will result in a greater speed in m/s, but not a greater frequency or angular speed. [See diagram above.]

EX. 3.9: (a) When an object is only increasing speed, its acceleration vector is in the **same direction** as its velocity vector. (b) When an object is only decreasing speed, its acceleration vector is in the **opposite direction** as its velocity vector. (c) When an object is only changing direction, its acceleration vector is in the perpendicular to its velocity vector. Note that a change in both speed and direction corresponds to an acceleration vector that has components both parallel to and perpendicular to the velocity vector.

EX. 3.10: Obviously the answer isn't always 3 + 4 = 7!

(a) $v_{\text{as seen from the river bank}} = v_{\text{river}} + v_{\text{rowing}} = 3\ \text{m/s} + 4\ \text{m/s} = 7\ \text{m/s}$

(b) $v_{\text{as seen from the river bank}} = v_{\text{river}} - v_{\text{rowing}} = 3\ \text{m/s} - 4\ \text{m/s} = -1\ \text{m/s}$ (upstream!)

(c) Oops! Can't just add or subtract the numbers ... so we call on Pythag.! (Or notice that this is a 3-4-5 right triangle!)

$$v_{\text{as seen from the bank}} = \sqrt{(3)^2 + (4)^2} = 5\,\tfrac{\text{m}}{\text{s}}$$

EX. 3.11: (a) If and ONLY if the tire doesn't slip, it will travel a distance equal to the circumference when it makes one revolution. So, 2.5 m. (b) Here, it rotates half a revolution, so it travels half the circumference. Again, this does NOT hold if the tire slides along the road.

EX. 3.12 : You really need to be able to figure this one out quickly! You know the constant acceleration equations, so put in the correct accelerations for the vertical and horizontal motion. See Ex. 2.25 if you need help.

EX. 3.13: This is just the magnitude and direction of a vector – the velocity vector.

$$\text{total speed} = v_{\text{total}} = \sqrt{v_x^2 + v_y^2}$$

angle of travel at an instant = angle of velocity vector

$$\theta = \tan^{-1}\left(\frac{v_y}{v_x}\right) \text{, measured from the horizontal}$$

EX. 3.14: Using θ as the angle between the velocity vector and the horizontal, basic trig. gives:

$$v_x = v\cos\theta \text{ and } v_y = v\sin\theta$$

EX. 3.15: Sketch it! Especially the paths and velocity vectors.

IMPLICIT DATA

"Dropped" $\Rightarrow$ $v_{\text{start, y}} = 0$, $v_{\text{start, x}} = 0$

"Thrown straight up" $\Rightarrow$ $v_{\text{start, y}} > 0$, $v_{\text{start, x}} = 0$, (if up is chosen as +)

"Shot straight down" $\Rightarrow$ $v_{\text{start, y}} < 0$, $v_{\text{start, x}} = 0$, (if up is chosen as +)

"Shot horizontally" $\Rightarrow$ $v_{\text{start, y}} = 0$, $v_{\text{start, x}} = v_{\text{start, total}}$, $\theta_{\text{start}} = 0°$

"Top of path" $\Rightarrow$ $v_{\text{top, y}} = 0$, $v_{\text{start, x}}$ = (never changes), $\theta_{\text{top}} = 0°$

Comparing 2 states at the same height $\Rightarrow$ $\Delta y = 0$, $v_{\text{moving down, y}} = -v_{\text{moving up, y}}$, v_x = (never changes), $\theta_{\text{moving down}} = -\theta_{\text{moving up}}$

"2 objects hit" (or pass) $\Rightarrow$ same place (x and y) at some time (say t_2)

EX. 3.16: The two factors that determine the acceleration due to a change in direction are

1. The **speed** at which the turn is taken. Greater speed $\Rightarrow$ greater rate of change of direction $\Rightarrow$ greater acceleration!

2. The **curvature** of the turn. Greater radius of the circle $\Rightarrow$ less sharp the turn $\Rightarrow$ less rate of change of direction $\Rightarrow$ less acceleration!

Greatest accel: case 3. Least accel.: case 1. I'd guess that case 4 has a greater accel. than case 2, but I'd need to know or measure the radius of the arc of the curve and then check the equation ...Hey, what *is* the equation? See the next exercise! ... It turns out that doubling the speed quadruples the acceleration!

EX. 3.17: Got the seconds right? Acceleration has units with s^2 in the denominator. R doesn't have units that include seconds, so use v. v has s in the denominator, so we'll have to square v to get s^2 in the denominator. ... but that gives

$$a = v^n \cdot R^m = v^2 \cdot R^m$$

Squaring v gives units of m^2/s^2. We need to cancel out one of those meters with the R. So just divide by R. This gives

$$a = v^n \cdot R^m = v^2 \cdot R^{-1} = v^2/R$$

EX.3.18: You've got to get this!

rad's ⇔ meters: Use the circumference equation to remind you that the angle ($\Delta\theta$) multiplied by the radius (R) gives the arc length (l); or faster yet, notice whether you need to put in a meter or take one out!

(a) rad · m = m, so $\Delta\theta \cdot R = l$ ... multiply by R!!

(b) m/m = rad, i.e. $l/R = \Delta\theta$ ... divide by R!!

(c) If you have $^{rad}/_{s}$ you need to *multiply* by m to get the m up top in the units and end up with $^{m}/_{s}$ ... so multiply by R. If you have v in $^{m}/_{s}$, you need to get rid of the m, so divide by R. You can also remember the circumference equation to see where the R goes. So, $v = \omega R$ and $\omega = v/R$ should be obvious!

rad's ⇔ rev's: You know that 2π rad = 1 rev, so just create a conversion factor. Faster yet, see if you need to put in a rad by multiplying by 2π or get rid of the rad units by dividing by 2π.

(d) rad's to rev's ... We want to get rid of the rad's so *divide* by 2π. Check this by multiplying by 1 rev/2π rad. So $\Delta\theta/2\pi$ = # rev's. (Right? If $\Delta\theta = 2\pi$ that's 1 rev, if $\Delta\theta = 4\pi$ that's 2 rev's, if $\Delta\theta$ = 0.9432 rad ... divide by 2π!)

(e) Converting from ω to f, we need to change those rad's to rev's, so divide by 2π. To convert f to ω, we need to end up with rad's on top, so multiply by 2π.

(f) A 2 step process. We want to go from f to v. We're starting with rev's in the numerator, so get rad's up there by *multiplying* by 2π. Then turn the rad's into meters by multiplying by R.

Make Sense? In Sum:

If you have rads, get meters by multiplying by R. If you have an arc length in meters, divide by R to get the rad's.

If you have rev's, get rad's by multiplying by 2π. If you have rad's, get rev's by dividing by 2π.

EX. 3.19:

(a) Valid. The B's effectively cancel.

(b) Valid. Just rearrange the terms so that you have $\vec{\mathbf{a}}_{1,3} = \vec{\mathbf{a}}_{1,2} + \vec{\mathbf{a}}_{2,3}$

(c) Invalid. First, don't just "cancel" those "cat" indices. The terms on the left are subtracted, not added. So rearrange the equation to get $\vec{\mathbf{v}}_{dog,\,cat} = \vec{\mathbf{v}}_{cat,\,rat} + \vec{\mathbf{v}}_{rat,\,dog}$. Now the "rat" subscripts cancel, but the order of the remaining subscripts is wrong. Oh well.

(d) Invalid. First, there are 4 references (you, me, N.Y., and S.F.), and that's not going to get it in this case. Second, there is only one "you" subscript (because you're special). So it just doesn't work!

(e) Valid. Check this out! $\vec{\mathbf{D}}_{you,\,me} = -\vec{\mathbf{D}}_{me,\,you} \Rightarrow \vec{\mathbf{D}}_{you,\,me} + \vec{\mathbf{D}}_{me,\,you} = 0$. But $\vec{\mathbf{D}}_{you,\,me} + \vec{\mathbf{D}}_{me,\,you} = \vec{\mathbf{D}}_{you,\,you}$. And certainly the displacement from you to you is 0. Furthermore, the displacement from you to me is just opposite the displacement from me to you. That makes sense, doesn't it?

(f) Valid. For the same reasons as in (e), although the physical explanation is different. Here, If I see you moving east, you see me moving west. Same speed, opposite direction.

(g) Valid! Turn $\vec{\mathbf{a}}_{3,\,4}$ into $-\vec{\mathbf{a}}_{4,\,3}$. Then the equation becomes

$$\vec{\mathbf{a}}_{5,\,3} = \vec{\mathbf{a}}_{5,\,4} - (-\vec{\mathbf{a}}_{4,\,3}) = \vec{\mathbf{a}}_{5,\,4} + \vec{\mathbf{a}}_{4,\,3}\,.$$

(h) Invalid. The "1" subscripts don't cancel since this is subtraction. The negative sign can reverse the order of the subscripts of the last term, but that's no good.

EX. 3.20: Here are two ways to get to the answer.

METHOD 1:

$$f = 2.0\,\frac{\text{rev}}{\text{s}} \Rightarrow \omega = \left(2\pi\,\frac{\text{rad}}{\text{rev}}\right)\left(f\,\frac{\text{rev}}{\text{s}}\right) = 4.0\,\pi\,\frac{\text{rad}}{\text{s}}$$

$$\Rightarrow v = \omega \cdot R = \left(4.0\,\pi\,\frac{\text{rad}}{\text{s}}\right)(0.12\ \text{m}) = 1.5\,\frac{\text{m}}{\text{s}}$$

$$\Rightarrow \Delta x = v \cdot t = \left(1.5\,\frac{\text{m}}{\text{s}}\right)(60\ \text{s}) = \mathbf{90\ m}$$

METHOD 2:

$$f = 2.0\,\frac{\text{rev}}{\text{s}} \Rightarrow \#\,\text{rev's} = \left(f\,\frac{\text{rev}}{\text{s}}\right)(60\ \text{s}) = 120\ \text{rev's}$$

$$\Rightarrow \Delta\theta = \left(2\pi\frac{\text{rad}}{\text{rev}}\right)\cdot 120\ \text{rev's} = 240\pi\ \text{rad}$$

$$\Rightarrow \ell = \Delta\theta \cdot R = 240\pi\ \text{rad} \cdot 0.12\ \text{m} = \mathbf{90\ m}$$

METHOD 3 OK, just one more. $\Delta x = v\,\Delta t = \omega\,R\,\Delta t = 2\pi f\,R\,\Delta t = \mathbf{90\ \ m}$!

Wow, that's almost a football field in one minute … then again, maybe that is pretty do-able. Let's see, the speed is 1.5 m/s, which is about 3 mi/h (really 3.3 mi/h), and that's standard walking speed. Perhaps a brisk walk, but certainly reasonable.

EX. 3.21: You know the plan, just get fast at putting it down. I'll fill in what's given, hidden, and wanted. A check, √, means that it's taken as known.

[MODEL] Projectile motion with negligible air resistance and no floating or self-propulsion.

[SETUP]

Diagram: Label the state just after launch as "1," and that just before landing as "2." Lands at same height so $\Delta y = 0$ (implicitly!).

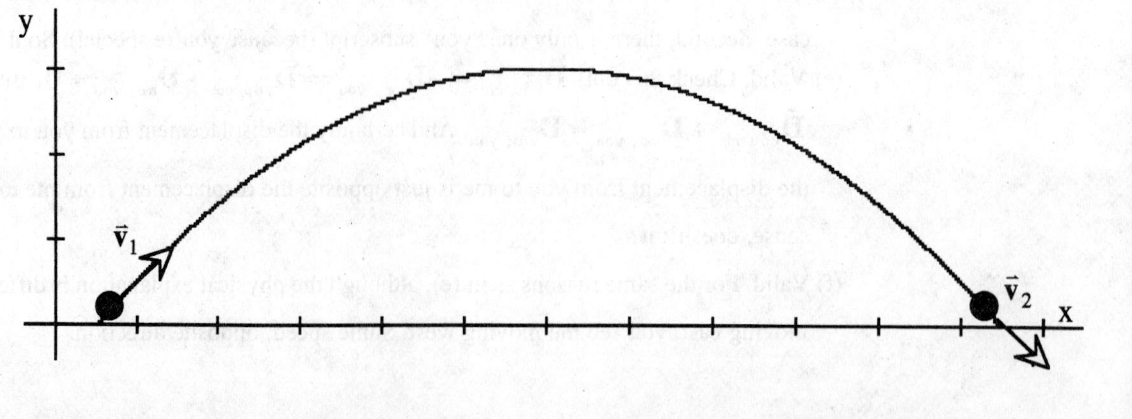

Data:

Organize the data according to the state. Data sought is listed as "= ?," while data assumed known has a check mark by it, "√."

State 1		State 2		...
$t_1 = 0$		$t_2 =$ "t" =		
$x_1 = 0$	$y_1 = 0$	$x_2 =$ "R" $- ?$	$y_2 = 0$	
$v_{1,x} = v \cos\theta = \surd$	$v_{1,y} = v \sin\theta = \surd$	$v_{2,x} = v \cos\theta = \surd$	$v_{2,y} = -(v \sin\theta) = \surd$	
$v_1 =$ "v" $= \surd$		$v_2 =$ "v" $= \surd$		
$\theta_1 =$ "θ" $= \surd$		$\theta_2 =$ (actually, we implicitly know this is - θ_1)		

*** Notice that "lands at same height as takes off" $\Rightarrow y_1 = y_2$ (i.e. $\Delta y = 0$), $v_{1,y} = -v_{2,y}$, $\theta_1 = -\theta_2$

Relations/Equations: Can you write them in less than a minute? That's your goal. There are 3 vertical equations, 1 horizontal equation. Also keep in mind the basic relationships for the velocity vector that relate the total magnitude and angle to the components.

[I'll choose up to be positive, so $a_y = -g$. Make sure that the equations that **you write** below reflect this choice.]

Horizontal Motion ($a_x = 0$)

1. $\Delta x =$

Vertical Motion ($a_y = g$, down)

2. $\Delta y =$

3. $v_{y,2} =$

4. $v_{y,2}^{\,2} =$

[SOLVE] (A quick outline.)

Method 1: We want R. We know that $R = \Delta x = v_{1,x}\, t = v \cos\theta\, t$. The only thing that we don't know here is t. So use another of the basic 4 equations.

- $\Delta y = 0$ & $\Delta y = v_{1,y}\, t - \tfrac{1}{2}gt^2 \Rightarrow 0 = (v \sin\theta)\, t - \tfrac{1}{2}gt^2 \Rightarrow 0 = t\,\{(v \sin\theta) - \tfrac{1}{2}gt\}$. Set each factor equal to zero to find two solutions for t. Clearly, $t = 0$ is no good. So $t = (2v \sin\theta)/g$. Now plug it back in to the Δx equation. Check with the text for the solution.

Method 2: Again, we want R. We know that $R = \Delta x = v_{1,x}\, t = v \cos\theta\, t$. The only thing that we don't know here is t. Now try an equation from the basic 4 equations that you didn't use in method 1.

- $v_{1,y} = -v_{2,y}$ & $v_{2,y} = v_{1,y} - gt \Rightarrow -v_{1,y} = v_{1,y} - gt \Rightarrow -2v_{1,y} = -gt \Rightarrow t = 2v_{1,y}/g = (2v \sin\theta)/g$, just as in method 1! As before, plug it back in to the Δx equation.

[ANALYZE] What are the extreme angles for launching the object upward? Check $\theta = 0$. Make sense? Check $\theta = 90°$. Make sense? Compare $\theta = 30°$ to $\theta = 60°$. Compare $\theta = 10°$ to $\theta = 80°$. Interesting! And how do these compare with $\theta = 45°$?

Ask your instructor whether you need to memorize this equation. I don't memorize it. I just derive it quickly whenever I need it. That way I only have to remember those 4 basic equations and fill in the details!

[Continued on next page →]

Find the angle that gives the maximum range for a set launch speed:

[See text Exercise 3.1, or check out the slightly different approach that I use below.] You have an equation for the range that depends on the angle of launch, the speed of launch, and the acceleration due to gravity. The latter two are assumed to be set, so there is only one variable, θ, that determines the range. Maximize the equation!

Take the derivative of the equation for R with respect to θ and set it equal to zero:

$$0 = \frac{dR}{d\theta} = \frac{d}{d\theta}\left(\frac{v^2}{g}\sin 2\theta\right) = \frac{v^2}{g}\frac{d}{d\theta}(\sin 2\theta) = 2\frac{v^2}{g}\cos 2\theta$$

Solve for the θ that makes this equation true (i.e. the θ that gives the max range). The only part that can be 0 is cos 2θ, so $0 = \cos 90° = \cos(2{\cdot}45°)$, so θ = **45°** gives the max range. If you want to know what the max range is, substitute $\theta = 45°$ back into the equation for R.

EX. 3.22: Now we don't have a lot that's given. I'm going to deviate from the solution template here, since I can get right to the solution. Remember, if you get confused, you need to sort things out carefully as I've shown. However, if you can get right to the solution, do it. If you get stuck, go back and take the slower route.

Diagram with states: ...... You sketch it. Just draw the first half of the diagram for the previous problem. DO NOT say "$v_{top} = 0$," $v_{top,y}$ is 0, but $v_{top,x} = v_x$ = constant.

Data:	Known	Implicit	Want
	v_1, θ_1	$v_{top,y} = 0$	t_{top} = ?, y_{top} = ?

Equations:

vertical

$\Delta y = v_{1,y}\, t - \frac{1}{2} g t^2$

$v_{2,y} = v_{1,y} - gt$

$(v_{2,y})^2 = (v_{1,y})^2 - 2g\Delta y$

horizontal

$\Delta x = v_x\, t$

Solve for t_{top}: Use $v_{top,y} = v_{1,y} - gt$, where $v_{1,y} = v_1 \sin\theta_1$ and $v_{top,y} = 0$.

This gives $0 = v_1 \sin\theta_1 - gt$. So the time to get from the start to the top is:

$$t_{top} = v_1 \sin\theta_1 / g$$

Solve for y_{top}: You could use the Δy equation, but that's a quadratic in t, so why not at least try bypassing it? Use $(v_{top,y})^2 = (v_{1,y})^2 - 2g\Delta y$, where $v_{1,y} = v_1 \sin\theta_1$ and $v_{top,y} = 0$, again.

This gives $0 = (v_1 \sin\theta_1)^2 - 2g\Delta y$ So the height, Δy, at the top is:

$$y_{top} = (v_1 \sin\theta_1)^2 / 2g$$

Check it out Dude! What do your results tell you? What happens when $\theta = 0°$ or 90°? When do you get the max time to the top and the max height? (First think conceptually, then notice that mathematically, the equation is max when $\sin\theta = 1$. If it makes sense, it just might be right.]

EX. 3.23: [Full Sol'n] How much time did it take you to set this problem up before moving on to the "solve" step? You *do* need to work on speed. Get the diagram, data, and equations/relations down quickly, but in an organized way. Here's one way:

[MODEL] Projectile motion with negligible air resistance and no floating or self-propulsion, but with rotations!

[SETUP]

Diagram: Label the state just after launch as "1," that at the top of the path as "2," and that just before entering the water as "3."

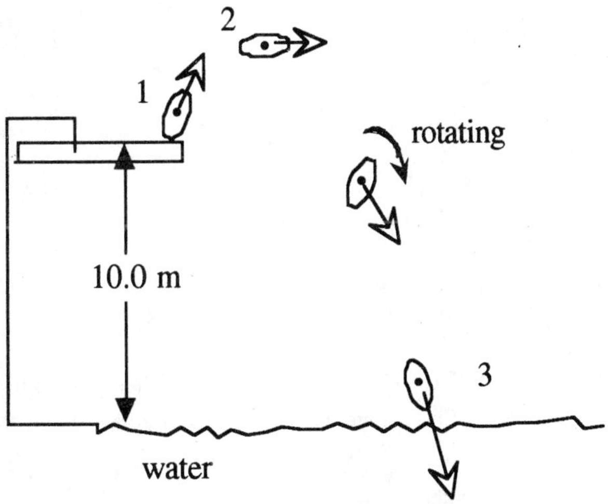

Data:

Organize the data according to the state. You may have or want any of the following data:

State 1

$t_1 = \mathbf{0}$

$x_1 = \mathbf{0}$ $y_1 = \mathbf{0}$

$v_{1,x} =$ $v_{1,y} =$

$v_1 = \mathbf{3.89}$ m/s

$\theta_1 = \mathbf{80^\circ}$, **up**

Not rotating yet.

State 2 (top) ...

$t_2 =$

$x_2 =$ $y_2 = \mathbf{?}$

$v_{2,x} = \boldsymbol{v_{1,x}}$ $v_{2,y} = \mathbf{0}$

$v_2 = \boldsymbol{v_{1,x}}$

$\theta_2 = \mathbf{0^\circ}$

Begin rotating. ω = ?

State 3 (enters water) ...

$t_3 =$

$x_3 =$? $y_3 = \mathbf{-10}$ **m**

$v_{3,x} =$ $v_{3,y} =$

$v_3 =$

$\theta_3 =$

Completion of 4.5 rotations.

Check this out. The data above came from "different places."

Chosen Data: $t_1 = \mathbf{0}$, $x_1 = \mathbf{0}$, $y_1 = \mathbf{0}$, I'll use up to be the positive y- direction.

Given Data: $v_1 = \mathbf{3.89}$ m/s, $\theta_1 = \mathbf{80^\circ}$, **up**, and a 10 m high platform.

Implicit Data: "Top" $\Rightarrow$ $v_{2,y} = \mathbf{0}$, $v_2 = \boldsymbol{v_{2,x}}$, $\theta_2 = \mathbf{0^\circ}$. (Note that $v_{2,x} = \boldsymbol{v_{1,x}}$ is always implicit, since doesn't change.)

From "up is +," "$y_1 = \mathbf{0}$," and "10 m platform" $\Rightarrow$ $y_3 = \mathbf{-10}$ **m**

Relations/Equations: [Again, I'll choose up to be positive, so $a_y = -g$.]

Proj. Equations:

vertical	horizontal
$\Delta y = v_{1,y}\, t - \frac{1}{2} g t^2$	$\Delta x = v_x\, t$
$v_{2,y} = v_{1,y} - gt$	
$2g\Delta y = (v_{1,y})^2 - (v_{2,y})^2$	

Velocity Vector Relations: $v_x = v\cos\theta \qquad v_y = v\sin\theta$

$$v = \sqrt{v_x^{\,2} + v_y^{\,2}} \qquad \theta = \tan^{-1}\left(\frac{v_y}{v_x}\right)$$

Circular Motion Relations: $\omega_{av} \equiv \frac{\Delta\theta}{\Delta t} = 2\pi f$, etc. ; $v_{tangential} = R\omega$, etc. ; $a_\perp = \frac{v^2}{R} = \omega^2 R$

[SOLVE]

(a) How high does she (actually her center) jump above the platform? We want Δy between states 1 and 2. Should we use the equation with time or the one without time? Since we don't have time, let's see if we can't bypass the need for it. Looking at the $2g\Delta y$ equation, we see that we only have one unknown. Yahoo!

$$2g\Delta y = (v_{1,y})^2 - (v_{2,y})^2 \Rightarrow \Delta y = \frac{(v_{1,y})^2 - (v_{2,y})^2}{2g} = \frac{(v_1 \sin\theta_1)^2 - (0)^2}{2g}$$

$$= \frac{(3.89 \sin 80^\circ)^2 - (0)^2}{2(9.80)} = \mathbf{0.760\ m}$$

(b) How far away, horizontally, does she land from beneath the platform? Find Δx between states 1 and 3. The only horizontal equation requires that we know the time between the states. Bummer. There are only 2 other projectile equations with time. Which do we know the most info. about? The easiest is the velocity equation, but we don't know the vertical velocity at state 3. So I guess we'll have to go with the Δy equation that's quadratic in t.

We know $\Delta y = -10$ m, $g = 9.80$ m/s^2, and $v_{1,y} = v_1 \sin\theta_1 = (3.89)(\sin 80^\circ) = 3.83$ m/s , so

$$\Delta y = v_{1,y} t - \tfrac{1}{2} g t^2 \Rightarrow \left(\tfrac{1}{2} g\right) t^2 - \left(v_{1,y}\right) t + \Delta y = 0$$

$$\Rightarrow \left(4.90 \tfrac{\text{m}}{\text{s}^2}\right) t^2 - \left(3.83 \tfrac{\text{m}}{\text{s}}\right) t + (-10.0\ \text{m}) = 0$$

→

Applying the quadratic equation gives:

$$t = \frac{-b \pm \sqrt{b^2 - 4ac}}{2a} = \frac{3.83 \pm \sqrt{(3.83)^2 - 4(4.90)(-10.0)}}{2(4.90)} = \frac{3.83 \pm 14.5}{9.80}$$

$$t = -1.09 \text{ s, or } 1.87 \text{ s}$$

Clearly she didn't hit the water 1.09 s before she jumped! So 1.87 s must be the physical solution. Thus,

$$\Delta x = v_x t = (v_1 \cos\theta_1)t = (3.89\cos 80°)(1.87) = \mathbf{1.26\ m}$$

(c) What is her average angular speed while she is spinning?

$$\omega = \frac{\Delta\theta}{\Delta t} = \frac{2\pi(\#\text{ rev's})}{\Delta t} = \frac{2\pi(4.50\text{ rev's})}{\Delta t_{\text{for rotations}}}$$

The time to complete the rotations is the time between states 2 (top) and 3 (at water). You can get this by:

Method 1: Use states 2 and 3 in $\Delta y = \ldots$ and find $t_{\text{from 2 to 3}}$.

Method 2: Use states 1 and 2 in $\Delta y = \ldots$, use the Δy found in part a, find $t_{\text{from 1 to 2}}$, then subtract this from the 1.87 s from 1 to 3 that we found in part b.

Method 3: Use states 1 and 3 in $v_{3,y} = v_{1,y} - gt$, since we can easily get $v_{1,y}$ and we know t = 1.87 s from 1 to 3 (found in part b). Now use $v_{3,y} = v_{2,y} - gt$, since $v_{2,y} = 0$. So $t = -v_{3,y}/g$ Note that this would be the time between 2 and 3, as desired.

Method 4: Can you outline another way? Probably.

Which looks best to you? I like Method 1. I'll show you what I noticed to make my decision.

Using states 2 & 3: $\Delta y_{\text{ between 2 \& 3}} = v_{2,y}\,t - \frac{1}{2}gt^2$. But 2 is at the top, so $v_{2,y} = 0$! This gives $\Delta y_{\text{ between 2 \& 3}} = -\frac{1}{2}gt^2$, so

$$t_{\text{between 2 \& 3}} = \sqrt{\frac{-2\Delta y_{\text{between 2 \& 3}}}{g}} = \sqrt{\frac{-2((-10) - 0.760)}{9.80}} = \sqrt{\frac{-2(-10.760)}{9.80}} = \mathbf{1.48\ s}$$

Finally,

$$\omega_{\text{av}} = \frac{\Delta\theta}{\Delta t} = \frac{2\pi(\#\text{ rev's})}{\Delta t} = \frac{2\pi(4.50\text{ rev's})}{1.48\text{ s}} = \mathbf{19.1\,\tfrac{rad}{s}}$$

(d) Approximate the tangential velocity of her head as she spins. [Hint: What is her approx. radius when curled up?] When I roll up in a ball, my diameter is about the length of my femur (thigh bone) ≈ 0.75 m. So I'll use a radius of 0.4 m (I don't give this number many significant figures).

$$v_{\text{tangential}} = \frac{\ell}{\Delta t} = \frac{\Delta\theta \cdot R}{\Delta t} = \omega \cdot R \approx (\mathbf{19.1}\,\tfrac{\text{rad}}{\text{s}}) \cdot 0.4\text{ m} = \mathbf{8}\,\tfrac{\text{m}}{\text{s}} \approx 16\,\tfrac{\text{mi}}{\text{h}}$$

(e) Using the approximation made in part (d), what is the acceleration of her head as she spins? [Ooh, you'll need to be careful to receive the gold medal on this part.] … Why is care needed? Hmmm.

Certainly, the acceleration from spinning at a constant speed is easily obtained as v^2/R, directed into the circle. So that's

$$a_\perp = \frac{v^2}{R} \approx \frac{(8\,\tfrac{\text{m}}{\text{s}})^2}{0.4\text{ m}} = 160\,\tfrac{\text{m}}{\text{s}^2} \text{ (or 200, to 1 sig. fig.)}$$

Here's the trick: her head has an acceleration due to the spinning *and* an acceleration due to gravity. After all, one would hope that their head would fall with their body! So,

$$\vec{\mathbf{a}}_{\text{head}} = (a_\perp,\text{ towards the center of rotation}) + (g,\text{ down})$$
$$= \left(160\,\tfrac{\text{m}}{\text{s}^2},\text{ towards the center of rotation}\right) + \left(9.8\,\tfrac{\text{m}}{\text{s}^2},\text{ down}\right)$$

Can I use the Pythag. th'm to get the magnitude? NO NO NO! The two parts are not always perpendicular. The rotational component changes direction as she spins around! So that's about all we can say, other than to show a few instances:

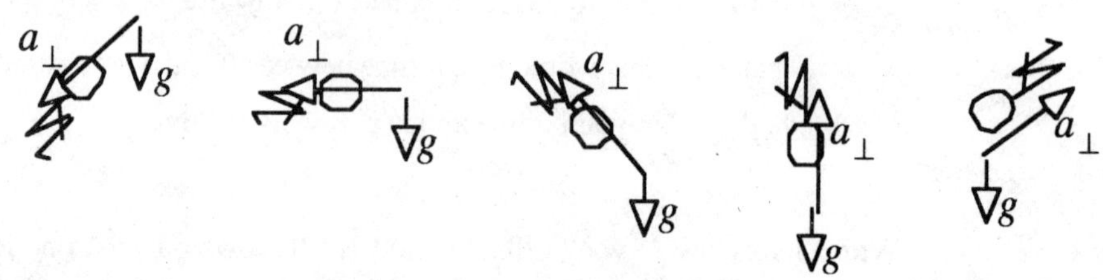

[ANALYZE] So are the numbers reasonable?

- $\Delta y = 0.76$ m above the board is actually quite a bit. In order to truly jump up 1 m, you would have to get your belly button above your head! Not many, if anyone, can do this. (High jumpers have a way of cheating by actually having their center of mass go below the high jump bar!) Still, this value doesn't seem too outlandish.

- $\Delta x = 1.26$ m seems about right for the distance out from the platform. They stay pretty close, but do drift out some.

- We found that it took 1.48 s to fall 10.76 m (that's correct), and in that time he spins 4.5 times. Top divers are able to do this, but it's the upper limit to date. Perhaps using $R = 0.4$ m is a little high. It's easier to spin when you are tucked up tightly, which may be one reason that divers don't tend to be very tall.

Ch. 4/5: Force and Newton's Laws/Using Newton's Laws

For perhaps obvious reasons, I've elected to treat these two chapters together. After reading the text, going to lecture, working through this tutorial, and working lots of problems, you will become quite the proficient force analyst! Vector addition, including getting vector components, will play a very big role in chapters 4 and 5. Be sure that you are up to speed on the techniques covered in chapter 1.

Overview

Now that you've learned how to use kinematics to *describe* the motion of objects, we turn our attention to the question of *why* things move the way they do. What is the natural motion of objects, what are externally imposed motions of objects, and what are the *causes* of the motions?

Most people's conception of why things move the way they do is rooted in so-called "common sense" derived from daily activities. To be sure, most people now *believe* that the Earth revolves around the sun and that things are made of molecules which are made of atoms which are made of electrons, protons, and neutrons. Still it's difficult to truly shake off the "common sense" that entrenches people in an Aristotelian worldview. It was Newton who clearly shook off the mental chains of shallow "common sense." You'll see that it's hard enough to shake these chains even if you have a text and instructor pointing the way. When you finally do – and you can – you will only then realize why physicists hold Newton in such high esteem.

In his "3 Laws of Motion," Sir Isaac Newton spelled out the underlying rules of nature that guide the motion of all things. These laws play a key role in describing and predicting the physical world. It is here that we need to consider the *mass* (AKA "inertia") of an object and the various types of *forces* (pushes and pulls) that can cause the object to *accelerate*. (We will need to modify Newton's laws to form quantum mechanics and relativity only when we begin to look at the very small – atoms and subatomic particles – and the very fast – say half the speed of light or faster –, respectively.)

In some cases we will need to use both Newton's laws and kinematics to fully analyze a process. How would we go about this? Well, what variable appears in both Newton's laws and kinematics?

Note: although most physicists won't admit it, we believe that Newton was Galileo reincarnated, since he was born the year Galileo died (1642). We also think that he can be traced back to Adam of Eden, because of the fig leaf – fig newton connection. Compelling evidence, don't you think? For those doubters, what about the apple connection, huh? Spooky! ... And he may go back further, since Galileo was born the year Michelangelo died.

Conceptual

A 2000 Year Conceptual Leap

To get at the subtlety here, let me lead you with a simple quiz-exercise. Go ahead and let your common sense guide you and you'll soon see the light.

EXERCISE 4/5.1: With which worldview do you observe the world? [For the most value and impact, physically do this. It'll only take a minute.] Hold a book in one hand and a pen in the other. Place your palms up so that the objects rest on your hands. Let's start with a simple, albeit obvious, question. Which object wants to fall to the earth more?

Aha!! Heretics! Aristotelians! Let's see if we can bring your worldview forward about 2000 years to the 1600's! First, let's understand an essential difference between the worldviews in question.

Aristotle (384 - 322 BC) asserted that things fall because it is *within their nature* to seek their natural or "proper" place. The more "earth" an object contains, the greater its tendency to seek the ground and the faster it falls. The more "air" it contains, the more it has a tendency to rise. Thus an object falls because of its own internal, self-contained properties–according to Aristotle and his followers.

Newton (1642 - 1727), aided by Galileo's work (1564 - 1642), asserted that objects fall to the ground because of an *external* cause; a separate, external agent (object) must be acting on the falling object. This, along with Kepler's analysis of Tycho Brahe's data of the orbit of Mercury, led to Newton's 3 laws and his law of universal gravitation.

Do you see how saying that an object "wants" to fall implies the perspective that it is an *internal* impetus that causes the fall? In fact, objects at rest *resist* being moved. A more massive object resists more. Neither the book nor the pen "want" to move, since their internal predilection, if you...*will*, is to resist acceleration. Furthermore, the book, having more mass, even resists falling more than the pen! Some *external* cause must be at work!

Something about that last paragraph should strike you as funny. Look again at the book and pen you're holding. Do you really believe that the book resists falling more than the pen? Maybe I'm trying to trick you. Let's defer to Newton, whose 3 laws reflect the "rules" that nature seems to closely abide by.

The Concepts

"Receptors" & "Sources"

These are two words that I use in a somewhat unusual way, but they can really help clear up some misconceptions fast and will continue to prove valuable throughout the course– especially when we get to electric and magnetic phenomena!

Receptor: the object whose motion we are focusing on, the one *receiving* the pushes and/or pulls, the *m* that accelerates in the $F = m\,a$ equation

Source: the object that is pushing or pulling the receptor, the *source* of the force!

Mass (m)

- an internal property of an object
- the more matter an object has, the more mass it has
- the resistance to accelerate (AKA "inertia")
- the property responsible for exerting and feeling gravitational forces
- a positive (+) scalar quantity

Force ($\vec{\mathbf{F}}$)

Every force has a source!

- received *by* an object (m, the "receptor") *from* another, external object (the "source")
- any push (away from the "source"), or
- any pull (towards the "source")
- a vector quantity!

Various Types of Forces

- any type of force may be added to any other type of force…vectorially!! That is, by components or graphically.
- look for forces in the following 2 categories. Notice that "contact" forces are those where the source touches the receptor, whereas "non contact" forces are those where the source doesn't even touch the receptor (weird!)

Contact Forces

- push or pull (from hands or whatever)
- surfaces - "normal force" ($\perp$ to surface)
  - friction (parallel to surface)
- tension (pull from a rope)
- spring force (push or pull)
- fluid forces (air resistance, etc.)

Non Contact Forces

- gravitational force ("weight")
- electric
- magnetic
- "weak" force (radioactive decay)
- "strong" force (nuclear physics)

The Laws

I've probably stated the laws a little differently than your text, but they are the same. Remembering these verbatim is fine, but not as effective as remembering them with some clear visual images of what they are saying.

> **Memory Tip:** ***Install a video file rather than a text (word) file in your head.***

(I) **Inertia (AKA Mass) & Natural Motion** - Any object's natural motion, i.e. when left alone, is $\vec{\mathbf{a}} = 0$ motion: it will remain at rest or move with constant speed and direction. ("Inertia" is the resistance to accelerate.)

(II) **Cause of $\vec{\mathbf{a}}$** - If an object's motion is not "natural," i.e. if $\vec{\mathbf{a}} \neq 0$, there must be an external cause (from one or more other objects)!

- The causes of $\vec{\mathbf{a}}$ are called "forces" – pushes or pulls of any type.
- Real, physical forces ($\vec{\mathbf{F}}$) are *always* interactions between 2 objects.
- Net External Forces ($\vec{\mathbf{F}}_{\text{total}}$) Cause (=) An Object ($m$) to Accelerate ($\vec{\mathbf{a}}$)

$$\underbrace{\vec{\mathbf{F}}_{\text{total}}}_{\text{from all sources}} = \underbrace{m}_{\text{of receptor}} \cdot \underbrace{\vec{\mathbf{a}}}_{\substack{\text{resulting accel.}\\ \text{of receptor}}}$$

or

$$\vec{\mathbf{a}}\,(\text{effect on receptor}) = \frac{\vec{\mathbf{F}}_{\text{total}}\,(\text{from sources})}{m\,(\text{of receptor})}$$

(III) **Equal & Opposite $\vec{\mathbf{F}}$ s** - When object A pushes or pulls on object B, object B pushes or pulls on object A…with equal magnitude and opposite direction. Always! Even with gravitational and electric forces! [Subtle exceptions may be noted later.]

$$\vec{\mathbf{F}}_{\text{on A from B}} = -\vec{\mathbf{F}}_{\text{on B from A}}$$

You will be amazed to see how many problems you can solve just based on these laws, as well as how easy it is to deviate from these seemingly simple laws. Your job is to learn to carefully apply these laws correctly and to avoid the misconceptions that "common sense" inevitably will lead you to. My job is to help you.

EXERCISE 4/5.2: Can only one force act on an object such that $\vec{a} = 0$ for that object (the receptor)?

EXERCISE 4/5.3: What is the net force on an object held in your hand? Identify the "receptor," the "source(s)," and the forces on the receptor.

EXERCISE 4/5.4: Look around you and list all of the *types* of forces that you see. What other types of forces are there? Is that all of the types of forces in the universe? How do you know when you've listed them all?

How much?!

Once again, let's get a feel for the magnitudes that you'll see when looking at these concepts. (The units of mass were also addressed in Ch.1 – see page I.1-4 and Ex. 1.7!).

Mass

S.I. unit of kilogram

1 kg = 1 (10^3 g) = 1000 g (where g means gram in this context)

1 kg ≈ 2 lb (+0.2 lb) [Ahhhh! Be careful here! Mass is NOT weight. In fact, an object with mass of 1 kg weighs 2.2 lb only on earth. More on this distinction soon.]

1 kg of water fills 1 liter (volume). [That's half of one of those plastic 2 liter soda bottles.]

Force

S.I. unit of newton

1 N = 1 kg·$^{m}/_{s^2}$

1 N ≈ 1/4 lb (really 0.224 lb), so ...

1 lb ≈ 4 N

200 N ≈ 50 lb (45 lb is closer)

I'm sure that you will never forget that $F = ma$, so you can easily use that equation to remind you that a newton (N) is equivalent to a kg·$^{m}/_{s^2}$. No excuses now.

EXERCISE 4/5.5: (a) Use the rough estimate that 1 N ≈ 1/4 lb to determine your weight in newtons. Then use the precise conversion of 1 N = 0.224 lb to obtain your weight in newtons. What is the percent error in using the rough estimate? [Notice that *weight* is a force–in N or lb, not a mass–in kg!] [See Hint for definition of % error.] (b) If a mass of 1 kg weighs 9.8 N on Earth, how much does it weigh in pounds when measured on Earth?

The Misconceptions (...well, some of them!)

A. N's 1st : Where's the throttle, Aristotle?

This is the misconception that I addressed in Ex. 4.1. Aristotle thought that things fall to the ground because of their own impetus – their own internal predisposition. Most people unconsciously see the world this way too. Even if someone says "things fall because of gravity," their intuition is probably based on the interpretation that heavier objects in some sense want to, or are predisposed to, fall to the ground more than lighter objects. That's simply not so, and Newton points the way with the necessary conceptual framework and logic to decipher what's actually going on. Newton pointed out that an object at rest would stay at rest if it weren't for an *external* force – call it "gravity" (from some other object – the Earth). In the Basic section of this chapter, we'll look at why two objects of different masses fall at the same rate (if air resistance is negligible).

B. Weight ≠ Mass

Visualize the difference. Imagine that you are floating in deep space with nothing in sight but a 5 kg lead ball and a 10 kg lead ball floating in front of you. Gently put your hand under each ball. How much do they weigh? Nothing! You can put your hand "below" them without feeling any pressure. Now grab the 5 kg ball and move it left, right , "up," "down," and notice that the object resists any attempt to change its motion. You feel this by the pressure felt in your hand. If you do this with the 10 kg ball, what do you feel? Twice the resistance! Twice the pressure. That's because it has twice the mass (not weight, since weight = 0 out in deep space). On Earth, the 10 kg objects weighs 98 N (22 lb). On the Moon it would weigh about 1⁄6 as much as on Earth. On Jupiter it would weigh about 3 times as much as on Earth. But on Earth, the Moon, Jupiter, or out in deep space, the object still has 10 kg of mass!

Objects don't have weight!

Weight is the pull (force) from some *other* object (Earth, Jupiter, etc.) due to gravitational attraction. Mass is proportional to the amount of matter–the more matter, the more the resistance to acceleration. Now mass *is* responsible for gravitational forces, but it is NOT equal to it. Mass is a scalar (number without direction), while weight, a force, is a vector. Furthermore, kg is not just another unit that can be used in place of N. Pounds and newtons are two units for the one concept of force. Kilograms, grams, and slugs (not used much), are different units of the one concept of mass. Mass ≠ Force, while Weight *is* synonymous with the force of gravity!!! Enough said? Let me give you one more image so that I know you know that objects have no weight! (I know you know, but do you know?)

(Here is a simplistic model that makes a point.) You and the 10 kg lead ball are floating out in space. In your palm, you have a very sensitive scale. You place your hand under the ball and the scale reads...that's right, zero. That's because the ball doesn't really *have*

Imagine all things attached to strings which the Earth tugs on. Can you see the strings? If so, it may to time to take a break!

weight. Now you notice that the ball has a string attached to it. All of a sudden the string is pulled between your fingers, the string pulls the ball into the scale, and you notice that the ball now "has" weight – that is, the scale no longer reads 0, but in fact it reads 98 N (22 lb). "Hey! Who's pulling that string? Someone or something must be. There must be a source (object) to that force. N's 2nd says so." So you follow the string down and you see that it is the Earth. The Earth pulls the invisible string (called gravity) which in turn causes the scale in your hand to register a "weight" for the ball. If you were on the Moon, the scale would read about 1/6 as much since the Moon isn't as strong as the Earth. So you see?! Weight is externally caused; it is the force of gravity from another object such as a planet. Granted, there is no "invisible gravitational string" that we know of, but the point is that weight is NOT an internal property of the object, mass *is*. NOW you know, objects don't really *have* weight. [Next time someone asks you your weight, tell 'em it's zero!]

C. N's 3rd : Give It Up, The Horse and Buggy Can't Move

Consider the assertion: " By Newton's 3rd law, if a horse pulls on a buggy, the buggy pulls back with an equal but opposite force. Consequently, they can never move." Okay, you know that this is supposed to be a misconception, but can you see how the assertion seems to make sense? Can you offer an explanation that unequivocally refutes it? For starters, horses *can* pull buggies. So what is it about N's 3rd that so many people are unclear on?

Which Is The Receptor, Which Is The Source?: That depends on which object's acceleration you want to find. In our $\vec{\mathbf{F}}_{total} = m\vec{a}$ equation, the "*m*" is for the object *chosen to be* the receptor, the "$\vec{a}$" is the acceleration of *that* object, and the total force is from all of the other objects that are pulling or puching on it. Push on a refrigerator. Which object is the receptor and which is the source? Your choice. You can view it whichever way suits your analysis:

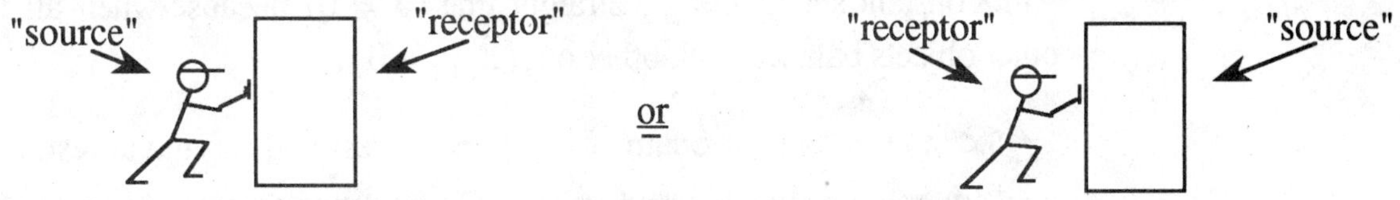

The Trick With N's 3rd: Of the equal and opposite force pair associated with N's 3rd, one force acts on object "A" while the other force acts on "B." They act on *different* objects!! So they do NOT NOT balance each other and the *effect* (acceleration) on one of the objects may be quite different than on the other!!

¡¡ N's 3rd does NOT deal with two equal and opposite forces acting on a single object !!

2 Factors Determine The Effect On Each Object In The Interaction:

1. The MASS of the object ("receptor") feeling the force
2. OTHER FORCES on that object that aren't on the other object

Note that these may be *very* different for the 2 objects creating the "equal & opposite" force pair. [Again, each object only feels one force from the N's 3rd law pair!!!]

EXERCISE 4/5.6: The Earth pulls on a large apple with a force of about 1N. With what force does the apple pull on the Earth? Which will have the greater acceleration? Why? [Hint: In each case, identify which you are viewing as the source and which as the receptor.]

EXERCISE 4/5.7: A noisy fly is buzzing around your head. You role up your newspaper and swat it out of the air. (a) Did *you* actually hit the fly? If you view the fly as the receptor of the force, who or what was the source of the force on the fly? (b) Now treat the object identified in (a) as the receptor. What is the source of the force on it? (c) How does the force on the fly compare to the force on the newspaper? (d) How does the effect on the fly compare to the effect on the newspaper? Explain.

EXERCISE 4/5.8: Hold a book on your palm. Doooo iiiiit. How many forces are on the book? Do they constitute a N's 3rd law pair? For the force of gravity on the book ("weight"), what other force makes up the N's 3rd force pair?

D. N's 2nd : Zero Net Force Means No Motion, Right? Wrong!

Just what does N's 2nd say when the net force is zero? Simply that there is no acceleration! The law says nothing about the velocity except the obvious inference that, if $a = 0$, the velocity is constant. Take a look at the possibilities, and what the law does and doesn't say, and remember the following terminology:

$\vec{\mathbf{a}} = 0 \Leftrightarrow \vec{\mathbf{F}}_{\text{total}} = 0$: In a state of *equilibrium*, the object of interest (m) is at rest or moves with constant speed along a straight line ($\vec{\mathbf{a}} = 0$) because/when all the forces from other objects balance each other out ($\vec{\mathbf{F}}_{\text{total}} = 0$).

- **Static** ($\vec{\mathbf{v}} = 0$) **Equilibrium** ($\vec{\mathbf{a}} = 0$) $\Leftrightarrow$ at rest and staying at rest.
- **Dynamic** ($\vec{\mathbf{v}} \neq 0$) **Equilibrium** ($\vec{\mathbf{a}} = 0$) $\Leftrightarrow$ constant speed and direction (straight line motion)

$\vec{\mathbf{a}} \neq 0 \Leftrightarrow \vec{\mathbf{F}}_{\text{total}} \neq 0$: If an object accelerates, i.e. changes speed and/or direction, there *must* be some net force from *other* objects. Conversely, if there is a *net* force from other objects, the object of interest will accelerate. Notice from Newton's 2nd law that the resulting $\vec{\mathbf{a}}$ is always *parallel* to the net $\vec{\mathbf{F}}$. Further notice that the value and direction of $\vec{\mathbf{a}}$ tells us nothing (Nothing!) about the value and direction of $\vec{\mathbf{v}}$ except how it is changing. [The 2nd law relates $\vec{\mathbf{F}}$ to acceleration, not velocity!]

E. N's 1st: Circles are Made With Preservatives

Consider this assertion: "Newton's 1st law says that objects tend to continue along the same path when left alone. Hence, an object moving along a circular path will continue to do so on its own accord. In fact, this is what satellites and the Moon do."

Now, is an object accelerating when it is moving along a circular path (even at constant speed)? ... Yes, because it's changing *direction*! Recall that at any instant $\vec{v}$ is tangent to the path, whatever the path may be, so clearly $\vec{v}$ must be changing.

But objects naturally resist acceleration. So some *external* agent must be present to *cause* the object to move in a circle. Get rid of the external agent, and the object will move along the line tangent to the circle with constant $\vec{v}$ (no $\vec{a}$). (Of course if gravity is present, the object will then become a projectile.) For circular motion, there must be a force pulling or pushing the object *into* the circle. This net inward force is what causes the object to change its direction from the straight line tangent path that it would take if left alone.

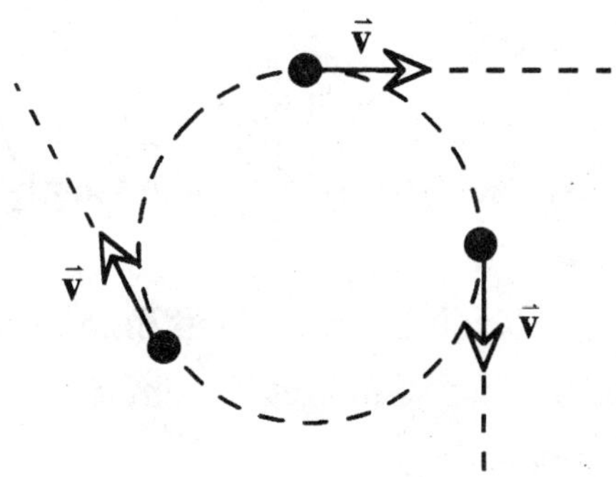

F. N's 2nd : Centrifugal Forces

Ready? You're about to get clear on something that most people are confused about. This little exercise will also help refine your newly acquired Newtonian way of seeing the world.

Attach a string to some object. I like to use a yo-yo. Spin the object above your head in a horizontal circle. Do you feel the force that is pulling the yo-yo outward, away from the center of the circular path? You do?! Hmmm... Well then, since every force must have a source, what is the object that is pulling the yo-yo outward? Come on, something's pulling it outward, no? Okay, get in a car and make a turn. You feel the outward force causing you to move to the outside of the circle, right? Or how about the Rotor carnival ride? (Figure 3.14 in *Physics: The Nature of Things*) What force is pressing you into the wall as you whirl around in a circle? What is the source object for that force? Hard to find the source, isn't it? That's because there is none!

The illusion comes from the confusion between force and inertia. Three examples will help separate reality from illusion!

(i) Bob is standing. He's just standing–minding his own business–when his friend, Freddy Force, pushes him in the back. Distinguish what happens from what it feels like happens.

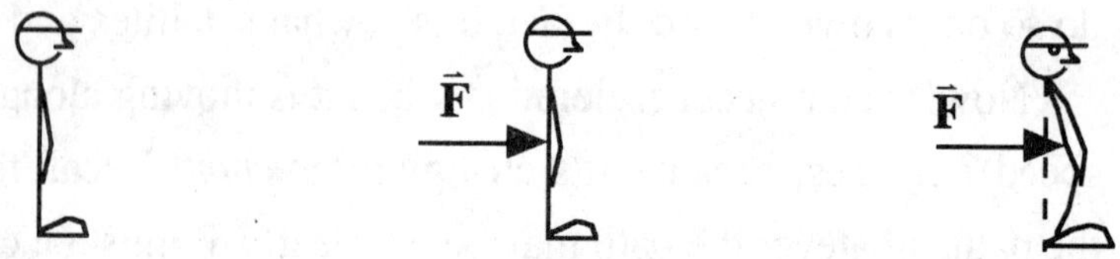

Bob says:

"Hey! I feel a push forward on my back! What's more, my head was thrown backwards!"

Freddy Force replies:

"I only pushed you forward. No one pushed or pulled your head back. Stop whining! It was just your fat head that tended to stay in place, inertia brain!"

I hate to admit it, Bob, but Freddy is right.

$\vec{F}_{real}$ is forward.

$\vec{F}_{ficticious}$ is presumed because it *feels like* your head is *thrown* back. But your head actually tends to stay put while your body is moved forward.

(ii) Bob is standing again. This time on a small throw rug that's placed on top of a wooden floor. Grinning Freddy Force sneaks up from behind and pulls the rug out from under Bob.

Bob says:

"Hey! You pulled the rug out from under me and made my head go forward!"

Freddy Force replies:

"I did pull the rug, which in turn pulled your feet back, but you can't sue me for whiplash because I didn't push your head forward. In fact, your head didn't even go forward…What is it with your head anyway?"

Once again, sly Freddy is right.

$\vec{F}_{real}$ is backward and from the rug.

$\vec{F}_{ficticious}$ is presumed because it *feels like* your head is *thrown* forward. Your head actually tends to stay put while your body is moved backward this time. Why don't you go out for a walk or something, Bob?

(iii) Bob is walking. It feels so good that Bob yawns and stretches his arms out to his sides. Freddy says "Don't mind if I do," and grabs one of Bob's arms and pulls him to the side.

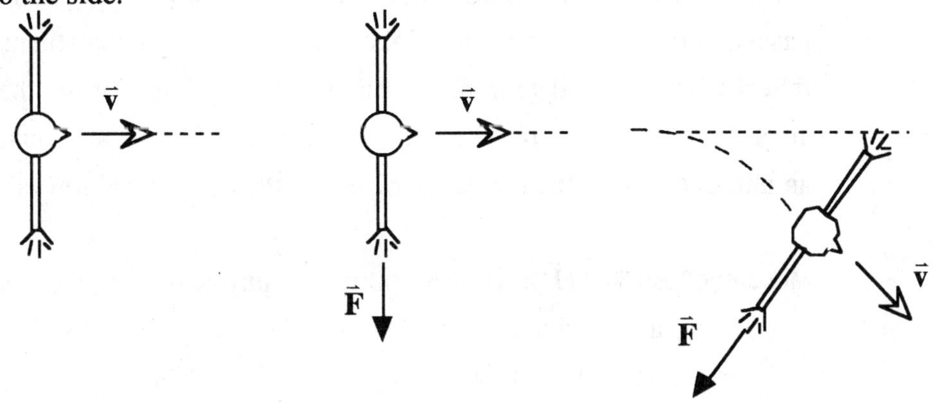

Bob says:

"This times it *is* whiplash! I was walking in a straight line, you pulled me to the side and my head was *thrown out* from the circle that you made me move in."

Freddy Force replies:

"(I love this guy.) I pulled you to the side while you moved forward. Consequently, your body moved in a circle. Your big ol' head tried to keep moving along the straight line path that it was on. I just pulled you to the side so that you would move in a circle. If you weren't so resistant to change, maybe your neck wouldn't hurt."

Yup.

$\vec{F}_{real}$ is to the side, i.e. *into* the circle.

$\vec{F}_{fictitious}$ - it just *feels* like there is an outward force. But there is no source creating an outward force! It is just your inertia (mass) that resists the change. Now there is an outward pull on Freddy–he pulls *in* on Bob, Bob pulls *out* on him–but there is no outward pull on Bob. The 1st law straight line path is viewed as out from the circle.

Got it? If you still insist that there is an outward pull on the yo-yo when you whirl it around, just tell me the source (external object) of the force and I'll concede. Terminology:

- **Centripetal**: "center-seeking;" any force that points into the center of the circular motion
- **Centrifugal**: "center-fleeing," where "fugal" is like "fugitive;" any force that points out away from the center of the circular motion

Inertia isn't something that acts on the object, it is a property of the object.

Note that "centripetal" and "centrifugal" are NOT new types of forces, they just designate the direction of the force. For the yo-yo, the tension force from the string pulls inward on the yo-yo, so tension *is* the centripetal force. There is no real outward (centrifugal) force in that case. There may be real centrifugal forces from strings pulling outward or whatever, but there must be a source to the force. Often the term is used to describe the apparent (pseudo, fictitious) force that is really the inertia of the object…but inertia ≠ force!

The Forces

There are forces from gravity (weight, attraction between masses), surfaces (friction & normal forces), strings (tension), springs (varying push & pull), hands, fluids (liquids & gases), and so on. For some of these forces, there are mathematical expressions which reflect what factors determine the magnitude and direction of the force. Usually, but not always, we will insert these expressions into the $\vec{\mathbf{F}} = m\,\vec{\mathbf{a}}$ equations. Below, try and get an intuitive feel for the factors involved for the mechanical forces.

Muscular Push/Pull ($\vec{\mathbf{F}}$): Unless you're into physical therapy or the like, you don't need to worry about what factors determine the strength of these forces. Just call them $\vec{\mathbf{F}}$ and let it go at that for now.

Tension Force ($\vec{\mathbf{T}}$): Imagine that you have a rope attached to your waist and someone is pulling on it. What do you notice? What can you say about the force from the rope? It should be apparent that the following points hold true.

1. The rope can only *pull*, it can't push!
2. The rope pulls in the direction of the rope itself (along the line of the rope).
3. The maximum possible pull from the rope is determined by the rope's strength.
4. Misconception: Is the person pulling on you? NO! NO! NO! The person is pulling on the rope. The *rope* is pulling on you. This distinction is very very important. Failure to make this distinction is one of the biggest errors in applying Newton's laws. More on this later. Just notice that the person isn't touching you, even though she *is* animating the process.

Common misconception

Normal Force ($\vec{\mathbf{n}}$): Push against some solid surface. Notice that it pushes back on you. Further notice the direction of the push. Push down on your desk top, it pushes up to support you. Sit on a chair. Gravity pulls you into the chair, the chair supports you (hopefully) by pushing up. Push against a wall, it pushes back with a horizontal force. Make sure that the following points become *obvious*.

1. The normal force is from the surface of a solid structure and $\perp$ to it.
2. The more an object is pressed into a surface, the greater the normal force.
3. The normal force is always *perpendicular* to and *away* from the surface. If the surface is horizontal (floor or desktop), $\vec{\mathbf{n}}$ is vertical and away from the surface. If the surface is vertical (wall), $\vec{\mathbf{n}}$ is horizontal and away from the surface, and if the surface is at an incline, $\vec{\mathbf{n}}$ is perpendicular to and away from that incline. It is never a force *into* the surface!

Common misconception

4. Misconception: n is *not* always equal to the weight of an object! Depending on the situation, you may have $n = w$, $n > w$, or $n < w$.. Don't carelessly jump to conclusions! Just put the variable n into the equations and see what you get!

Friction Force ($\vec{\mathbf{f}}$): Right this minute, push some things across the table in front of you and notice the friction force at play. What do you notice? How can you increase or decrease the friction? Try it! What's the direction of the friction? Can there be friction when an object isn't sliding across a surface? Can there be no friction at all on a rough surface?

1. The rougher the two surfaces, the more friction possible. (roughness → μ)
2. The more the surfaces are pushed together, the more friction possible. Try this out by sliding a book across the desk and then sliding it while pushing down on the book. More friction, no? (n tells how much they are pushed together)
3. Friction forces are always along (parallel to) the surface.
4. For *kinetic* friction (when one object slides across another), the force is always in a direction *opposite the motion* – it impedes the sliding.
5. For *static* friction (the object is being pushed, but isn't moving), the force is directed to *oppose the onset* of motion.

Common misconception

6. Misconception: Kinetic friction is constant (in most cases), but static friction ranges from 0 to some maximum value. The more you push a box that won't budge, the greater the static friction force must be to keep the box static. At some point, the ground is pushing as much as it possibly can $\left(\vec{\mathbf{f}}_{s,\,max}\right)$ and the object is at the "breaking point." If you push harder, you will overcome the static friction, the object will start to slide, and then you have kinetic friction.

Spring Force ($\vec{\mathbf{F}}_{spring}$): Imagine a spring. A slinky is okay for stretching a spring, but you can't compress it. So imagine one that you can both stretch and compress.

1. The stiffer (stronger) the spring, the stronger the force. (stiffness →k)
2. The more the compression or stretch (x), the more the force from the spring. The spring force is *not constant.*
3. When you *compress* a spring, the spring pushes out. When you *stretch* a spring, it pulls back. Springs exert a force in a direction *opposite* to the displacement of the spring's end. The spring force is always towards the equilibrium point, hence it is called a "restoring force".

Common misconception

4. Misconception: Some students (not you) try to use the constant acceleration equations for things attached to springs. But if the spring force is the only force, and since the spring force isn't constant, the acceleration isn't constant – it's greater when the force is greater, i.e. when there is more compression or stretch.

Gravitational Force ($\vec{\mathbf{W}}$ or $\vec{\mathbf{F}}_G$): Newton posited that every object attracts every other object by a gravitational force – "universal gravitation." We distinguish two different situations, the gravitational force from the Earth (or other planet or moon) when near the surface, and the completely general gravitational force between any two objects.

Case 1: Near the surface of something big (called *weight* in this context, $\vec{\mathbf{W}}$)

1. The more *mass* the receptor (thing being pulled) has, the more the pull on it. It's kind of like the object has more handles for the source to grab onto.
2. The greater the acceleration when in free fall ("free flight"), the greater the net force. Since gravity is the only force, a greater acceleration while in free flight reflects a greater pull from gravity. (Acceleration in free flight, i.e. with only gravity acting on the object is specially labeled as $a_{\text{free flight}}$ = "g")
3. Misconceptions: Remember, objects don't really "have" weight, they have mass. Weight *is* the grav. force on an object from another object (Earth, Moon, Mars, Jupiter,...). Weight is a force (in units of N or lb); mass is not a force (it has units of kg, grams, slugs).

Common misconception

Case 2: Between any two objects (completely general, $\vec{\mathbf{F}}_G$)

1. Again, the more *mass* the *receptor* (thing pulled) has, the more the pull on it.
2. Similarly, the more *mass* the *source* (thing pulling) has, the stronger the pull. Think of a more massive puller as being stronger.
3. The *further apart* the centers of the objects are, the *weaker* the force. Think of it as though the source calls out to the receptor. The call gets fainter with distance.
4. Remember N's 3rd: we can choose which object we view as the receptor and which the source. Either way the forces on each are always equal and opposite.

Other Forces: There are certainly other forces around us. There are the non-mechanical electric and magnetic forces, the "strong force" that holds a nucleus together, and the "weak force" involved in some radioactive decay processes. Can you think of others? (I can think of one!) Well, they are all just pushes or pulls and can be added freely to one another. We'll leave a detailed study of the electric and magnetic forces for later. For now, let's work with the list of mechanical forces above.

The Applications (using Newton's laws)

An important component of doing a force analysis is to notice and determine the type and direction of the forces on an object. Of course, we need to decide which object (the "receptor") we are interested in. With practice and attention, you can always know that you've got *all* of the forces on an object. In fact, there are three simple questions that you can ask yourself to check to make sure that you've gotten all of the forces on the object of interest:

1. "What is the object that I am interested in, i.e. the 'receptor'?"
2. "What 'non contact' forces are present?" [Hint: only gravity until later chapters!]
3. "What is in contact (actually *touching*) the receptor?"

To help you see and keep track of the forces, you will always want to make a "Free-body Diagram." This will also prove essential when it comes time to break the force vectors down into components. The technique is simple *if* you can see the forces, Luke.

FREE-BODY DIAGRAM METHOD

1. Take a **snapshot** (freeze!) of the system at the instant (state) of interest.
2. Identify the **receptor** (object you're focusing on).
3. Make a simple **schematic sketch** of the receptor as a dot, block, circle, rod, etc.
4. List all **sources** (all objects pushing or pulling on the receptor).
5. Draw in the **force vectors** for all of the forces acting on the receptor.
   (i) Draw in the gravitational force (AKA weight) vector.
   (ii) Draw in all contact forces. Get the directions right! See the list from the previous section.

You absolutely need to be fast and accurate here. In some ways it is harder than it seems–as evidenced by the many students that have trouble–and in some ways it is easier: where's the weight and what's touching the thing?! Get up to speed with the following exercises. It will pay big dividends. Expand your awareness of forces. What kind and direction are they? Keep N's 3rd in mind, but learn to stay clear of those misconceptions.

EXERCISE 4/5.9: Bob's walking. Take a snapshot of him in mid stride (one foot on the ground) and show the forces on him. State the source of each force. How can he walk if he and the floor are exerting equal and opposite forces on one another? Who animates the process?

EXERCISE 4/5.10: You're sitting in the passenger seat (right side) of a car in the U.S. and the driver makes a left hand turn. Consider a snapshot moment of you as you round the turn. What are the forces on you? ...The *real* forces! Identify the source of each force.

EXERCISE 4/5.11: For the designated "receptor," draw a free-body diagram. Label the forces by type.

(a) Receptor: the projectile in free-flight. Consider each of the 3 stages shown.

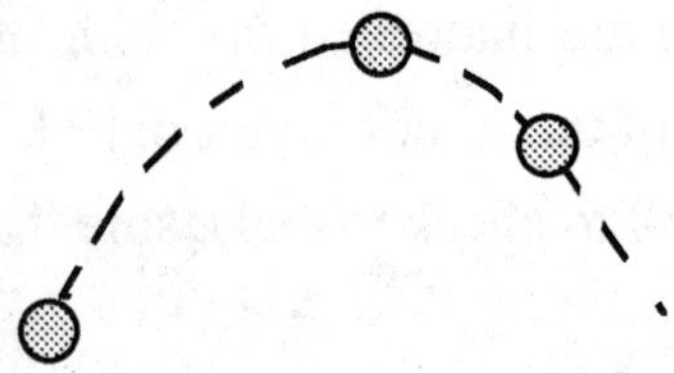

(b) Receptor: the pendulum bob at the end of a string.

(i) Does it matter what $\vec{\mathbf{v}}$ is?

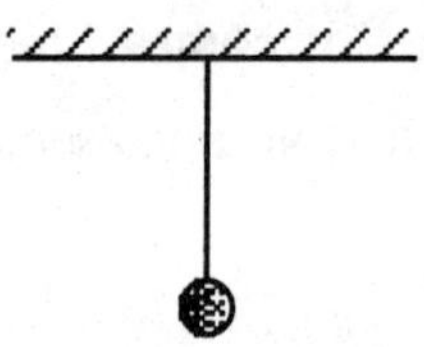

(ii) Does it matter which way it's going or if $\vec{\mathbf{v}} = 0$? (It could be moving up, down, or at rest in the picture below.)

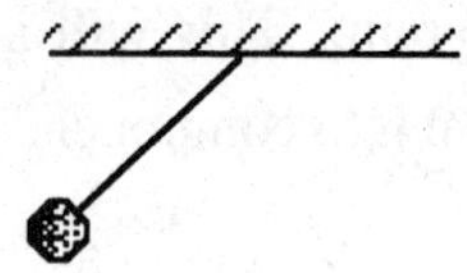

(c) Receptor: the block; surfaces are *not* frictionless. Does the direction of motion matter?

(i)

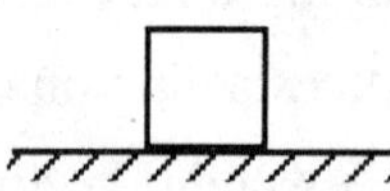

(ii)

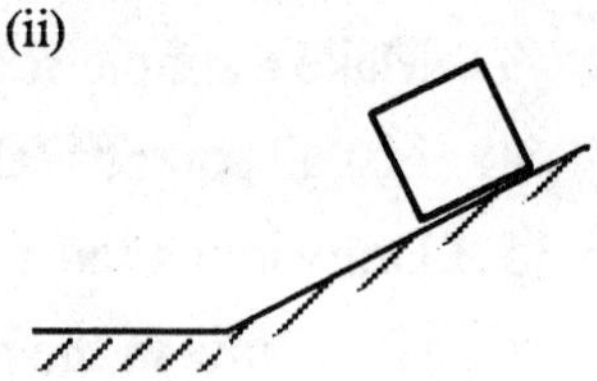

(d) Receptor: the rectangular sign

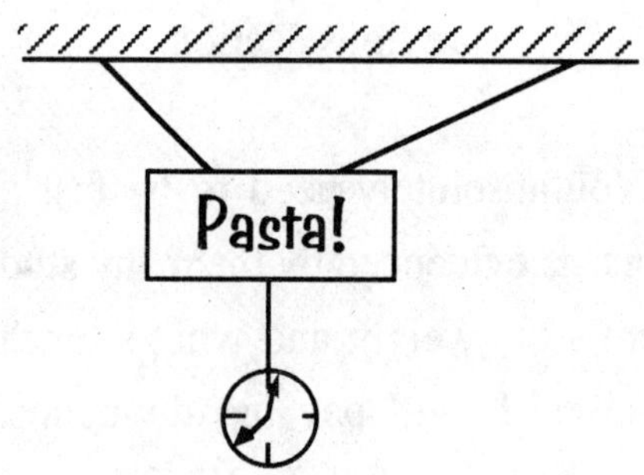

(e) Receptor: the volleyball held under water

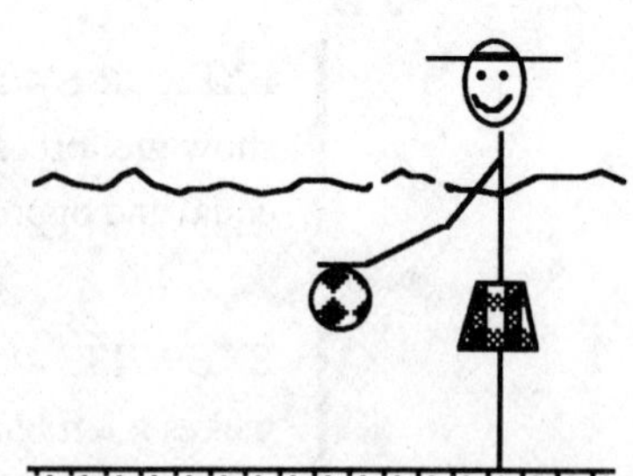

EXERCISE 4/5.12: More fun with free-bodies. See the force. Feel the force! Become the force.

(a) Receptor: the round object connected to the string & spring

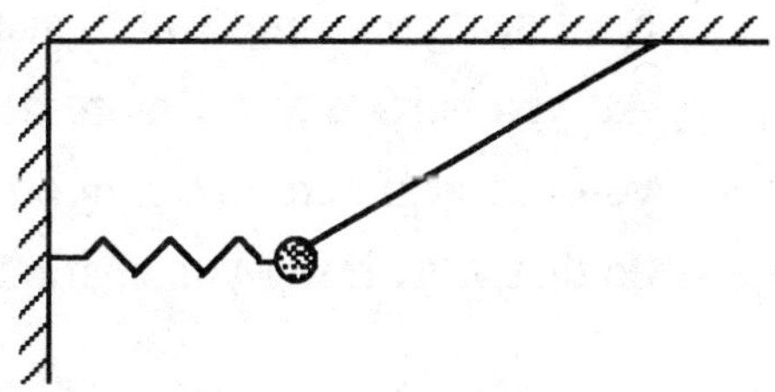

(b) Receptor: the bead on a frictionless wire

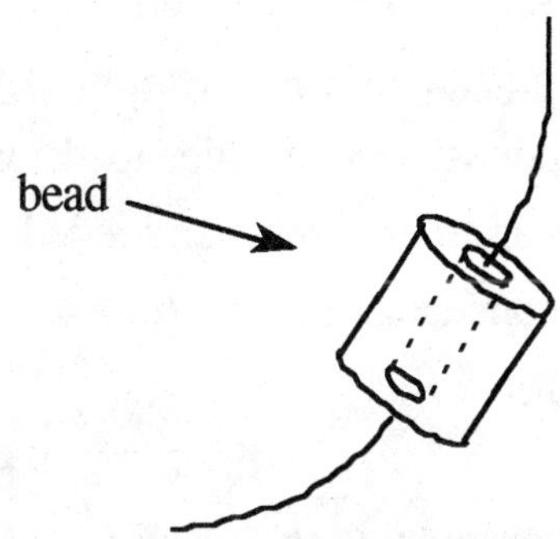

(c) Receptor: the box at the end of the rope that is strung through an ideal pulley system

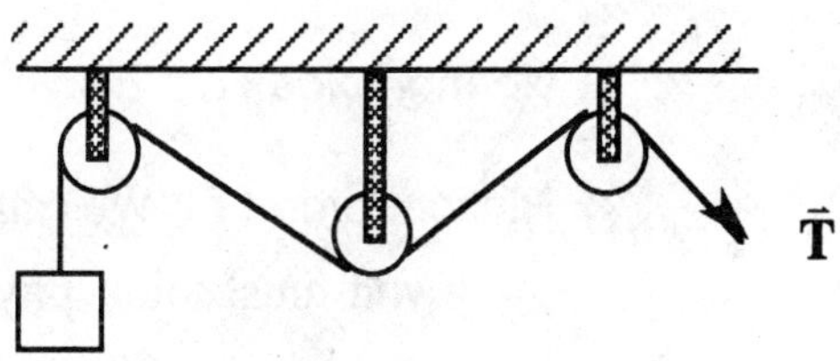

(d) Receptor: first treat each block separately, then treat all 3 as a single receptor

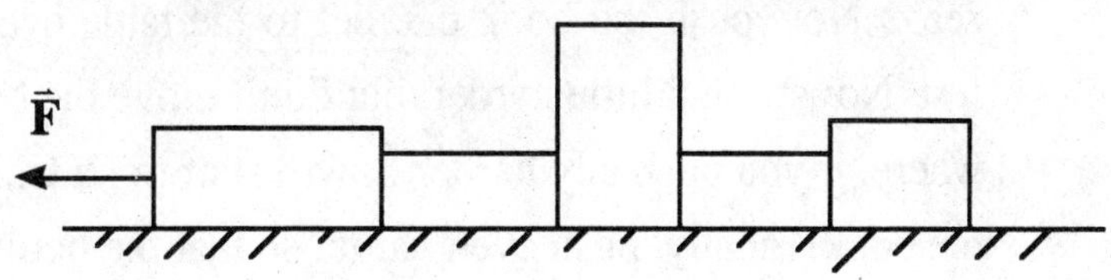

(e) Receptor: first treat each of the 3 stacked objects separately, then together as a single object

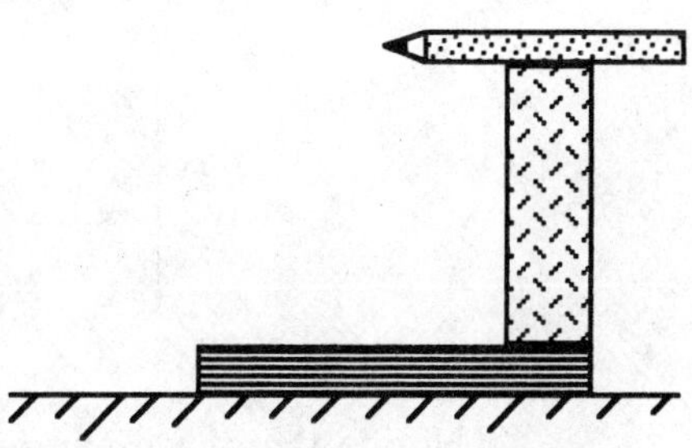

Basic

The Forces

So, I hear you can't wait to start calculating. You know the template. It's just $F_{net} = m\,a$. All you need is a few more details about the ingredients that go into that template so that you can begin *constructing* equations that are appropriate to specific situations. Once you do that, it's off to Mathematic Land to solve for the unknowns.

The Basic Relations

As you know, the different types of forces have different factors which contribute to the magnitude and direction of the force. Now we must address just how much each factor contributes in order to determine mathematical expressions to use in place of the force variables whenever needed. Be sure to review the list of forces and related factors in the conceptual section (p. I.4/5 - 12) anytime that you feel unclear about a particular type of force.

Muscular Push/Pull ($\vec{\mathbf{F}}$): Leave it as is; no more detail is usually given.

Tension Force ($\vec{\mathbf{T}}$): Leave it as is.

Normal Force ($\vec{\mathbf{n}}$): Leave it as is. See, this isn't so bad, is it?

Friction Force ($\vec{\mathbf{f}}$): We can (sometimes) break friction down into its determining factors;but you must think physically about the situation because friction can be expressed in three forms. You must decide which is appropriate. A little experiment will help. Put a book on a horizontal table in front of you. What's the friction on it? (Hint, it's zero.) Now push the book parallel to the table ever so slightly so that it remains at rest. Now push a little harder, but don't move the book yet. Find the "breaking point" where, if you push any harder, it will start to slide. Now push a little harder and slide the book. Finally, push even harder so that the book accelerates even more. That's all you need in order to physically reason out what form of friction is appropriate!

$$f = \begin{cases} f_s = 0 & \text{; no other forces parallel to the surface} \\ f_s & \text{; static, before breaking point} \\ f_{s,\,max} = \mu_s n & \text{; static, at breaking point} \\ f_k = \mu_k n & \text{; sliding over a surface} \end{cases}$$

Notice that these expressions give the magnitude of the friction force for particular situations (**the direction is parallel to the surface so as to oppose the motion or the**

Common misconception

onset of motion). Further notice that static friction can only enter the equations as "f" if the object is not at the breaking point. We cannot write $f_S = \mu_S n$ unless the object is at the breaking point!

Spring Force ($\vec{\mathbf{F}}_{\text{spring}}$): A spring always pushes or pulls so as to return the object to the equilibrium position. That's why it is called a "restoring" force. Sometimes this force speeds the object up (as it returns to the equilibrium position), sometimes it slows the object down (as it moves away from the equilibrium position), but it is always directed back to the equilibrium. That's why there is a minus sign in the expression. Furthermore, the spring strength (k) and amount of compression or stretch (x) both affect the magnitude of the force

$$F_{\text{spring}} = -kx$$

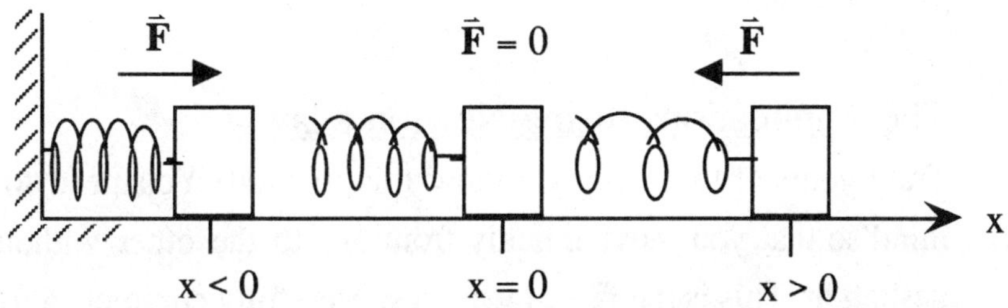

Gravitational Force ($\vec{\mathbf{W}}$ or $\vec{\mathbf{F}}_G$): see page I.4/5 - 14 for further discussion.

Case 1: Near the surface of something big (called *weight* in this context, $\vec{\mathbf{W}}$)

$$W = mg$$

Where g is the acceleration when in free-flight while on the planet specified (usually Earth).

Case 2: Between any two objects (completely general, $\vec{\mathbf{F}}_G$)

$$F_G = G\frac{m_1 \cdot m_2}{r^2}$$

Where G = 6.67×10^{-11} N m^2/kg^2 is the universal gravitation constant (it never changes), and r is the distance between the centers of the two objects. (Really between the centers of gravity, but we'll wait to make that distinction.)

Does it make sense to you that the gravitational force gets weaker according to r^2? It can. Recall that the area of a rectangle is length X width and the surface area of a sphere is $4\pi r^2$, so area is in units of a distance squared (m^2, ft^2, mi^2, etc.). Now consider a couple of analogies. The strength of a balloon depends on the thickness of its rubber walls. As you blow up the balloon, the rubber is spread over a larger area and the balloon walls get thinner and weaker. Sound is similar. If you call out to a friend, the sound you put out spreads out over a sphere (roughly) and gets weaker with distance as the area, $4\pi r^2$, increases. Finally, think of one of the masses as putting out a gravity call (the gravitational "field") which gets weaker with distance in the same way. In each case a reasonable relation would be:

$$\text{strength} \approx \frac{1}{\text{area}} \approx \frac{1}{r^2}$$

Sure, these are just analogies, but they give you some feel for the plausibility of the relation. By no means is this a proof. The force could have turned out to be very different.

The Applications (using Newton's laws)

Train yourself to carefully follow this method. You need to ingrain these steps in your mind so that you move quickly from one to the other without skipping any and without hesitation. This is the fastest way to successfully complete a force analysis.

METHOD FOR FORCE ANALYSIS OF 1-D PROBLEMS

0. Take a **snapshot** (freeze!) of the system at the instant (state) of interest. Think about what is physically going on.
1. Identify the **receptor** (m)–the object of interest.
2. Identify the **forces** ($\vec{F}$) by listing the **sources** and drawing a **free-body diagram.**
3. Identify the **acceleration** ($\vec{a}$). What do you know? Is $a = 0$? What direction is $\vec{a}$? Is it completely unknown?
5. Choose a **convenient direction** to be the + direction. If you know the direction of $\vec{a}$, use that.
6. **N's 2nd**: Fill in the particular forces on the left side of $F = ma$. (Don't forget the correct sign for each force.) Put in what, if anything, you know about the acceleration. (Often you will have $a = 0$, which simplifies the algebra a bit.)
7. **Need more equations?** That is, do you have enough equations to solve for your unknowns? If no, think about the kinematics equations or anything else that is pertinent. If yes, go to Mathematic Land and **solve** away!

Okay, practice makes perfect! Remember, you are *not plugging into* an equation, you are *constructing* an equation based on the 2nd law template which points out just what you need to be aware of:

$$\sum F \quad = ma$$

$$\underbrace{\qquad\qquad} = m\_\_\_$$

And don't try to solve *until* you've got this equation set up. On the left side you've got your causes, on the right, you've got the effect. Once you've put in the details, you can algebraically move things from side to side and mix up the logic as much as you need to in order to solve for the unknowns.

EXERCISE 4/5.13: In each of the 3 cases below, find the magnitude of the normal force from the table. The mass of the block is 10.0 kg.

(a) (b) (c)

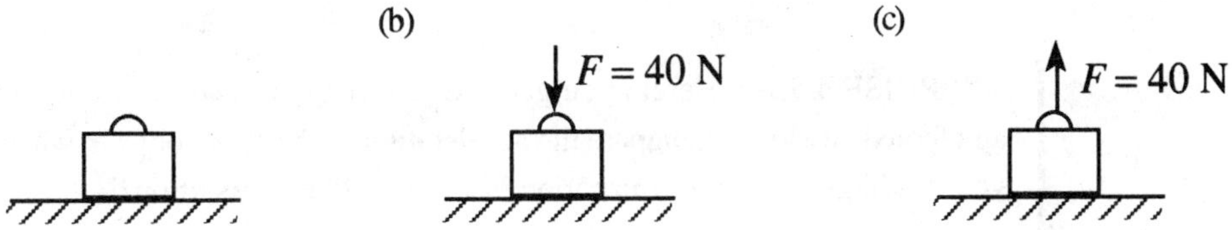

EXERCISE 4/5.14: Simple friction. For your pleasure, I've set this one up to work out "just right." Find the friction force in each case. The weight of the box is 4.00 x 10^2 N (roughly 100 lb), the coefficient of *static* friction is $\mu_s = 0.200$, and the coefficient of *kinetic* friction is $\mu_k = 0.100$. [See Hint section!]

(a) (b) $F_{applied} = 20.0$ N (c) $F_{applied} = 40.0$ N

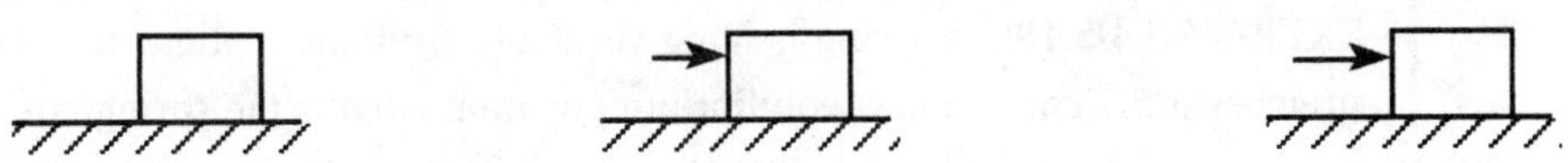

(d) $F_{applied} = 80.0$ N (e) $F_{applied} = 120$ N (f) $F_{applied} = 160$ N

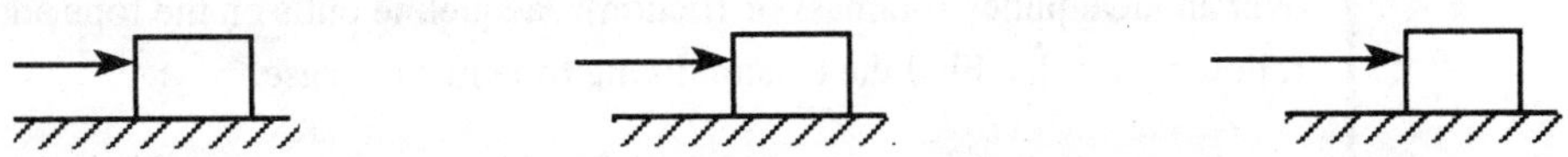

Now put a 2.00 x 10^2 N object (≈ 50 lb) on top of the 4.00 x 10^2 N box and quickly repeat parts a - f.

EXERCISE 4/5.15: Vertical free-flight. Find the acceleration in each case. Justify with a free-body diagram and N's 2nd law.

(a) Before release:

(b) After release when dropped:

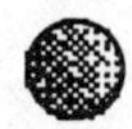

(c) After release when thrown up at an angle:

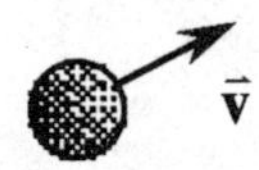

(d) After release when thrown down at an angle:

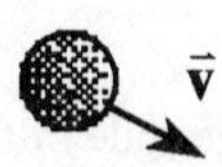

(e) At the top of the path:

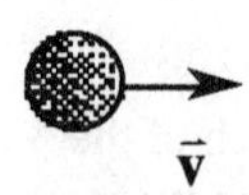

(f) Just before hitting the ground:

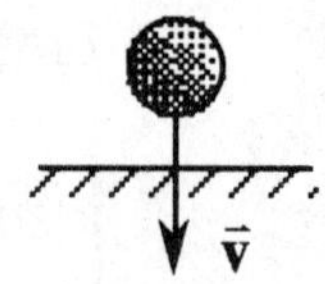

EXERCISE 4/5.16: Herein you get to discover the answer to a question that often bugs most students and fooled Aristotle. Compare the acceleration of the three objects when they are in free-flight. Explain your findings, i.e. what explanation justifies Galileo's assertion?!

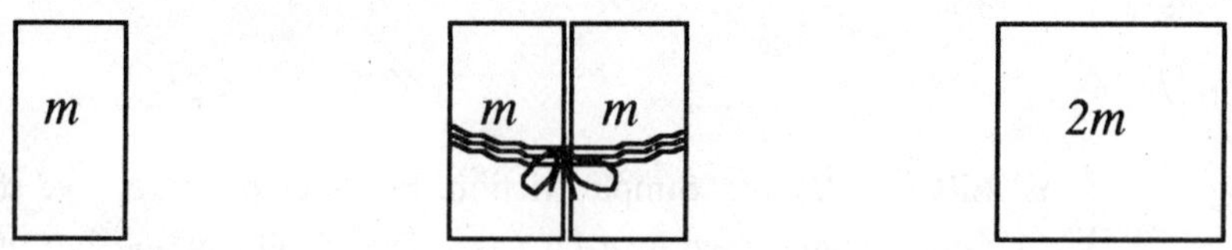

EXERCISE 4/5.17: An apple (about 0.1 kg) near the surface of the Earth pulls on the Earth with what force? Use both expressions for the gravitational force and compare results.

EXERCISE 4/5.18: A spring is hung vertically from the ceiling. If a 1.0 kg object is placed on it and stretches it 5.1 cm to a new equilibrium position, what is the spring constant, k, of the spring?

EXERCISE 4/5.19: Don (m = 1.00×10^2 kg) has a rope tied around his waist and the rope is strung over an ideal pulley (no mass or friction). Jacqueline pulls on the rope and at different times achieves the following results. Find the tension in the rope in each case.

Don

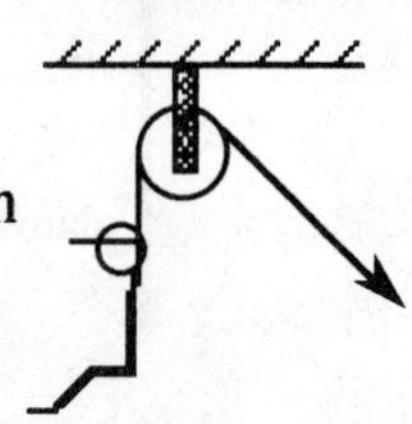

(a) $a = 0,\ v = 0$
(b) $a = 0,\ v > 0$
(c) $a = 0,\ v < 0$
(d) $a = 0.20\ \mathrm{m/s^2}$, up
(e) $a = 0.20\ \mathrm{m/s^2}$, down

Intermediate

The Applications (using Newton's laws)

Now we up the ante a bit. The new variations will include circular motion, force analysis combined with kinematics, hidden constraints, 2-D analysis, and multiple objects. This will often require selecting convenient coordinates and breaking force vectors down into components, but the essential technique is the same. You'll also want to start making sure that your numerical results make physical sense, if you haven't already been doing that.

F = ma Combined With Kinematics

Take the A-train! [All right, that one was even before my time.] But notice that the acceleration occurs as a variable in both methods, so acceleration is the link. N's 2nd won't tell you anything about time, displacement, or velocity directly, but if you get a from N's 2nd, you can go to the kinematics equations to find that information. But be careful. The constant acceleration equations are only good if a is constant, which in turn means that the net force must be constant. Alternatively, you may need to start with the kinematics equations to find a and then go to N's 2nd to find some unknown force or factor of a force.

2-D Analysis

The difference here is that you will have forces that are not just parallel and anti-parallel. You will employ standard vector techniques to handle this situation, but rather than adding the force vectors up at the start, you do the following:

- choose convenient coordinate directions to be reference directions
- for each force, get the components in each direction
- write N's 2nd for each direction separately
- if necessary, put vectors "back together" to find the net magnitude (Pyth. th'm) or angle (using the inverse tangent)

Circular Motion

Here you will find that one convenient direction for a coordinate axis is radially into the circle. We know the acceleration into the circle at any instant is given by $a_c = v^2/r$, where v is the magnitude of the velocity at that instant. We'll take a closer look in just a bit.

Multiple Objects

When there are 2 or more objects for analysis which are pushing on one another, write equations for each object separately. Often their motions are related – for example, they may have the same magnitude acceleration. Once you've written the equations, you're back to solve-mode and cranking the mathematical wheel. Again, we'll take a closer look in a bit.

Implicit ("Hidden") Constraints

Sometimes you may find that you've written the 2nd law equations, put in the correct expressions for the forces, and maybe even used the kinematics equations and you still can't get what you're looking for. What's missing is probably some condition that is implicit in the wording of the problem or the physical situation at hand. Those implicit conditions must be made explicit, i.e. they must be expressed mathematically as an equation or by saying $n = 0$, or whatever. You'll learn the common constraints when you work through the exercises.

General Method For Force Analysis

Once you have decided that a force analysis is an appropriate method (or worthy of a try), proceed through the following steps exactly as listed below.

Remember! A force is a push or pull on one object from another object. Thus it is necessarily an interaction between two objects.

[Setup]

1. **Free Body Diagram**
   - Select the **state (instant!)** that you are interested in.
   - Select the object/system **("receptor," *m*)** of interest.
   - If confused about the forces, **list the objects ("sources")** pulling or pushing on the receptor (i.e. all objects touching the receptor and all objects, such as the Earth, exerting non contact forces).
   - Draw a **simple schematic** of m (receptor) and show all of the forces on m as vectors ($\vec{\mathbf{F}}$'s).

2. **Choose Coordinate Directions**
   - The objective is to make the math as easy as possible by the following steps.
   - Choose one + direction to be in the **direction of the acceleration** ($\vec{\mathbf{a}}$).
   - If you need a second direction, choose it to be $\perp$ **to the first** and is appropriate to the problem. [This way you may have a_x = "a" and $a_y = 0$, $a_x = 0$ and $a_y = -g$, or something nice like that instead of $a_x = a\sin\theta$ and $a_y = -a\cos\theta$ ... yuk!]
   - If $\vec{\mathbf{a}} = 0$, either choose the reference directions according to how m will accelerate in other parts of the problem, or choose reference directions that make it easier to get the force components.

3. Get Force Components

- **Leave numbers out** until the 2nd law equations are completely set up!! Don't worry if you don't know a variable, you don't have to solve for the variables first. You'll get to that in the "solve" step.
- Quickly draw the following table:

Forces $\vec{\mathbf{F}}$	x-components $(0, +, -?)F_x$	y-components $(0, +, -?)F_y$

- **Column 1**: very quickly (5 sec.!) list all the forces that you see on your free body diagram. Just put in the vector variables without further ado.
- **Columns 2 & 3**: take each force one at a time and get the components in the directions you chose to be the coordinate directions.
  - * 1st: check if each component is **0** or in the + or – direction.
  - * 2nd: use the "Getting Components" subroutine as follows
    - Draw the **vector** that you're working on.
    - Choose/sketch the **coordinate directions.**
    - Draw a **GRID** (parallel to coord. directions and through the tip and tail of the vector).
    - Draw in the **components**: ***from*** the ***tail*** along one grid line, then draw an arrow head, make a 90° turn and go along the next grid line ***to*** the ***tip***, and draw the arrow head of that component.

4. N's 2nd Law Equations Constructed

- I strongly recommend writing these down exactly as I do (vertically), keeping the forces (causes) on the left and the "ma"s (effects) on the right. Once constructed, you can algebraically mix the equations up as needed.

$$\sum F_x = ma_x \qquad\qquad \sum F_y = ma_y$$

$$\underbrace{\text{from column 2}} = m\_\_\_ \qquad\qquad \underbrace{\text{from column 3}} = m\_\_\_$$

- Be sure you put in the correct signs for the components!!

– If you know that one or both accelerations are 0, then $m\,a$ is 0. Put that information in on the second line! If one direction is into the center of a circular path, then you know that the acceleration takes the form "v^2/r."

5. **SOLVE**

– Put a check mark over the known quantities in your equations.
– Circle the unknown variables.
– Check that you have enough equations to solve for the unknowns. If not, do you need to use kinematics, momentum, energy, or some implicit information?
– Now you can finally do what you've been dying to do!! Chug away!

6. **ANALYZE**

– **Reality Check**: Do your results make physical sense? Check extreme limits. Check units. Check your intuition! Compare magnitudes with the "How Big?!" sections.

¡ Fin !

Now practice until you don't have to think about what step comes next. When you can quickly look at a physical system and the forces involved, break it down in the mathematical manner outlined, and employ any math techniques (algebra, geometry, trig., calculus) necessary, you can call yourself a master of Newton's 2nd Law … well, until you get to upper division work!

N's 2nd: A Useful Alternate Form

As noted, $\sum F = ma$ has a cause = effect form. If, however, you know that $a = 0$, you can write this in a balanced forces form:

$$\sum F_{\text{in one direction}} = \sum F_{\text{in the other direction}}$$

For instance, when a book rests on a table and there are no other forces besides the normal force from the table and the weight from the Earth, you can say that $n = mg$ straight-away. This is rooted in the $\sum F = ma$ form, but may be faster in obvious situations. Just be careful. When in doubt, draw the free-body and go through all of the steps in the analysis.

Problem Solving Tip: *Don't look for the answer, the answer will find you. More Zen? No, I just mean that it is common to fixate on getting the answer to the degree that one loses sight of the process. With the answer-person frame of mind, these problems are going to look insurmountable. With the detective's frame of mind, you work steadily through the clues and apply the method step by step. Each step is a manageable chunk. Before you know it, the solution is either staring you in the face or you've uncovered some missing information that you need to obtain. So, proceed by asking the answerable questions given in the method above.*

EXERCISE 4/5.20: More difficult free-body diagrams.

(a) A box rests on a moveable incline. A spring is attached to a wall and the incline. The box/incline system is pushed so as to compress the spring and then released. For a state while the spring is expanding, draw a free-body diagram for the box and one for the incline. Are these diagrams applicable to all later times? Explain.

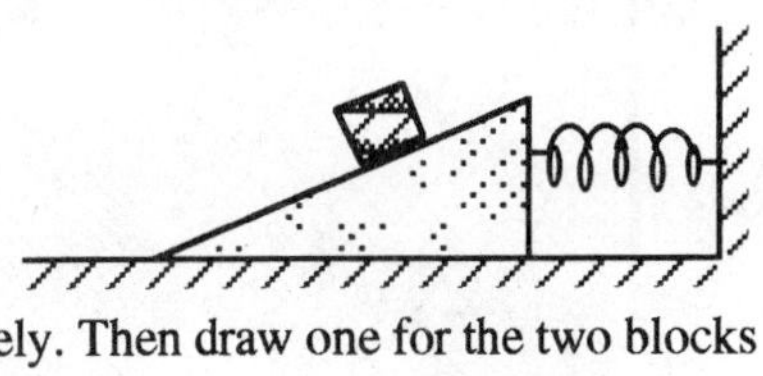

(b) Draw free-body diagrams for each box separately. Then draw one for the two blocks taken as one object.

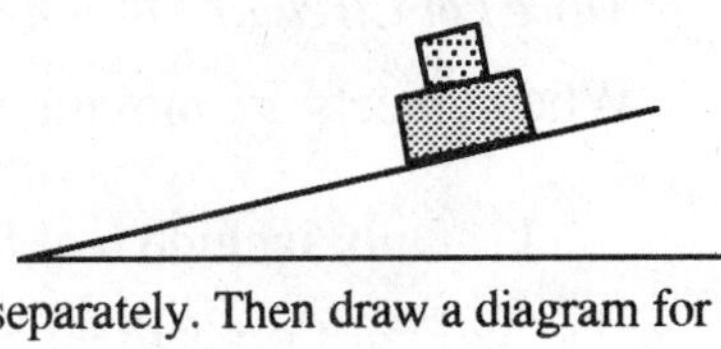

(c) Draw free-body diagrams for the box and ball separately. Then draw a diagram for the two treated as one object.

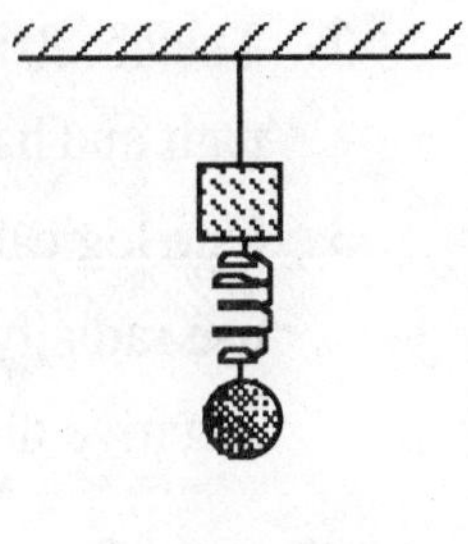

Choosing coordinate directions based on the *direction of the acceleration* is often an essential ingredient in doing a force analysis. If there is no acceleration, look to the *directions of the forces* for guidance. The objective is to make the mathematics of the problem easier. This step needn't take long at all, but you should be able to reason the choice out for yourself. Then you can move on to getting the components of the forces.

EXERCISE 4/5.21: Choose convenient coordinate directions. First base your choice on the direction of the acceleration. If the acceleration is clearly zero, base your choice on the direction of the forces. For cases where there is more than one object of interest, find the relationship between the accelerations of each object.

(a)

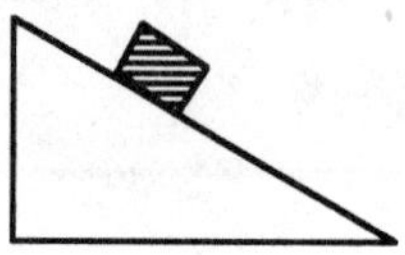

(b)

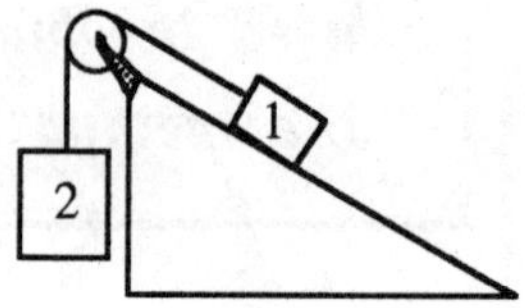

(c)

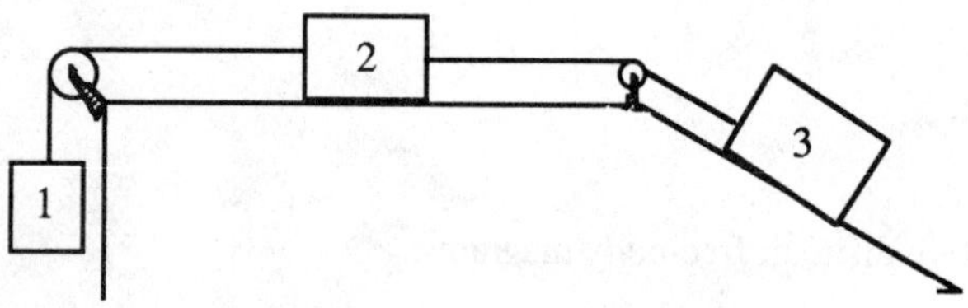

(d) Ball rolling on an incline with a changing slope.

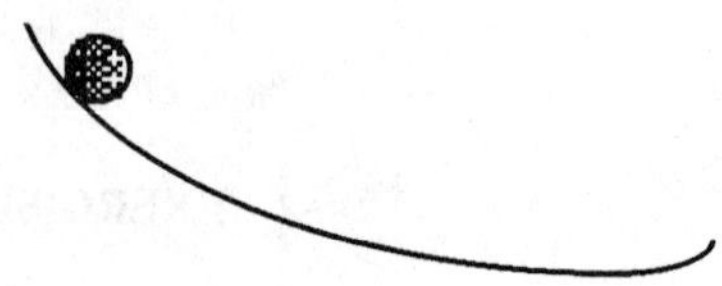

(e) A vertical pole (perhaps a telephone pole) in the ground with two taut wires attached.

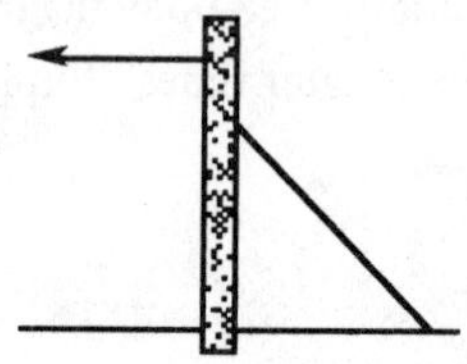

More On Circular Motion

When objects are moving in circular motion, keep in mind the following:

1. Only include real forces (from some source object) in your free-body diagram. Don't draw in an outward force if there is nothing pulling outward!
2. The acceleration vector for changing direction is in towards the center of the circular path and has a magnitude of $a_c = v^2/r$ (which can also be written $\omega^2 r$).
3. One logical coordinate direction is then radially into the circle. (Some people like to use radially outward. That's okay too, if you then say that the acceleration is in the negative direction.)

4. Sometimes additional coordinate directions are necessary. Two common convenient choices are:
   (i) tangent to the circle (where $a_{tangential}$ may or may not be zero), and/or
   (ii) the vertical direction (since that's the direction of the gravitational force)
5. Remember to take a snapshot to write N's 2nd law. Hence, the coordinate directions may change from moment to moment (state to state).

EXERCISE 4/5.22: Draw the free-body diagrams, note the acceleration directions, and determine the most convenient choices for the coordinate directions.

(a) Block on a table top, attached to a rope, and moving along a horizontal, circular path :

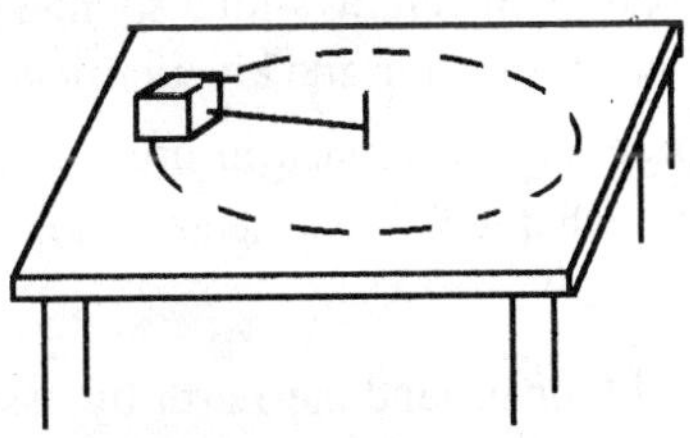

(b) Block sliding along a vertical, circular track:

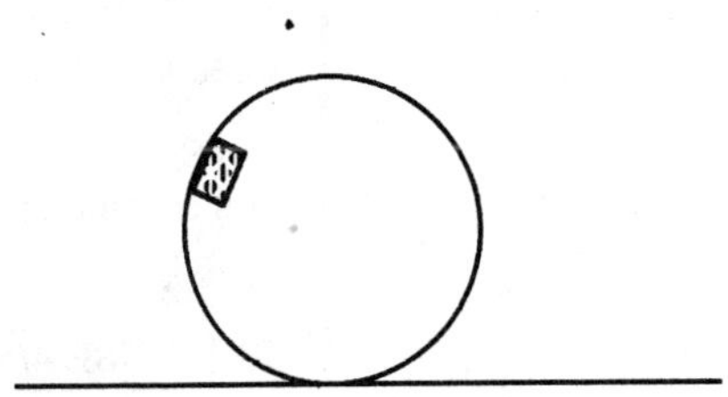

(d) Ball rolling on the inside of a cone. Assume that the speed & height remain constant.

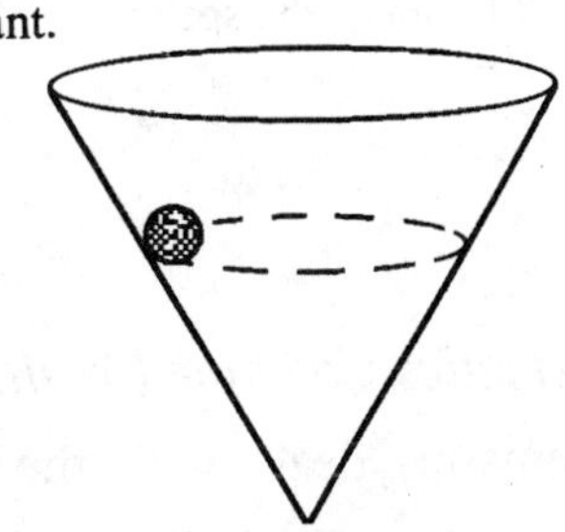

Now work a few from start to finish. Follow the steps carefully and don't worry about solving until you're at that step. If you get stuck, try to verbalize or write down what exactly is bothering you. You may surprise yourself and work through your confusion! That's what it means to be like a detective.

EXERCISE 4/5.23: Find the magnitude and direction of the force that the ground must exert on the 220 kg pole in order for the pole to be in static equilibrium. [See the Hint section.]

$T_1 = 440$ N

$T_2 = 330$ N

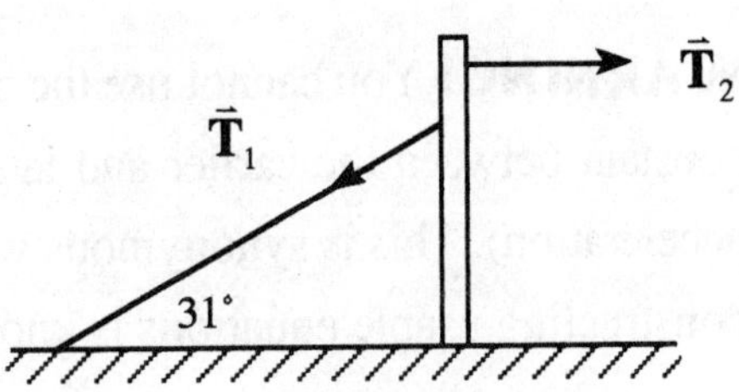

EXERCISE 4/5.24: Kym ($m_K = 52.2$ kg) and Jason ($m_J = 63.3$ kg) are ice skating. They face each other and Kym pushes Jason's chest with a force of 24 pounds. Find the acceleration of each.

EXERCISE 4/5.25: Find the acceleration of each block and the forces of block 1 on block 2 and block 2 on block 3. The coefficient of kinetic friction is $\mu_k = 0.123$, and the masses are $m_1 = 11$ kg, $m_2 = 33$ kg, $m_3 = 22$ kg. Assume that the blocks are sliding from the external push of 160.5 N. [Do you see how this could be very useful to a neurosurgeon?]

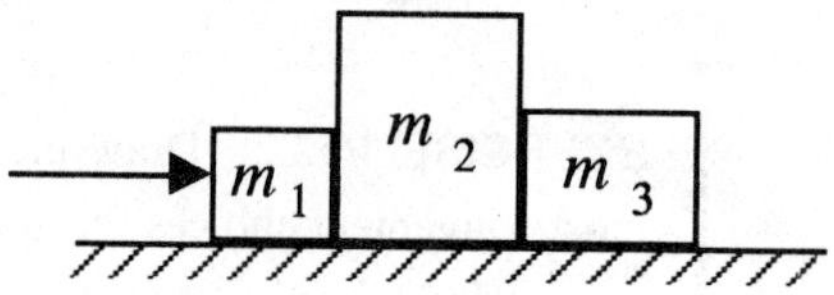

EXERCISE 4/5.26: While cleaning the bathroom ceiling with a sponge ($m = 0.045$ kg), you push in a direction 25° from the vertical. The coefficients of static and kinetic friction are $\mu_s = 0.41$, and $\mu_k = 0.21$, respectively. (a) With what minimum force must you push in order to get the sponge moving? (b) If you continue pushing with that same force, what will be the acceleration of the sponge? [See the Hints!]

EXERCISE 4/5.27: Find the speed of the Moon around the Earth by assuming a circular orbit. Get the quantities that you need from a table in your text. What fraction of the speed of light (3×10^8 m/s) is this? Estimate the speed in mi/h. You might also try this for the speed of Earth around the Sun!

Kinematics Combined With N's 2nd Law

Kinematics deals with the concepts of time, position and displacement, velocity, and acceleration. For 1-D motion, the list of variables is (t, x, v, a). Newton's 2nd uses the concepts of force, mass, and acceleration, so the 1-D list is (F, m, a). The algebraic connection is obviously the acceleration. There are two typical solution plans:

1. Find the acceleration via kinematics and use it to find a force, any factor of a force (such as the coefficient of friction), or a mass.

2. Find the acceleration from a force analysis and then use kinematics to find the time interval, displacement, or velocity.

WARNING! You cannot use the constant acceleration equations unless the acceleration is constant between the earlier and later states that you are relating (or you use the average acceleration). This is synonymous with saying that the net force must be constant. Part of constructing usable equations is knowing when the methods do and do not apply.

Constraints On The Motion

A few common constraints come up time and time again. You need to develop your physical intuition and observation so that you notice these sorts of things. Lack of awareness of a constraint is often the problem when someone says "I don't have enough information!" So here is a starter kit. Keep thinking physically!

In general, there are **data constraints** – for example, is $v = 0$, $a = 0$, or some force = 0? and there are **relationship/equation constraints** – like the conditions between linear and rotational motion when rolling without slipping (see chapter 3), or the conservation of string. How would you classify the following constraints?

- Movement *along a surface* restricts the possible directions for acceleration.
- $f_s = \mu_s n$: only used when the object is *at the breaking point*, i.e. about to slide. Physically determine the direction of f_S (to oppose the onset of motion) based on the other forces present.
- $\vec{\mathbf{n}} = \mathbf{0}$: Set the normal force equal to zero if given the condition that the object *just loses contact with the surface.*
- *Conservation of String*: (see page 168 of *Physics: The Nature of Things*)

  Objects attached to the same taut rope must move together with the same magnitude (not necessarily direction) of displacement, velocity, and acceleration.

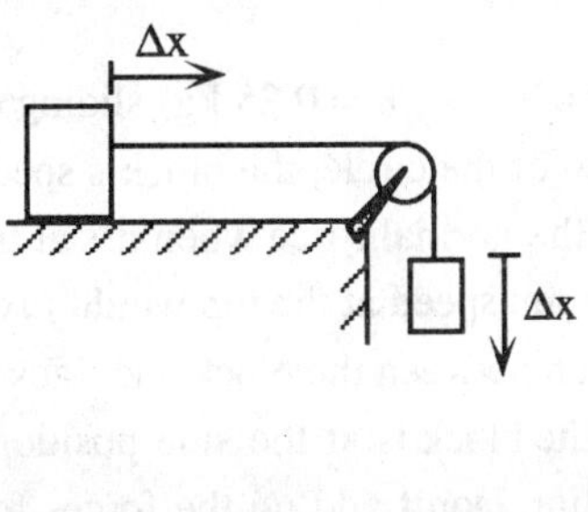

or

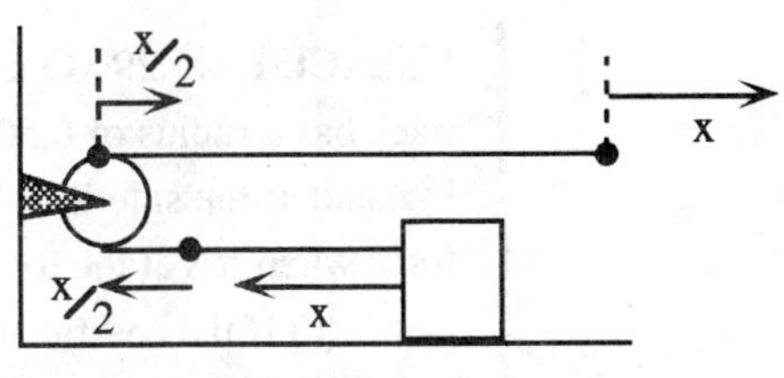

And now for an interesting and useful twist. ...let's see...

Objects attached to a free pulley – not to the rope itself – behave differently. If a rope goes around the pulley once, and one end is pulled, the pulley and object attached will move 1/2 as far, at 1/2 the speed and 1/2 the acceleration as the end pulled. If the rope is wrapped twice, as in the text fig. 5.25, the block moves 1/4 the distance, speed, and acceleration of the pulled end.

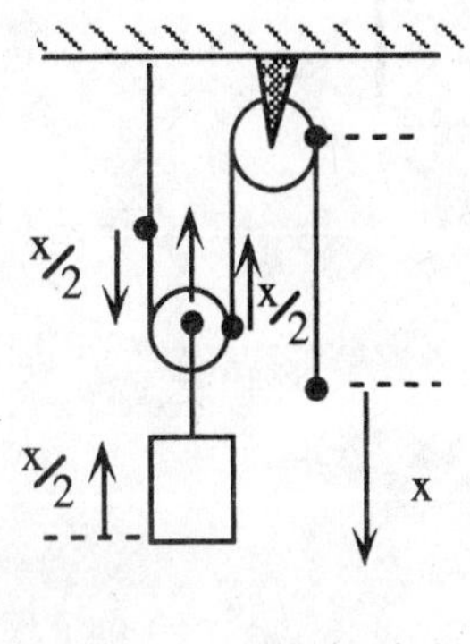

EXERCISE 4/5.28: Measure the coefficients of friction! Use the apparatus shown below. Assume that the pulley and rope are close enough to ideal for the accuracy desired, and that the blocks and table are made of rough wood.

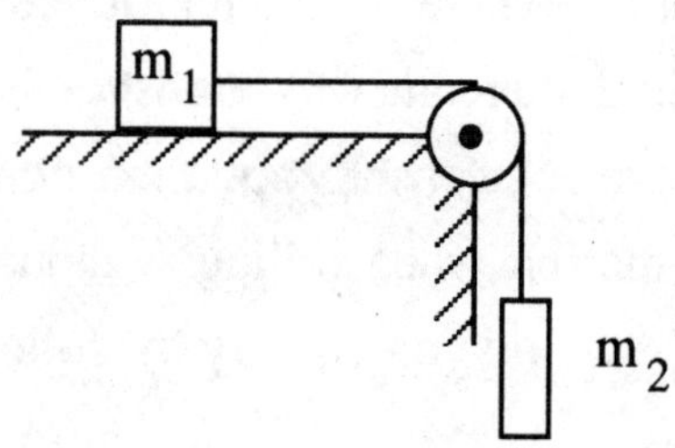

(a) If $m_1 = 1.0$ kg, and you find that the greatest mass that you can hang over the edge and still keep the system static is 2.0 kg, what is the coefficient of static friction between the wooden block (m_1) and the wooden table?

(b) Keeping the same m_1, you double m_2 and measure the acceleration of m_1 to be 7.4 $^m/_{s^2}$. What is the coefficient of kinetic friction between the wood block and the wood table?

(c) Bonus points: If the objects start from rest, how fast are they going after 1 second? How far has the block dropped?

EXERCISE 4/5.29: Once again, consider a block ($m = 0.25$ kg) sliding on a vertical circle. The circular track has a radius of 0.50 m. At the bottom of the circle, the block's speed is 6.0 $^m/_s$, at the top it is 3.0 $^m/_s$, and at the side it is 4.0 $^m/_s$. (a) Find the normal force when it's at the bottom. (b) Find the normal force when it's at the top. (c) What minimum speed at the top would *just* keep the block on the circular path? (d) If the coefficient of kinetic friction between the block and track is 0.10, what is the magnitude and direction of the *net* acceleration when the block is at the side position shown? What is the net force magnitude and direction at that instant? [Hint: Don't add up the forces for the last part. It's much easier than that!]

EXERCISE 4/5.30: The old block on an inclined surface! If the coefficient of kinetic friction is 0.15 and the slope of the incline is 24°, (a) what *vertical* height does the block drop in 2.0 seconds? and (b) how fast is the block going? Assume that the block starts with a velocity of 1.0 m/s.

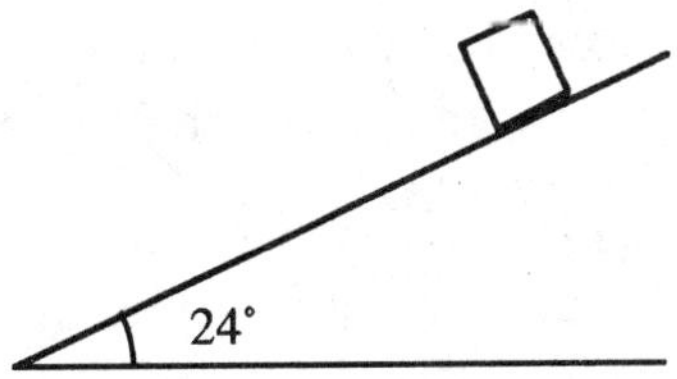

STUDY TIP: ***You've come a long way baby! So it's probably a good time to make your own short summary of what you've learned. It will help you pull the material together by forcing you to organize the information in your own way. After doing this, you might find yourself asking "Why didn't they just say this?" The answer is that summaries don't make much sense until you've gotten your hands dirty. They are also much more useful and comprehensible if you do them yourself.***

Quiz

1. True or False? "An apple is 1 m above the ground and feels a certain gravitational pull from the Earth. When you move the apple to a height of 2 m, the force is $1/4$ of that when it was at 1 m. This is because F_G is proportional to $1/r^2 = 1/(2r_o)^2 = 1/4r_o^2 = 1/4 \cdot 1/r_o^2$, i.e. the distance is doubled, so the force is decreases by a factor of $2^2 = 4$."

2. Two people on ice skates are at rest and standing face to face. They push off of each other and each moves away from the other. In this situation we can definitely say that:

   (a) the acceleration of each is the same during the pushing stage.
   (b) the accelerations are related by the ratio of their masses.
   (c) by N's 3rd law, the forces "cancel out." Therefore there must be an additional external force for them to separate.
   (d) because of their accelerations, during each successive 1 second time interval, they separate by greater and greater amounts. That is, Δx for the 1st second of separation is less than Δx for the 2nd, and so on.

3. True or False? "Initially, the distance between the centers of two baseballs is 1 m (roughly arm length). When the baseballs are moved closer so that the distance between their centers is 10 cm (roughly the width of your palm), the gravitational force of attraction increases 100-fold!"

4. Neglecting any effects due to the air, the diagram (ball swung above your head in a horizontal circle such that the string goes *up* from your hand to the ball) shows a situation that

   (a) is impossible: it disobeys N's 2nd
   (b) is only possible at low speeds
   (c) is only possible at high speeds
   (d) is easily achieved (especially if you have cowboy boots or your name is David)

5. True or False? "When you attach one end of a rope to your instructor's belt and then you pull on the other end of the rope, you are exerting a force on your instructor."

6. True or False? "Springs always exert forces given by the Hooke's law equation: $F_{spring} = -k\,x$."

7. A person on the "Rotor" (carnival ride in the text, fig. 3.14) experiences an outward force pushing their back into the cage (wall) that is

   (a) equal to zero
   (b) proportional to a_{rotor}
   (c) dependent on the radius of the rotor according to $1/r$
   (d) proportional to $\omega^2\, r$

8. True or False? "There is no gravitational force in a vacuum."

9. True or False? "In order for an object to keep moving, there must be a net external force on it."

10. True or False? "A pound of lead has more mass than a pound of feathers."

11. True or False? "High heel shoes are bad for floors because they exert a greater force on floors than flat shoes."

12. True or False? "First, an 8 N force from a spring and a 6 N force from rope are applied to an object. Then a single force of 14 N is applied by a person. Either way, the net effect will always be the same."

13. In the diagram below, the blocks have the same mass, the surfaces are assumed frictionless, and the pulley and rope are assumed ideal. The slopes of the inclines are different, but otherwise unspecified. Pick the statement which is unquestionably true.

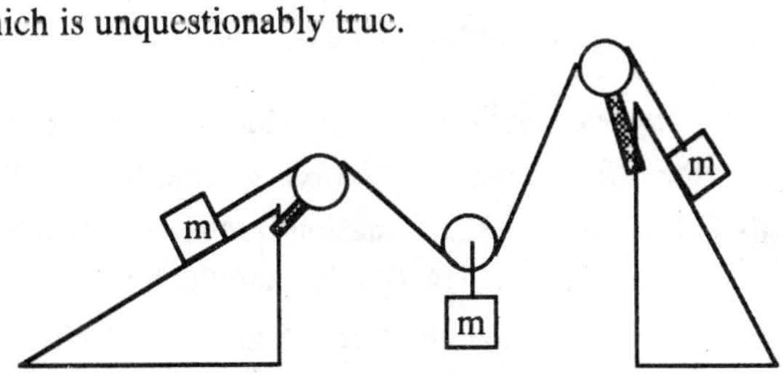

(a) Equilibrium is absolutely impossible.
(b) If one slope is 30° and the other is 60°, there will be equilibrium. Similarly, 20° and 70°, 40° and 50°, and so on will work.
(c) The system will always be in equilibrium.
(d) The tension in the rope is independent of the mass of the blocks.

14. Astrology in question: Using the average distances between the planets and the Sun, when the Sun, Earth, and Jupiter form a line, Jupiter is about 4.2 Au = 6.3×10^{11} m from Earth. How close would a person need to be to exert a gravitational force of the same order of magnitude as Jupiter? [Note 1 astronomical unit, Au, is the ave. distance between the Sun and Earth, i.e. 1.5×10^{11} m $\approx 10^8$ mi.]

15. What's wrong with the following free-body diagram?

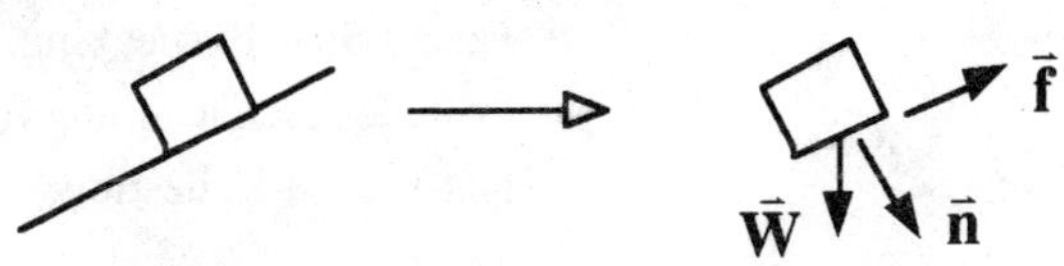

16. True or False? "The acceleration of the 100 N box will be the same whether you hang a 200 N box on the other side or you pull with a force of 200 N."

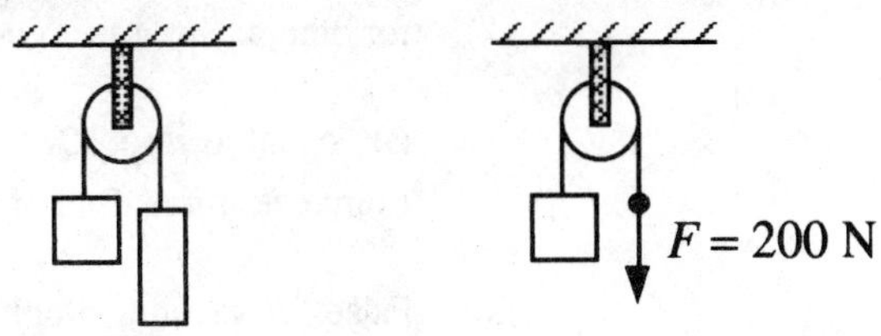

17. Object A weighs 10 N on Earth. Object B weighs 10 lb on Earth. Which has more mass?

18. Person A weighs 120 N on Earth. Person B weighs 120 N on the Moon.

(a) They have the same mass. (b) $m_A > m_B$ (c) $m_A < m_B$
(d) It's impossible to tell, since the laws on Earth differ from those on the Moon.

19. In which case of circular motion is there an *outward* (centrifugal) force?

(a) A car driving along a flat (horiz.) circular track.
(b) A yo-yo spun so that it makes a horizontal circle (as in lecture)
(c) A yo-yo spun in a vertical circle when at the bottom of the circle.
(d) Liquid in a test tube spun in a horiz. circle (centrifuge).
(e) both a & b
(f) both b & c
(g) all of a through d

Quiz Solutions

1. False! The distance above the ground doubled, but that is not the distance in the gravitational force equation. The distance there is the distance between their centers (of mass). Thus the distance to focus on has changed from (radius of Earth + 1 m) to (radius of Earth + 2 m). A pretty negligible change, wouldn't you say?

2. These are worth studying one by one.

   (a) "The acceleration of each is the same during the pushing stage." – By N's 3rd, the force on each is the same, so what if the masses are different? Different accelerations!

   (b) (TRUE). "The accelerations are related by the ratio of their masses." – An objects acceleration has a magnitude of F_{net}/m, and the forces are the same by N's 3rd.

   (c) "By N's 3rd law, the forces 'cancel out.' – Therefore there must be an additional external force for them to separate." But the forces act on different objects!

   (d) "Because of their accelerations, during each successive 1 second time interval, they separate by greater and greater amounts. That is, Δx for the 1st second of separation is less than Δx for the 2nd, and so on." – The force only acts during the pushing stage. After that we enter the coasting stage. Don't mix-up stages!

3. True. If the distance goes down by a factor of 10, the force goes up by a factor of 10^2.

4. (a) is impossible: it disobeys N's 2nd. Notice that there is a downward force on the object (gravity, AKA weight!) Now what is balancing this force? Nothing! The only other force also has a vertical component that is downward. Thus the object must fall until there is a force that can stop it. Hence the rope must angle downward so that the tension has an upward component.

5. False. Strictly speaking, the rope is pulling on them. This is a very important distinction that tends to elude students doing force analyses. Sure, you animate the process by pulling on the rope, but that's not the question.

6. False. This is an ideal model equation which works well for some types of springs and very poorly for others. But it is a good starting point in the study of springs.

7. (a) equal to zero. Or better yet, non existent. What object is pushing them outward? None. No source, no force. End of story.

8. False. A vacuum simply means there is no air or gas. This means no air or gas pressure, so objects with internal pressure may explode from the lack of external pressure to keep them in equilibrium. But gravity has nothing to do with air pressure! It is the attraction between two objects due to their masses. Furthermore, you can't block the gravity of one object with another object. If you could, you could do some pretty neat things!

9. False! A net external force will cause an acceleration (N's 2nd law). Otherwise, an object will remain at rest or keep coasting in a straight line at a constant speed (N's 1st). When the net force is zero we either have static ($v = 0$) or dynamic ($v \neq 0$) equilibrium ($a = 0$).

10. False. Same weight (assumed on same planet or moon) → same mass (the volumes, and hence densities, differ.

11. False. If the same person is wearing them, the same force is applied. (The pressure, i.e. force/area, is very different. The damage is caused by the force all being applied to a small area.)

12. False. The two forces may not be exerted in the same direction. What if they are anti parallel or not collinear?

13. Which one is totally supported in the vertical direction? Which of the remaining two is supported more? Remember, the surfaces are assumed to be frictionless. So how is equilibrium possible?

14. Use Newton's general law of Universal Gravitation: $F_{grav} = Gm_1m_2/r^2$. In order for the gravitational pull on you from Jupiter to be about the same as the gravitational pull on you from the person next to you, we need $m_{Jupiter}/(r_{to\ Jupiter})^2 \approx m_{person\ next\ to\ you}/(r_{to\ person\ next\ to\ you})^2$. Why? Because G and your mass are the same in both cases. If we use $m_{person\ next\ to\ you} = 75$ kg ($\approx$ 150 lb), and we look up the mass of Jupiter to be 1.9×10^{27} kg, we can solve to find $r_{to\ person\ next\ to\ you} \approx 13$ cm. So a person standing close to you pulls with about as much force as Jupiter! (Feel it? I didn't think so.) And what about the houses, buildings, trees, and so on? So what do you think? If astrology is a dependable system, do *you* think that gravity is at the heart of it? (I certainly haven't *proved* anything. But a bit of healthy (and open) skepticism is a good thing in my book. ... By the way, just how strong is that force anyway? There is an easy way to find out!

15. The normal force vector should be in the opposite direction.

16. False!! Ooooh, interesting, huh? Why do most people think that this is true? Because they, at least subconsciously, still perceive the object as *having* weight. Remember, weight is the pull from the Earth. So in the case on the left, with the second mass attached, there is an *external* force of 200 N applied by the Earth. In the case on the right, without the second mass, there is also an external force of 200 N applied. In which case is there more mass to accelerate? Yes. With the extra mass, the system is harder to accelerate. Thus the acceleration is greater for the case on the right (no 2nd block).

17. They must have the same mass.

18. (c) $m_A < m_B$ If you brought B back to Earth they would weigh about 6 times as much ($\approx$ 720 N $\approx$ 180 lb) as on the moon. Person A weighs 120 N $\approx$ 30 lb on Earth. (Person A is a baby!) "Weight" depends on the thing exerting the gravitational force.

19. (c) A yo-yo spun in a vertical circle when at the bottom of the circle. What is the source of the real outward force? The Earth! The Earth pulls the yo-yo down with a gravitational force (weight). When the yo-yo is at the bottom, the weight is out of the circle; when it is at the top, the weight is into the circle; when it is halfway between the top and bottom, the weight is tangent to the circle. That's about all I have to say about that.

Exercise Hints & Solution Plans

EX. 4/5.5: "% error" is used to compare two numbers. One of the numbers is the reference number. It is either known exactly, or is considered to be the better known of the two. The other is the number that we are checking – "How good is the approximation?" "How accurate was the measurement?" I'll call one number the "reference," and the other the "approx." Take a look at these three ways of citing how much one number differs from another:

- Difference: (approx.) – (reference)
- Fractional Error: $\dfrac{(\text{approx.}) - (\text{reference})}{(\text{reference})}$
- Percent Error: $\dfrac{(\text{approx.}) - (\text{reference})}{(\text{reference})} \times 100\%$

For all three ways a positive result implies that the approx. is greater than the reference, while a negative result implies that the approx. is less than the result. Further note that the latter two are generally more informative than the first. A difference in speed of 2 m/s is enormous for a snail and negligible for a car. Try it out. For a snail use approx. = 2.1 m/s , exact = 0.1 m/s. For a car use approx = 30 m/s and exact = 32 m/s. Notice that the differences are the same while the % errors are vastly different.

EX. 4/5.13: [Hint] First look at all of the possible friction forms and magnitudes of each. Can you see which friction should be used in each case? Sure, I knew you could.

EX. 4/5.22: [Solution Plan] For force analyses, you will do well to follow the "general method for force analyses" solution plan step by step. Notice that key questions that you should ask yourself are about the mass ("What's the receptor?"), the acceleration ("What direction is the acceleration, and what direction is the acceleration zero?"), and the forces ("What are the sources of the forces?" Draw the forces on a free body diagram. And get the components of the forces.) These key ingredients are easily remembered since they are the variables in $F = ma$.

EX. 4/5.25: [Hint] The solution for part (a) is strange. It will take some thinking to figure out what is going on. Most importantly, make sure that you are confident in your setup. Then take a peak at the solution if you can't see how to interpret the result. For part (b) try $F = 4.0$ N and see if you get a reasonable result.

Exercise Solutions

EX. 4/5.1: It *seems* like the book wants to fall more. But that is viewing the world through Aristotelian eyes. It is much more powerful to view this as Newton does. Can you make the leap? In the one worldview, the book's own internal nature is the cause of falling, while in the other worldview it is an *external* agent pulling on the book that causes the fall. (Which view is Newton's?)

EX. 4/5.2: No! With only one force, there is no other force (or forces) to balance it. Thus $\vec{\mathbf{F}}_{net} \neq 0$ and the object must accelerate in the direction of $\vec{\mathbf{F}}_1$.

EX. 4/5.3: Since it's static for an extended period, $\vec{\mathbf{a}} = 0$. This can only be so if $\vec{\mathbf{F}}_{net} \neq 0$.

Receptor: "object"

Sources: your hand, the Earth

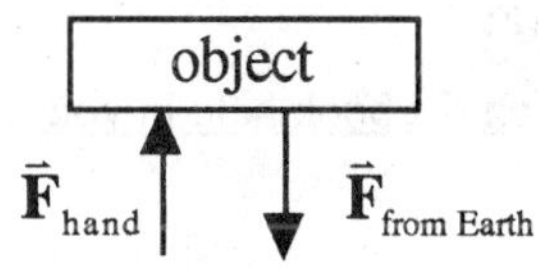

EX. 4/5.4: You know that you've accounted for them all when you include all of the forces from objects touching each other (contact forces) and from the Earth (gravity) ... OK, electric, magnetic, weak, and strong forces could be considered, but we'll leave those alone for now.

EX. 4/5.5: [See Hint for definition of % error.]

(a) I'll use 160 lb. Then 160 lb ≈ 160 ($^1/_4$ N) = 640 N. More carefully,

$$160 \text{ lb} = 160 \text{ lb}\left(\frac{1 \text{ N}}{0.224 \text{ lb}}\right) = 714 \text{ N (or 710 N, to 2 sig fig's)}$$

$$\text{Percent Error} = \frac{640 \text{ N} - 710 \text{ N}}{710 \text{ N}} \times 100\% = -9.9\%$$; The estimate is low, but just within 10%.

(b) Roughly, 1 kg weighs 9.8 N = 9.8 ($^1/_4$ lb) = 2.5 lb. More carefully, 9.8 N = 9.8 (0.224 lb) = 2.2 lb.

EX. 4/5.6: The apple pulls the Earth toward it with a force of 1 N (≈ $^1/_4$ lb), by N's 3rd law. However, a 1 N tug on the Earth won't give the Earth much of an acceleration, whereas a 1 N tug on the apple will give it an acceleration greater than any race car. (In fact, the apple accelerates at 9.8 $^m/_{s^2}$.) Why the difference in accelerations from the same magnitude forces? Because the mass of the Earth is about 6 x 10^{25} times greater than that of the apple. In an exaggerated view we can draw this as below.

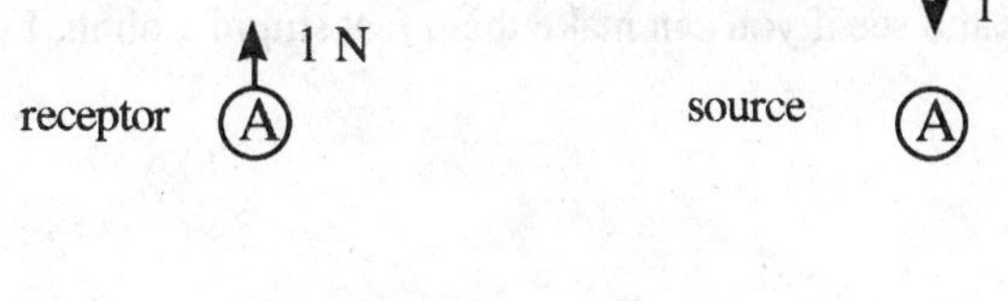

EX. 4/5.7:

(a) Strictly speaking, *you* did not hit the fly, the newspaper was the source of the force on the fly. You animated the process by pushing on the newspaper. In many cases, this distinction is critical.

(b) The newspaper receives a force form the fly. Hence the fly is the source of the force on the newspaper – not that the fly *wanted* to hit the newspaper!

(c) Equal and opposite!

(d) Not equal!! First, the fly has less mass than the newspaper and so will tend to be affected more. Second, the fly only feels the force from the paper, while the paper also feels the force from your hand holding it. Thus, even if the paper would recoil from the fly hitting it, your hand keeps this from happening. In sum, different masses and different (other) forces acting on each object.

EX. 4/5.8:

- Forces on the book held in your hand:

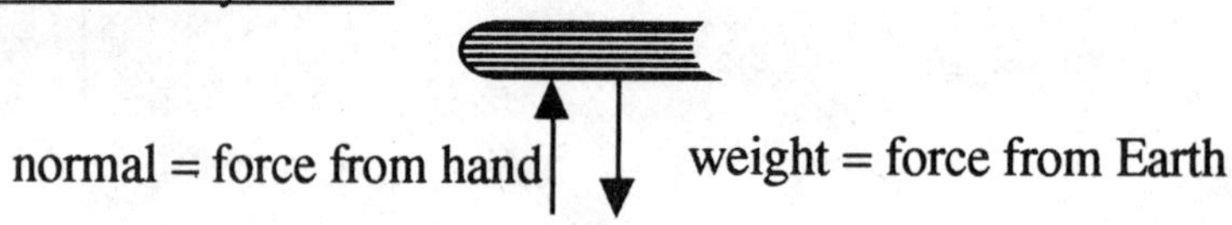

- Are these a N's 3rd law pair? NO WAY, JOSÉ!!! Why not? After all, they certainly are equal and opposite forces. But rememberrrrrrrr, N's 3rd is NOT simply about any old equal and opposite forces!! The two forces ON THE BOOK are equal and opposite by N's 2nd law – they balance each other out. Did you catch that? The normal force from your hand and the weight of (really "on") the book both act on the same object, they are NOT from the same interaction. Rememberrrrrr, the two forces of any N's 3rd law pair necessarily act on different objects: the force on A from B is equal and opposite the force on B from A! The subtlety is important, elusive, and the point of this exercise. Let's take a closer look.

- N's 3rd Law Pairs:

  Receptor: book

  Sources of forces on book: hand, Earth ... nothing else is touching the book!

Forces ON BOOK		Force ON SOURCE
$F_{\text{normal, on book from hand}}$	⇐ 3rd Law Pair ⇒	$F_{\text{normal, on hand from book}}$
$F_{\text{gravity, on book from Earth}}$	⇐ 3rd Law Pair ⇒	$\mathrm{F}_{\text{gravity, on Earth from book}}$

Try some of your own to make sure that you really do understand Newton's 3rd Law. Then go quiz other people and see if you can make them feel stupid...ohhh, I was just kidding!

EX. 4/5.8:

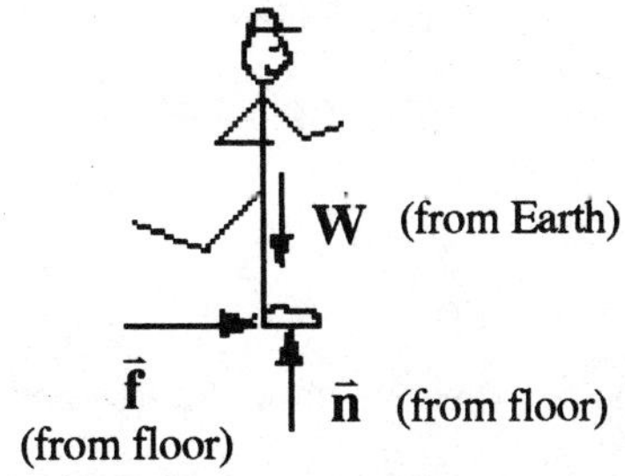

Bob animates the process. His foot pushes backward and the floor pushes forward. The Earth pulls Bob down into the floor and the floor resists by pushing up. Notice that the floor pushes both up and forward. Also notice that these two forces are different types.

Although the forces between the floor and Bob are equal and opposite, it is the forces *on Bob* that move him. The forces on the floor would move it, except that the floor is attached to the Earth. Got it? **The equal and opposite forces of N's 3rd act on different objects and so usually result in very different effects.**

EX. 4/5.9:

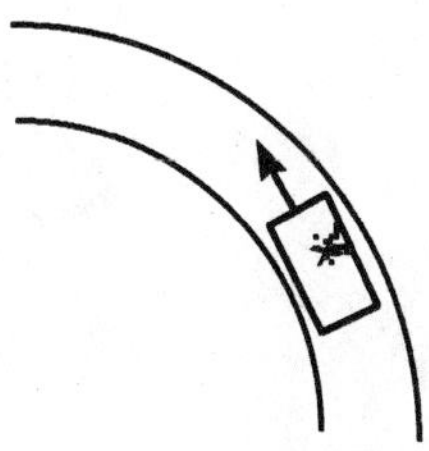

source	force
Earth	weight (down)
seat	$\vec{n}_1$, (up)
back of seat	$\vec{n}_2$, (forward)
car door	$\vec{n}_3$, (into circle)

The forces on you *must* be from non contact forces (gravitational, electric, magnetic, weak, or strong) or contact forces. The predominant non contact force is clearly from the Earth pulling you down toward it ("gravity"). Now what's touching you? The seat pushes up from below and forward from behind. (Feel it?) Additionally, the car door is probably touching your right shoulder and pushing you to your left, i.e. into the circle causing you to turn.

EX. 4/5.10:

(a) If we neglect air resistance, then the only force on an object in free fall is the force of gravity, AKA "weight." This is true at every instant and every point of the path. Thus, no matter what the velocity or position, the free-body diagram is as shown.

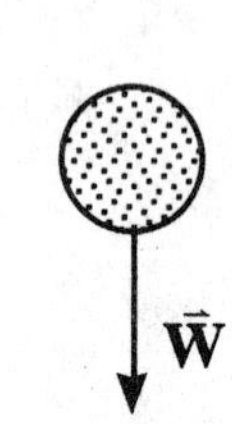

(b) In both cases neither the speed nor direction matter when we draw the free-body (force) diagram. The only forces on the pendulum bob are tension from the string and gravity from the Earth.

(i)

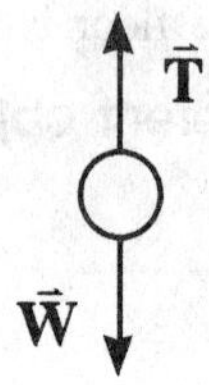

(ii)

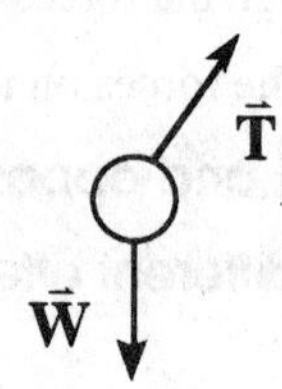

(c) Here the direction of the velocity *does* matter, since the direction of the friction force depends on it. It is instructive to look at all possible cases.

(i) Not Moving

Moving Left (←)

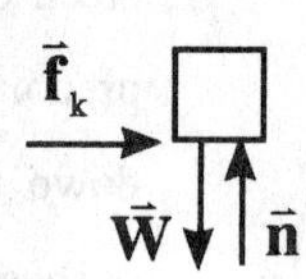

Moving Right(→)

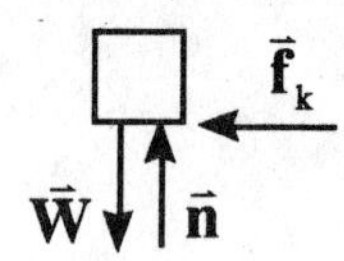

(ii) Notice a few subtleties in the diagrams below. First, the normal force vector is drawn above the box for convenience. We know that the surface is really touching the bottom of the box, but that's OK. Second, I've drawn the weight force in a place that is convenient for finding the components. Physically, we might prefer to put it at the center (of mass) of the object. Third, notice when static friction is used and when kinetic friction is used.

Not Moving

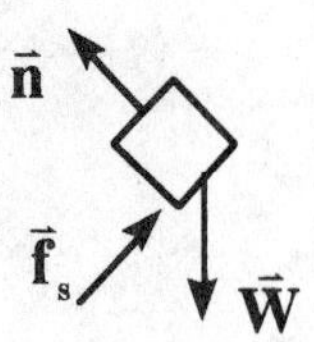

Moving Down

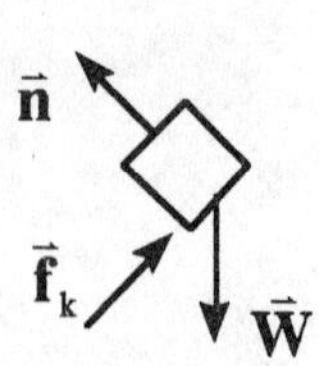

Moving Up

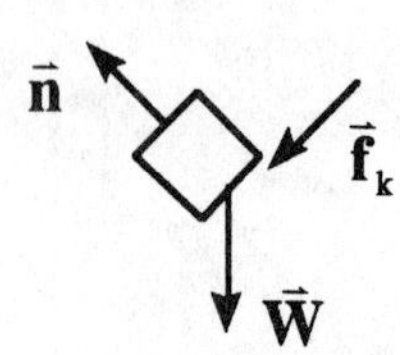

(d) Basta pasta! ("Basta" means "enough" in Italian.) So where's the force of gravity from the Earth and what's touching the sign? That's all you need to worry about. Does the clock pull on the sign? No, not directly! That can be very important in a force analysis. We'll get to that later.

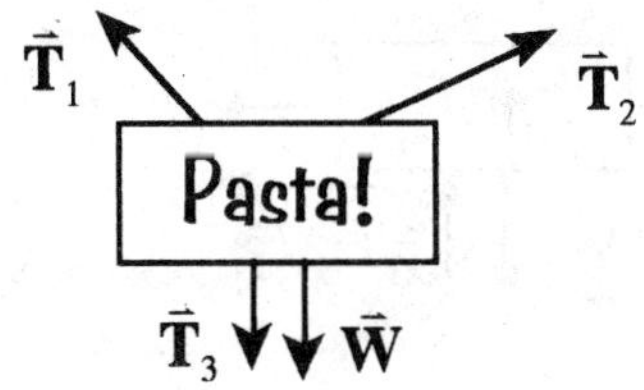

(e) A different type of force! The water "buoys" the ball up with a "buoyant force." And even though the ball would rise up if you let go, gravity is still acting on the ball. Notice that the force of your hand + the force of gravity must be exactly equal in magnitude to the buoyant force.

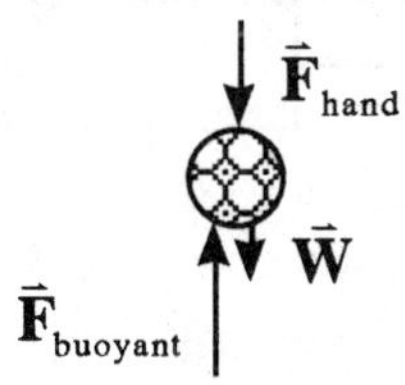

EX. 4/5.11:

(a)

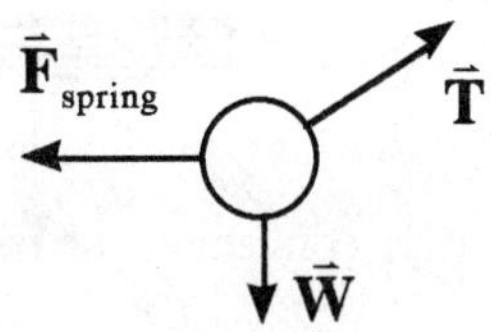

(b) The wire can only push in directions ⊥ and parallel to it. Since the wire is frictionless, we have:

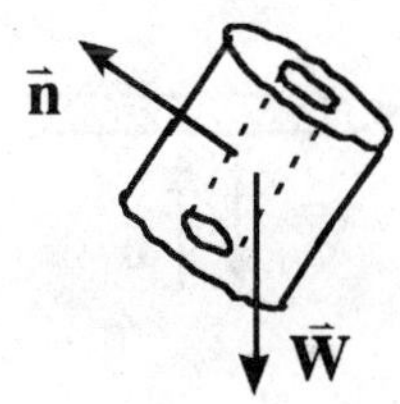

(c) For an ideal rope and pulley system, the tension is the same everywhere in the rope.

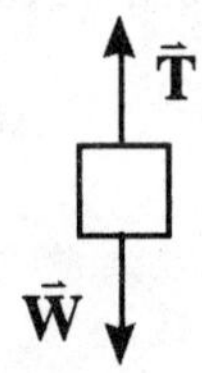

(d) Does $\vec{F}$ pull on the block on the right? No! Be careful only to include those things touching the receptor of interest. Notice that the tension for the same rope is the same magnitude at both ends, but may be different in different ropes.

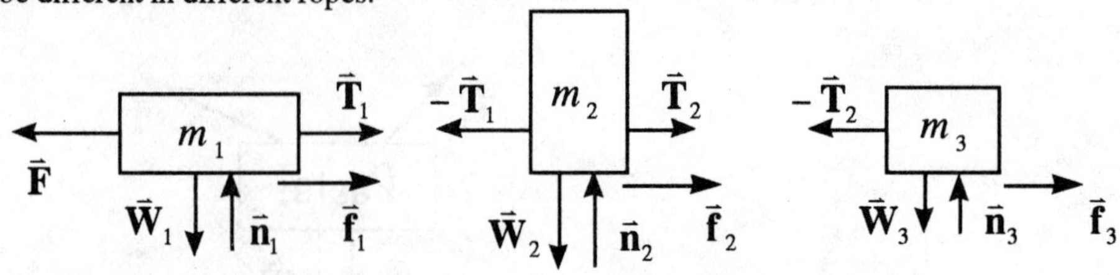

If we treat the three blocks (and two ropes) as one object (like a train), we don't include the "internal" forces (here the tensions) on the free-body diagram. The dashed line is the imagined boundary of the "single" object.

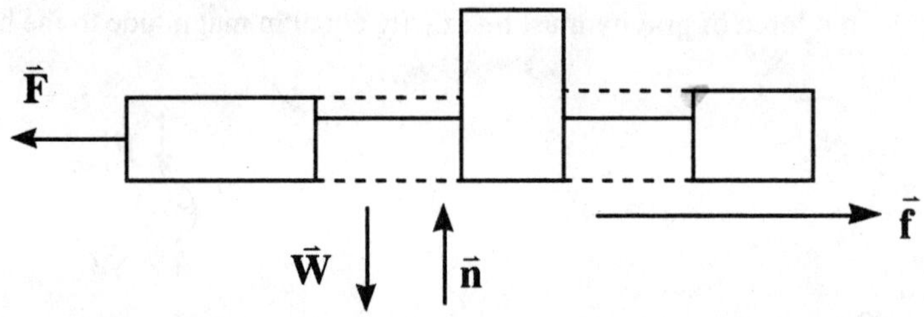

The total weight is the sum of the weights, the total normal force is the sum of the normal forces, and the total friction is the sum of the frictions.

(e) Notice the equal and opposite forces.

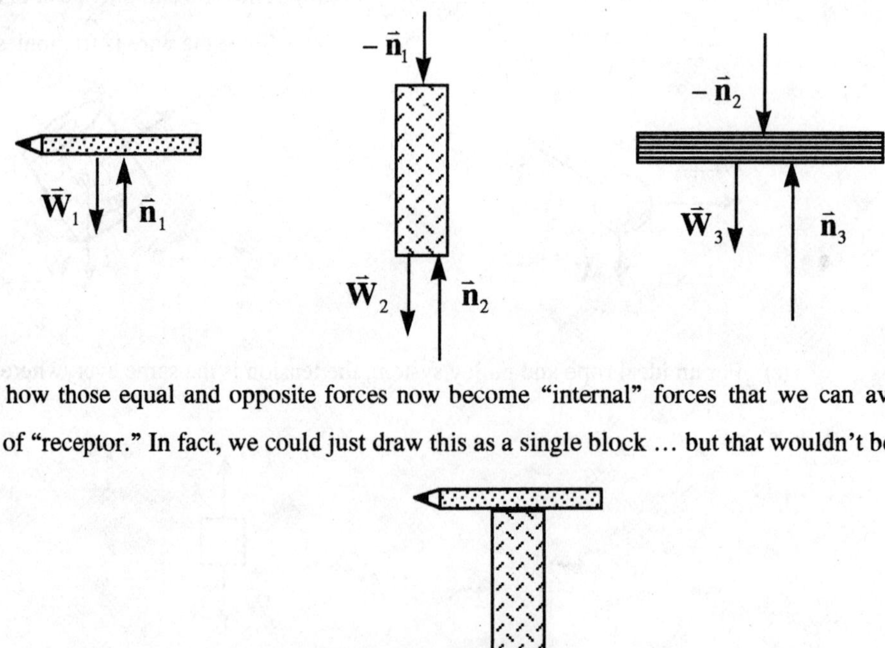

Notice how those equal and opposite forces now become "internal" forces that we can avoid by this choice of "receptor." In fact, we could just draw this as a single block … but that wouldn't be as pretty.

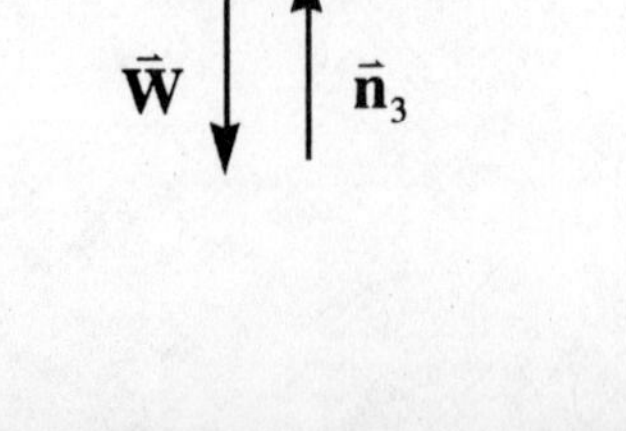

EX. 4/5.12: You could do this one in your head, but it's good to take a close look when you are first learning this stuff. Pay attention to the method so that you can apply it to more difficult problems. First, let's take a look at the free-body diagrams.

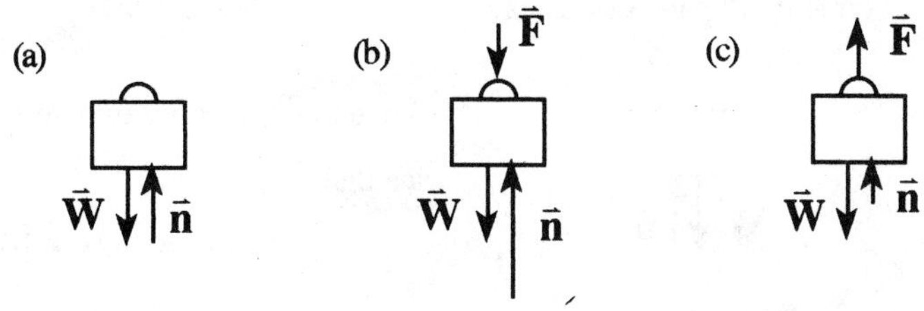

My drawings already show insight into the problem, but I could have just drawn in the vectors without concern for length and let the math guide me to this awareness. Next, let's take a look at the problem by carefully using the 2nd law template: $\sum F = m\,a$.

- Data: Given: Hidden: Want:

Given	Hidden	Want
$m = 10.0$ kg	$a = 0$ (Make sure you see why.)	$n = ?$
$F = 40.0$ N		

- What is the magnitude of the weight? 10.0 kg? NO NO NO. Mass ≠ weight. $W = m\,g = (10.0\text{ kg})(9.80\,) = 98.0$ N (For a quick check of the weight in newtons, just multiply the kilograms by 10.)

- Acceleration: It should be clear that $a = 0$ in each case. 40.0 N isn't enough to lift a 98.0 N object.

- Choose coordinates: I'll choose up to be positive. There is no need to look at the horizontal direction, since there is no horizontal force component.

- Newton's 2nd law template:

(a)

$$\overbrace{\sum F}^{n - W} = m\underline{a}_{0}$$

$$n = W$$

$$= 98.0\text{ N}$$

(b)

$$\overbrace{\sum F}^{n - W - F} = m\underline{a}_{0}$$

$$n = W + F$$

$$= 138\text{ N}$$

(c)

$$\overbrace{\sum F}^{n - W + F} = m\underline{a}_{0}$$

$$n = W - F$$

$$= 58.0\text{ N}$$

Them Signs, Oh Them Signs: I could have written "$n + W = 0$" and "$n + W + F = 0$," and then put in –98.0 N for W, and ± 40.0 N for F. That is, I could "suppress" the sign. There are advantages to *not* putting in the sign right away. For starters, if you don't know the direction of a variable, putting it in as positive will allow you to discover the direction by finding out the sign in the end. In the case above, we clearly knew the directions and I wanted to emphasize when the forces were adding to or reducing the normal force required to balance the block.

EX. 4/5.13: [SETUP] A nice approach here is to look at what is the same in all cases and look at the four possible forms of the friction force at the start. First, I'll choose the coordinates to be positive in the horizontal-right, and vertical-up directions in all cases. Second, notice that vertical forces are the same in *all* parts of this exercise:

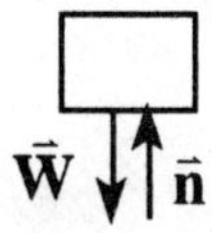

Since there are only two vertical forces, it is easy to see that:

$$n = W = 4.00 \times 10^2 \text{ N}$$

And third, I'll evaluate the possible friction forces:

$$f = \begin{cases} f_s = 0, \text{ with no other forces parallel to the surface} \\ f_s = \ldots \text{ it depends on the other forces or components that are parallel to the surface} \\ f_{s,\,max} = \mu_s n = (0.200)(4.00 \times 10^2 \text{ N}) = 80.0 \text{ N, at the breaking pt.} \\ f_k = \mu_k n = (0.100)(4.00 \times 10^2 \text{ N}) = 40.0 \text{ N, when sliding over a surface} \end{cases}$$

Now that we have set this problem up nicely, we can quickly go through each part.

(a) Not moving, no forces with components parallel to the surface → $f = 0$.

(b & c) Not moving, before the breaking point. The vertical forces have been done, so we only need:

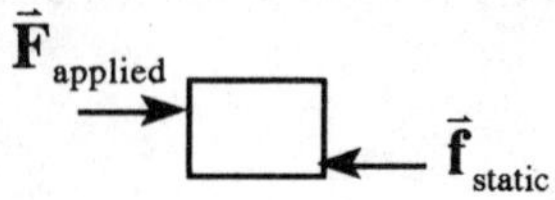

$$\overbrace{F_{applied} - f_s}^{\sum F_x} = m\underline{\overset{a_x}{0}} \Rightarrow f_s = F_{applied}$$

Thus, (b) $f_S = 20.0$ N, and (c) $f_S = 40.0$ N

(d) This is at the breaking point since this is the maximum force that the friction can supply to resist the onset of motion. Thus the block is not moving/just about to move and $f = f_{S,\,max} = 80.0$ N.

(e & f) Object is sliding since $F_{applied} > f_{S,\,max}$. Thus $f = f_K = 40.0$ N.

Finally, by placing a 2.00×10^2 object on top of the first box, we increase n to 6.00×10^2 N, which in turn increases $f_{S,\,max}$ to 120 N and f_K to 60.0 N. The friction for each part then becomes: (a) $f = 0$; (b-d) $f_{S,\,max} = F_{applied}$; (e) Breaking pt., $f = f_{S,\,max} = 120$ N; (f) $f = f_K = 60.0$ N.

There, you've mastered friction!!

EX. 4/5.14: Remember? The force diagram doesn't depend on the velocity when in free fall. So (b) through (f) are all the same. (Don't draw the velocity vector...or at least don't confuse it with a force vector.)

$$\sum F = ma$$

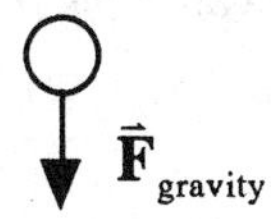

$$\overbrace{F_{gravity}} = ma$$

$$mg = ma \quad \Rightarrow a = g \quad \text{at all times while in free fall .}$$

But what about part (a)? If the object is just dropped, then the acceleration *before* release is 0. The object holding the ball exerts a force which balances the force from the Earth. If the ball is thrown, then it accelerates from the applied force from whatever projects it.

EX. 4/5.15: When air resistance is negligible, why is it that objects with vastly different masses fall with the same acceleration? (Drop a pen and a book at the same time, from the same height!) Let's see if this fact, which seems odd to most people at first, can become intuitively clear ... or at least understood. Recall that the force of gravity (expressed as mg or GMm/r^2) is proportional to the mass of the object being pulled. This is illustrated by the following

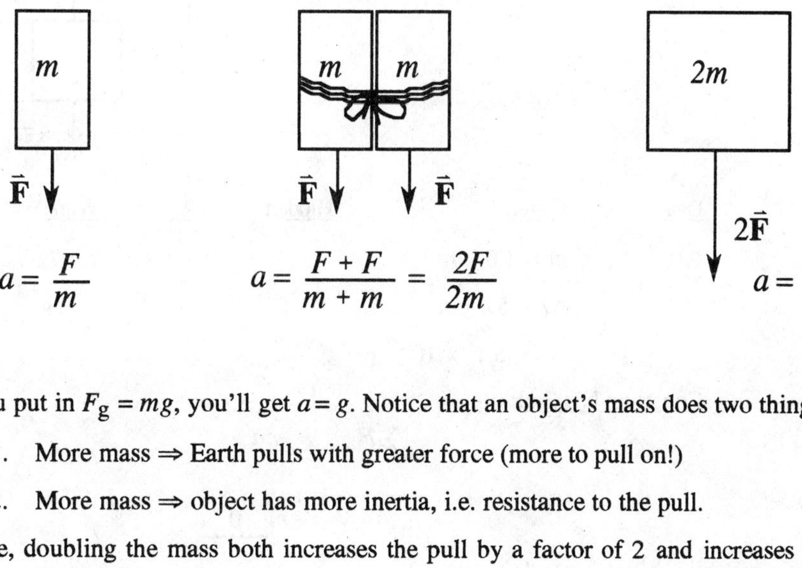

$$a = \frac{F}{m} \qquad a = \frac{F+F}{m+m} = \frac{2F}{2m} \qquad a = \frac{2F}{2m}$$

If you put in $F_g = mg$, you'll get $a = g$. Notice that an object's mass does two things here.

1. More mass $\Rightarrow$ Earth pulls with greater force (more to pull on!)
2. More mass $\Rightarrow$ object has more inertia, i.e. resistance to the pull.

Hence, doubling the mass both increases the pull by a factor of 2 and increases the resistance by a factor of 2. The net result is that all objects (neglecting air resistance) fall with the same acceleration. This dual role of mass (1. grav. force, 2. inertia) is known as Einstein's equivalence principle. (It could have been that there were different properties responsible for each role.)

EX. 4/5.16: From Ex. 4/5.6, we know that the apple pulls on the Earth with a force equal and opposite to the force on the apple from the Earth...OK, OK, Ex. 6 isn't the law, it's N's 3rd.

i. Force on apple from Earth when near the surface of the Earth (AKA "weight") = $m_{apple}\, g$ = (0.1 kg)(9.8 $^{m}/_{s^2}$) = 0.98 N.

ii. Using Newton's force of universal gravitation between any two objects, we need the mass of each (m_{apple}, M_{Earth}), the distance between their centers (≈ radius of the Earth = R_{Earth}), and the universal gravitation constant (G).

$$F_G = G\frac{M_{Earth}\, m_{apple}}{R_{Earth}^{\ 2}} = \left(6.67\times10^{-11}\,\mathrm{N\cdot m^2/kg^2}\right)\left(\frac{(5.98\times10^{24}\mathrm{kg})(0.1\ \mathrm{kg})}{(6.38\times10^{6}\ \mathrm{m})^2}\right)$$
$$= 0.98\ \mathrm{N}\ !!!$$

EX. 4/5.17:

[MODEL] Assume that we have an ideal spring so that $F_{sp} = k$ (amount stretched or compressed).

[SETUP]

Free Body Diagram With Coord. Choice:

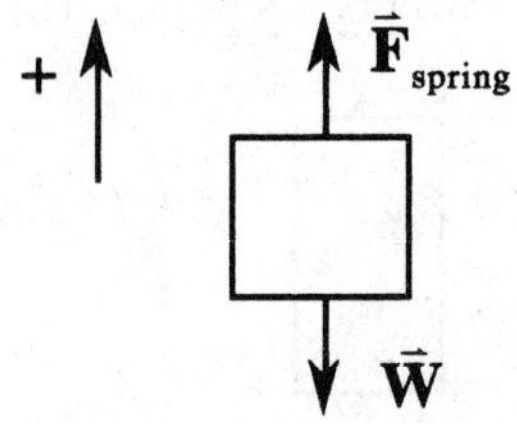

Data:	Given	Hidden	Want
	m = 1.0 kg		k = ?
	Δx = 5.1 cm		
	= 5.1 x 10^{-2} m		

Construct N's 2nd:

$$\sum F = ma$$
$$\overbrace{F_{sp} - W} = m\ \underline{0}$$

[SOLVE]

$$F_{sp} = W \Rightarrow k\,\Delta x = m\,g \Rightarrow k = \frac{m\,g}{\Delta x} = \frac{(1.0\ \mathrm{kg})(9.8\ \frac{\mathrm{m}}{\mathrm{s}^2})}{5.1\times10^{-2}\ \mathrm{m}} \Rightarrow k = 190\ \frac{\mathrm{N}}{\mathrm{m}}$$

[ANALYZE] This is a good way to measure the spring constant of a spring. You must, however, be careful to take a number of measurements to determine whether the spring is linear or not (i.e. whether you get the same k for different Δx's). Also notice that it is easy to remind yourself of the unit for k – since $k\ \Delta x$ must be in newtons, k must be in N/m.

EX. 4/5.18:

[MODEL] Assume an ideal pulley and rope. Then T is the same at both ends of the rope.

[SETUP]

Free Body Diagram With Coord. Choice:

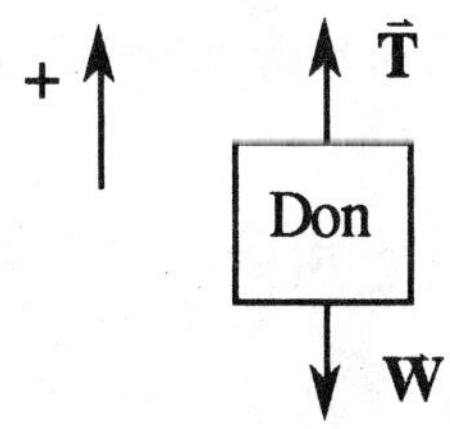

WARNING! Do not make the all too common, but sorely mistaken, assumption that $T = W$. This is only true if $a = 0$. If there is an acceleration, these are NOT NOT NOT equal.

Data:	Given	Hidden	Want
	$m = 1.00 \times 10^2$ kg		$T = ?$

Construct N's 2nd:

$$\sum F = ma$$

$$\overbrace{T - mg} = ma \Rightarrow T = ma + mg = m(a + g)$$

NOTE: The velocity does NOT matter! If $a = 0$, the tension is the same whether moving at constant velocity or not moving. What matters is the acceleration.

[SOLVE] Now it's easy!

(a - c) $a = 0 \Rightarrow T = 9.80 \times 10^2$ N

(d) $a = +0.20$ m/s^2 $\Rightarrow T = 10.0 \times 10^2$ N

(e) $a = -0.20$ m/s^2 $\Rightarrow T = 9.60 \times 10^2$ N

[ANALYZE] When $a = 0$, the tension does equal the weight ($T = mg$). When the acceleration is up, T must be greater than mg. When the acceleration is down, T must be less than mg to allow Don to downward acceleration.

EX. 4/5.19: Draw in the force of gravity and the contact forces and you're done.

(a) No matter what the motion is, the free-body diagrams will be the same, although the magnitudes of the forces may change. Exception: if and when the box losses contact with the incline, $n_1 = 0$. Notice that the block on the incline does not directly fell the force from the spring, so the spring must not be on block's free-body diagram.

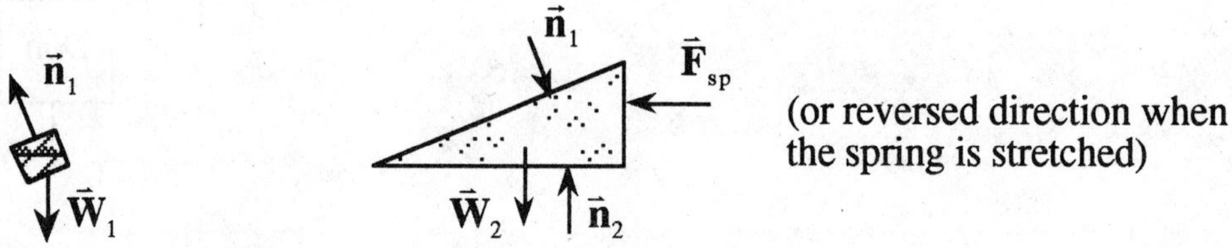

(b) Notice that the internal force, here n_1, is not considered when you treat the objects as one object.

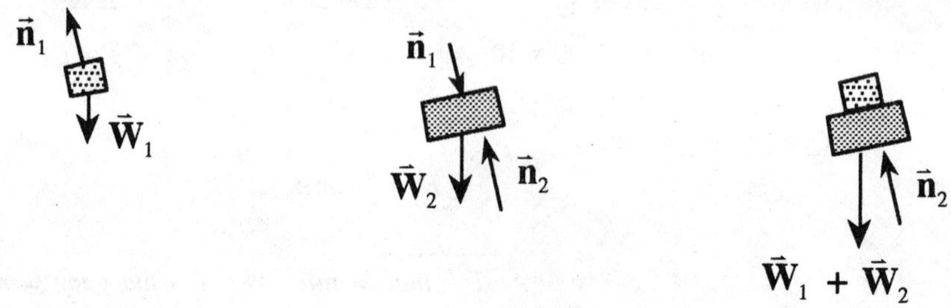

(c) Again, we only include forces from direct contact with the object (receptor)...and the gravitational force. Here we could jump to the conclusion that $F_{sp} = W_2$ and $T = W_1 + W_2$, since $a = 0$.

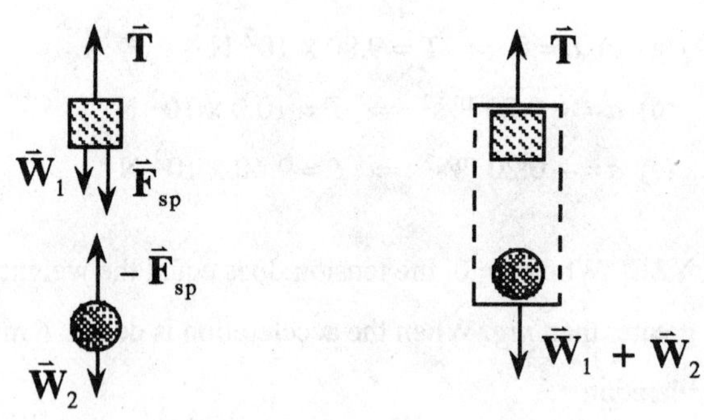

EX. 4/5.20:

(a) The acceleration is either down the inclined plane or 0 if the friction is large enough. (In terms of the forces, there is one vertical force, the weight, one force perpendicular to the surface, the normal force, and one force along the surface, the friction. Thus with the chosen coordinate system, there is only one force that has components in both the x and y directions.

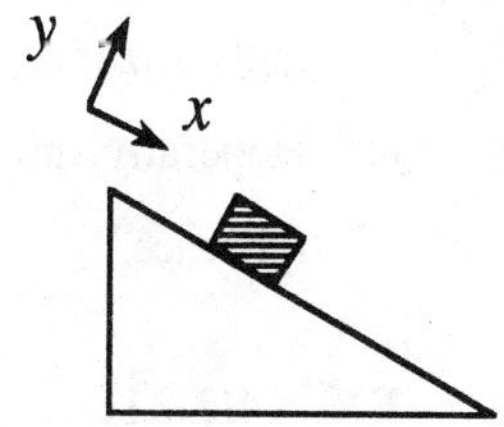

(b) Notice that, when the larger (hanging, "2") block moves down, the smaller block moves up the incline. Further notice that a taut rope requires that the speed and the magnitude of the acceleration of the two blocks are the same. For this reason you might choose:

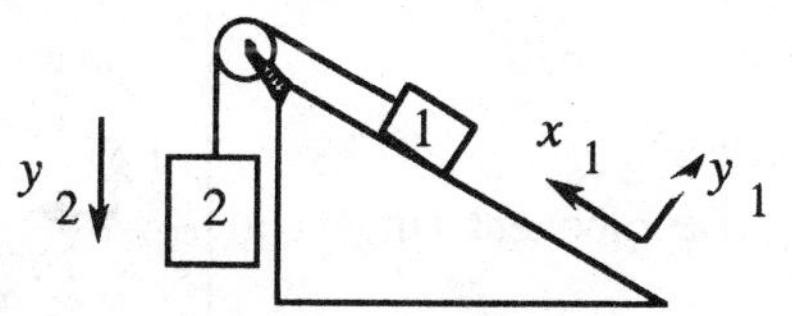

With this choice we have

$$a_{1,x} = +a_{2,y} = \text{"}a\text{"}$$

We could also choose:

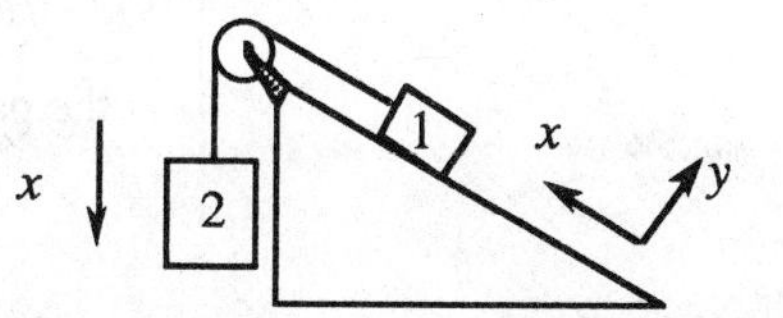

This is my favorite. Then I just write

$$a_1 = +a_2 = \text{"}a\text{"}$$

If you thought that object 2 would go up and 1 would go down the plane, you might want to reverse the directions. If it turned out that you guessed wrong, you would find $a < 0$. You could also make the following choice, but I'm not sure why you would want to overlook the fact that the motion of the two blocks are tied together.

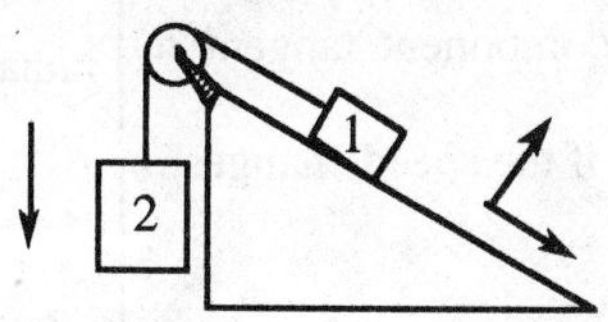

Yuk. You would then relate the acceleration variables by $a_1 = -a_2$. This increases the likelihood of making a sign error, but it'll work if you're careful.

(c) I would guess which way the blocks will move if I knew the masses and angle. Guessing that the top block moves to the right(assuming block 3 is more massive than block 1), I'll choose

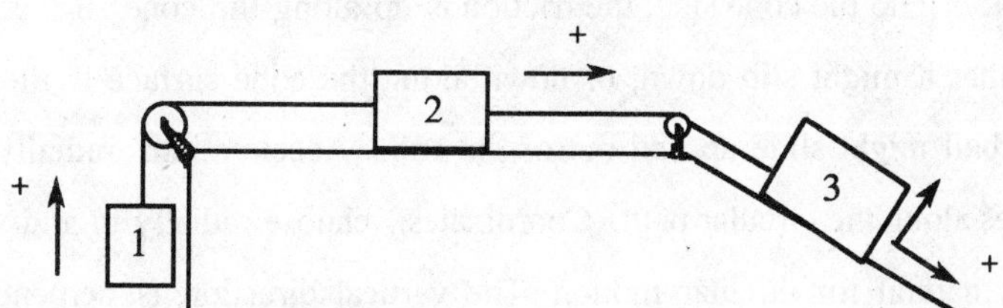

Here we have:

$$a_1 = a_2 = a_3 = \text{"}a\text{"}$$

(d) The acceleration here is a bit more difficult to describe than the others. We have the acceleration component parallel to $\vec{v}$ for the changing speed, and the acceleration component into the radius of curvature for the changing direction. And since the slope changes, so does the acceleration vector. As for the forces, the normal force ($\vec{n}$) changes direction (it is ⊥ to the surface), while the gravitational force ($\vec{W}$) is always vertically down. ... Might as well choose the old standby of vertically up for $+y$ and horizontal and right for $+x$.

(e) Hopefully, no acceleration here. Again, I'll choose the vertical and horizontal directions for my reference.

EX. 4/5.21:

(a)

<u>Free-body</u>

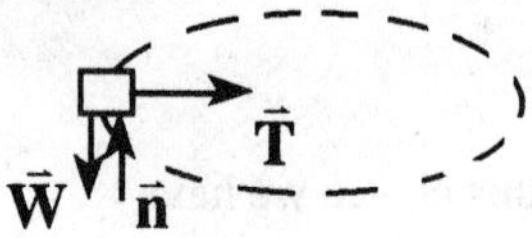

If there is friction, it is kinetic and tangent to the path, opposite the motion.

<u>Accel.</u>

$\vec{a}_r$: radially in for changing direction.

<u>With friction</u>:

$\vec{a}_{tangent}$: a component tangent to the path if the speed changes.

<u>Coord.'s</u>

Without friction: choose radially in and vertical

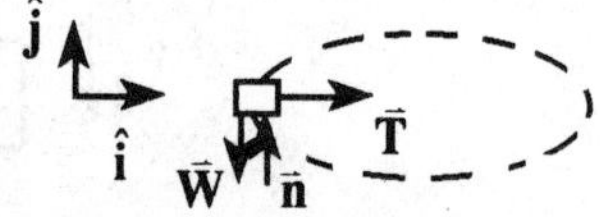

With friction: choose tangent to the path to be a third direction.

(b)

<u>Free-body</u>

(Assuming block moves ccw.)

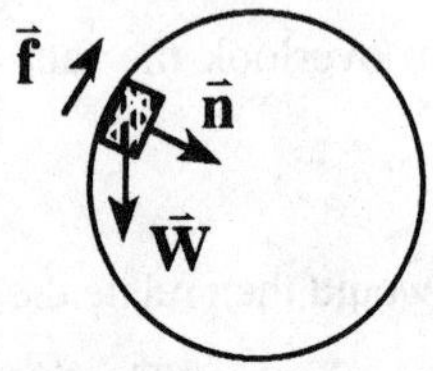

<u>Accel.</u>

$\vec{a}_r$: radially in for changing direction.

<u>with friction</u>:

$\vec{a}_{tangent}$: a component tangent to the path if the speed changes.

<u>Coord.'s</u>

Choose radially in and tangent to the path. Notice that the vertical direction is not always ⊥ to the radial direction

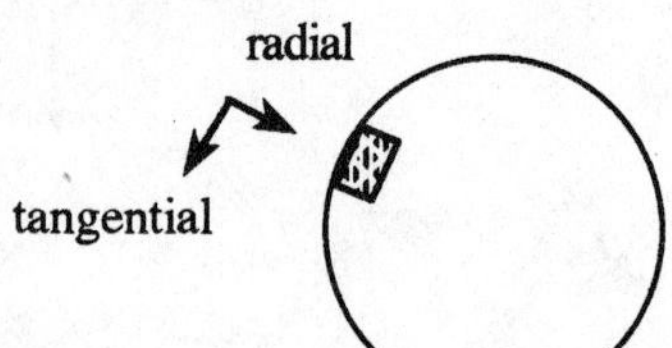

(c) I'll describe this one and you draw it. Free-body: the weight is vertically down; the normal force is up at an angle so that it is ⊥ to the cone side; the friction is up along the cone surface if the ball is going slowly enough so that it might slip down, or down along the cone surface if the ball is rolling fast enough so that the ball might slide up and out of the cone. Acceleration: radially in due to changing direction as it moves along the circular path. Coordinates: choose radially in and vertical. The radially inward direction is natural for circular motion. The vertical direction is perpendicular to the radial direction in this case.

EX. 4/5.22:

[MODEL] "Equilibrium" ⇒ $a = 0$, so $F_{net} = 0$, $F_{net,x} = 0$, and $F_{net,y} = 0$. "Static" ⇒ $v = 0$.

The ground is assumed to be able to push in any direction., thus we write the ground's force as the vertical and horizontal components H_y and H_x, respectively. (I used "*H*" for lack of a more compelling choice.)

- Diagram: See the problem statement and the free-body diagram below.

[SETUP]

- Data:

Given	Hidden	Want
$m_{pole} = 220$ kg	$a_x = 0$	$\vec{\mathbf{H}}$
$T_1 = 440$ N	$a_y = 0$	
$T_2 = 330$ N		
$\theta = 31^\circ$ (angle shown)		

- Object of interest: the pole
- Free-body Diagram: I will guess the directions of H_x and H_y. It should be obvious that H_y is correct. If the direction of H_x is incorrect, I will get H_x = a negative number. (The negative tells me I guessed the wrong direction.)

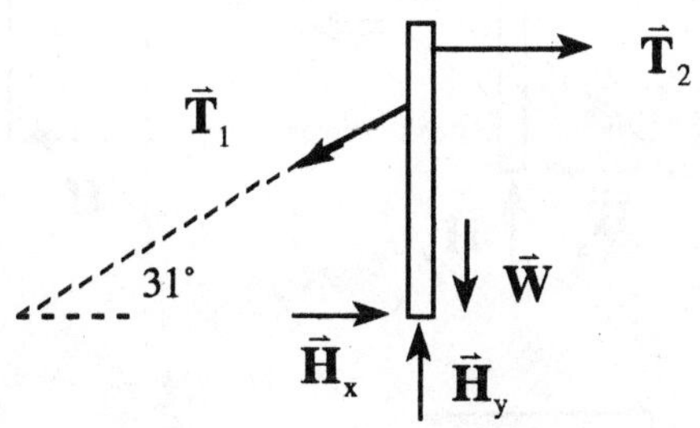

- Constraints: none really, just that it is in static equilibrium.
- Accel.: $a = 0$ (since static)
- Coordinate Axes: Might as well choose the vertical (↑) and horizontal (→) directions. $a_x = 0$, $a_y = 0$.
- Force Components: Start by quickly filling in the first column with the forces on your free-body diagram. Here we have $\vec{\mathbf{W}}$, $\vec{\mathbf{H}}$, $\vec{\mathbf{T}}_2$, $\vec{\mathbf{T}}_1$ (Do the easiest ones first to get rolling!)

Force	F_x (include the sign)	F_y (include the sign)
$\vec{\mathbf{W}}$	0	$-W = -mg$
$\vec{\mathbf{H}}$	$+H_x$	$+H_y$
$\vec{\mathbf{T}}_2$	$+T_2$	0
$\vec{\mathbf{T}}_1$	$-T_1 \cos 31^\circ$	$-T_1 \sin 31^\circ$

NOTE: In the component table of any force analysis, there are usually some vectors that only have one component and one or more that need to be broken down into two components. For vectors that need to be broken down, as $\vec{\mathbf{T}}_1$ does above, it is usually a good idea to make a separate sketch so that you can make sure that you get the angles correct. I've skipped this step because the components of $\vec{\mathbf{T}}_1$ are pretty easy to see in this case.

- Construct Equations With N's 2nd: Just add the columns in the force component table and put in what you know about the acceleration components.

$$\underbrace{\sum F_x}_{H_x + T_2 - T_1\cos 31^\circ} = ma_x \qquad\qquad \underbrace{\sum F_y}_{H_y - W - T_1\sin 31^\circ} = may$$

$$H_x + T_2 - T_1\cos 31^\circ = m\cdot 0 \qquad\qquad H_y - W - T_1\sin 31^\circ = m\cdot 0$$

[SOLVE]

$$\Rightarrow \; H_x = T_1\cos 31^\circ + T_2 = +710 \text{ N} \qquad\qquad \Rightarrow \; H_y = mg + T_1\sin 31^\circ = +2400 \text{ N}$$

Sketch $\vec{\mathbf{H}}$: Since you have the components of $\vec{\mathbf{H}}$, you can make a quick sketch to see the component triangle and the direction of $\vec{\mathbf{H}}$. This sketch is NOT to scale, but it does show that $H_y > H_x$.

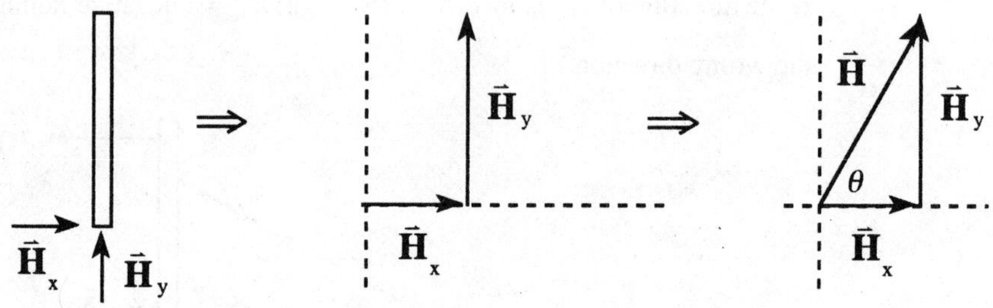

So, $H = \sqrt{H_x^{\,2} + H_y^{\,2}} = 2500 \text{ N}$, and $\theta = \tan^{-1}\left(\frac{H_y}{H_x}\right) = 74^\circ$

[ANALYZE] Let's see. If $T_1 = 0$, it makes both conceptual and mathematical sense that $H_x = T_2$ and $H_y = W$. The result we found is $H = 2500$ N ≈ 625 lb. The weight of the pole is mg = (220 kg) (9.8 m/s²) ≈ 2200 N, so we expect H to be a bit more, and it is. Looks OK to me.

EX. 4/5.23: Hey, something looks familiar here. Hmmm, could there be a method to the madness?

[MODEL] Neglect friction from ice. The acceleration occurs only when she is pushing (so that $F_{net} \neq 0$). After the acceleration stage, there is a coasting stage which is *not* a part of this problem.

- Diagram: (Hey, it works!)

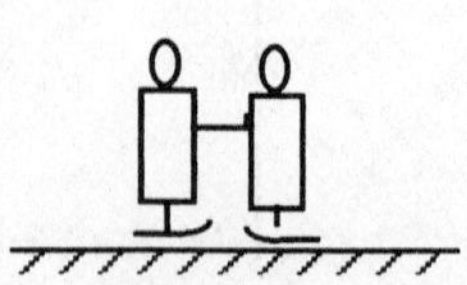

[SETUP]

- Data:

Given	Hidden	Want
$m_K = 52.2$ kg	$\vec{\mathbf{F}}_{\text{K on J}} = -\vec{\mathbf{F}}_{\text{J on K}}$	$a_K = ?$
$m_J = 63.3$ kg		$a_J = ?$
$F_{\text{K on J}} = 24$ lb		
$= 24\ (4.448\ \text{N})$		
$= 107$ N		

- Objects of interest: Kym and Jason taken separately.
- Free-body Diagrams: One for each object.

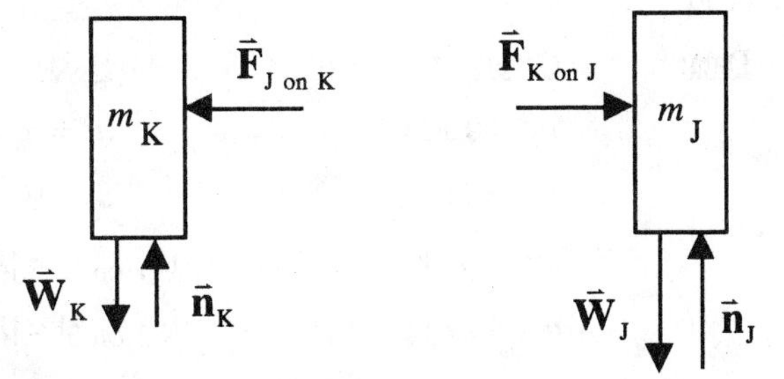

- Constraints: By N's 3rd, the magnitudes of the F's satisfy $F_{\text{J on K}} = F_{\text{K on J}}$.
- Accel.: All in the horizontal direction. Kym's is to the left, Jason's to the right.
- Coordinate Axes: Might as well choose the vertical (↑) and horizontal (→) directions, so $a_y = 0$ for both. In this case we know that the directions of the horizontal forces are the directions of the respective accelerations. Thus we don't really need to worry about the signs here.
- Force Components: All forces are either vertical or horizontal. The vertical forces on each person balance out and since we do not need $\vec{\mathbf{n}}$ for either, we only need to look at the horiz. forces and accel's.
- Construct Equations With N's 2nd: Just add the columns in the force component table and put in what you know about the acceleration components.

Kym

$$\underbrace{\sum F}_{F_{\text{J on K}}} = ma \qquad F_{\text{J on K}} = m_K a_K$$

Jason

$$\underbrace{\sum F}_{F_{\text{K on J}}} = ma \qquad F_{\text{K on J}} = m_J a_J$$

[SOLVE]

$$a_K = \frac{F_{\text{J on K}}}{m_K} = \frac{107\ \text{N}}{52.2\ \text{kg}} = 2.05\,\frac{\text{m}}{\text{s}^2} \qquad a_J = \frac{F_{\text{K on J}}}{m_J} = \frac{107\ \text{N}}{63.3\ \text{kg}} = 1.69\,\frac{\text{m}}{\text{s}^2}$$

[ANALYZE] $a_J < a_K$ is to be expected, since $m_J > m_K$. The acceleration is less that "*g*," *which* is good since "*g*" is a pretty big acceleration. For a relatively simple problem such as this you might be able to get away with skipping some of those steps. But skipping steps will generally lead to more confusion and wasted time in more difficult problems. Applying the steps methodically even for the simpler problems helps improve your speed with force analyses.

EX. 4/5.24:

[MODEL] Since we want the force between the blocks, we'll need to break this group up and look at the free-body diagram for each block separately. However, we do know that they will move together so that $\vec{\mathbf{a}}_1 = \vec{\mathbf{a}}_2 = \vec{\mathbf{a}}_3$. If we just want to find $\vec{\mathbf{a}}$ of the group, why not try treating the three blocks as a unit?

- Diagram: Notice that the darkened lines show surfaces in contact!

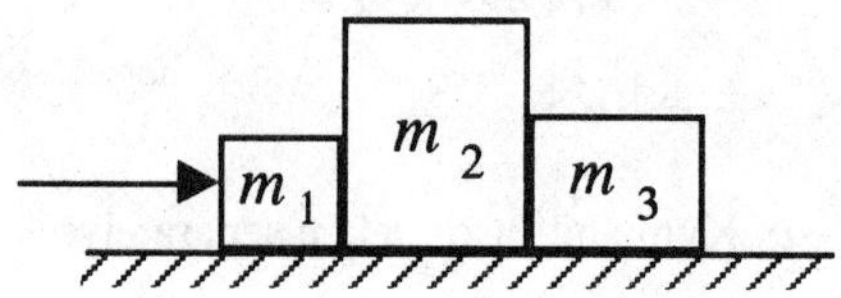

[SETUP]

- Data:

Given	Hidden	Want
$\mu_k = 0.123$	$a_1 = a_2 = a_3$	$a = ?$
$m_1 = 11$ kg	$a_y = 0$	$F_{1\text{ on }2} = ?$
$m_2 = 33$ kg	$\lvert F_{1\text{ on }2}\rvert = \lvert F_{2\text{ on }1}\rvert$	$F_{2\text{ on }3} = ?$
$m_3 = 22$ kg	$\lvert F_{2\text{ on }3}\rvert = \lvert F_{3\text{ on }2}\rvert$	
$F_{\text{applied}} = 160.5$ N		

- Receptor (object of interest): First, take the three blocks as one single receptor. This way we can avoid the "internal" forces of $F_{1\text{ on }2}$ and $F_{2\text{ on }3}$. Make sure that you see why.
- Sources of Forces: The only things pushing or pulling on the group as a whole are the Earth, the surface it slides on, and whatever is pushing from the right.
- Free-body Diagram: Note that the weight is the combined weight, as are the normal and kinetic friction forces.

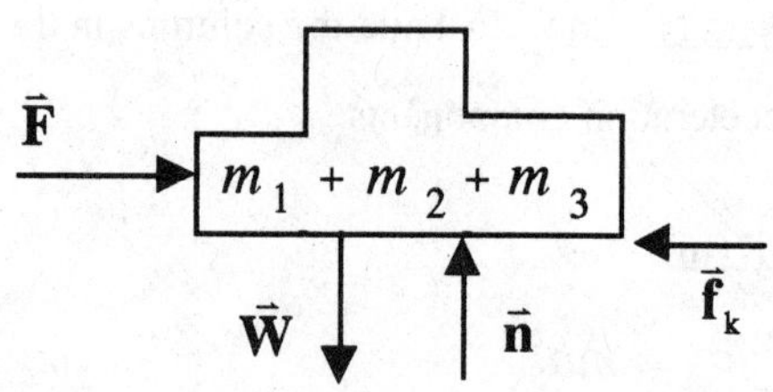

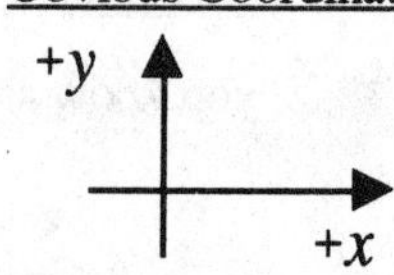

- Constraints: They move in unison.
- Accel.: $a_1 = a_2 = a_3 =$ "a" and is to the right, $a_y = 0$.
- Coordinate Axes: Shown above. Based on the direction of the acceleration and forces.
- Force Components: None of the forces need to be broken into components, so we can move straight to the 2nd law.
- Construct Equations With N's 2nd: Just notice the directions of each force. You could skip the vertical components, since it is obvious here that $W = n$. Again, this is NOT always the case.

$$\overbrace{\sum F_x}^{} = ma_x \qquad\qquad \overbrace{\sum F_y}^{} = ma_y$$

$$F - f_k = m_{total} a_x \qquad\qquad n - W = m \cdot 0$$

$$F - \mu_k n = m_{total} a_x$$

[SOLVE] Glancing up at the data table we see that the only unknown is a. Go for it. (We are restricted to two sig. fig.'s by the mass data.)

Vertical $\Rightarrow$ $n = W = (m_1 + m_2 + m_3)\,g$

Horizontal $\Rightarrow$ $a_x = \dfrac{F - \mu_k n}{m_{total}} = \dfrac{F - \mu_k (m_1 + m_2 + m_3)\,g}{m_{total}}$

$= 1.2\ \dfrac{\text{m}}{\text{s}^2}$

[ANALYZE] These blocks weigh about (66 kg) (10 m/s^2) = 660 N ≈ 660 (1/4 lb) = 165 lb. and are pushed by a 160.5 N ≈ 40 lb force which is countered by friction. This gives a substantial, but not enormous acceleration. Seems reasonable enough.

[SETUP - part 2]

- Receptor (objects of interest): Now we take each object separately. This "forces" us to look at the forces that one block exerts on another. But that's exactly what we're after! (Said another way, the forces which were "internal" when we grouped the blocks now become "external" forces on the blocks.) Remember: select the object of interest ("receptor") and ask yourself "What objects are touching it?" That, along with the pull from the Earth ("weight"), will give you all of the forces on your receptor.
- Sources of Forces: Each block has the Earth pulling on it, the table's normal and friction forces, and one or more other blocks touching it. Only the first block has F from something on the right.
- Free-body Diagrams: Notice the N's 3rd law force pairs. If you don't notice these, you will introduce way too many variables. Also notice that the weight, normal, and friction forces on each depend on the mass of each. Thus we subscript like mad to keep it all straight.

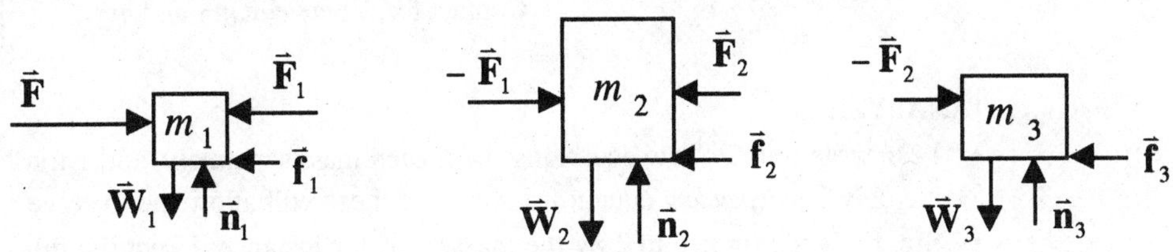

- Constraints: Just the equality of the N's 3rd law force pairs.
- Accel.: We still have $a_1 = a_2 = a_3 =$ "a" and is to the right, $a_y = 0$. It is, after all, the same physical situation that we are analyzing.
- Coordinate Axes: Same as before.
- Force Components: Again, there are no forces to break down.

- Construct Equations With N's 2nd: Since there are no vertical forces other than the weights and normal forces, and there is no vertical acceleration, we can see straight-away that $n_1 = W_1$ (i.e. $m_1 g$), $n_2 = W_2$ (i.e. $m_2 g$), and $n_3 = W_3$ (i.e. $m_3 g$). THE SIGNS!! ... Since I know the directions of the forces, I will put the force variables into N's 2nd with the signs written explicitly. That is, the variables represent the positive magnitudes. If I happen to guess a direction wrong, I will solve for that variable and find a negative number. That will tell me "You guessed the direction wrong. It's the other way!!" So when I put F_1 in the equation for m_1, I subtract it since $\vec{\mathbf{F}}_1$ is in the direction I called negative. When I put F_1 in the equation for m_2, I add it since $-\vec{\mathbf{F}}_1$ is in the direction I called positive. Keep it simple. The vectors are in opposite directions, so one is the negative of the other. But when it comes to putting them into N's 2nd, look at their directions relative to your chosen coordinates.

[SOLVE] I'll set 'em up and solve here. Recall that we've already handled the vertical components.

$$m_1:\ F - F_1 - f_1 = m_1 a \ \Rightarrow\ F_1 = F - f_1 - m_1 a$$
$$= F - \mu_k m_1 g - m_1 a \qquad \text{(since } n_1 = m_1 g\text{)}$$
$$= F - m_1(\mu_k g + a) \qquad \text{(unknowns: just } F_1\text{!)}$$

$$m_2:\ F_1 - F_2 - f_2 = m_2 a \ \Rightarrow\ F_2 = F_1 - f_2 - m_2 a$$
$$= F_1 - \mu_k m_2 g - m_2 a \qquad \text{(since } n_2 = m_2 g\text{)}$$
$$= F_1 - m_2(\mu_k g + a) \qquad \text{(unknowns: just } F_2\text{!)}$$

m_3: $F_2 - f_3 = m_3 a \ \Rightarrow$ unknowns: we know everything here! This is a fourth equation which is redundant since we used the three blocks as a unit to start with. Had we not used the three blocks as a unit, we could have used this for the third equation.

Using $a = 1.2\ ^{m}/_{s^2}$ from the first part, I find F_1 from the m_1 equation and F_2 from the m_2 equation (or the m_3 equation). I get

Contact force between m_1 and m_2 : $\left|\vec{\mathbf{F}}_1\right| = F_1 = 130\ \text{N}\ \ (134\ \text{N})$

Contact force between m_2 and m_3 : $\left|\vec{\mathbf{F}}_2\right| = F_2 = 51\ \text{N}\ \ (50.6\ \text{N})$

[ANALYZE]

- I can write four 2nd law equations: 1 for each mass separately, and 1 for the group as a whole. But there are only 3 *independent* equations. Any 3 of these will allow me to solve for a, F_1, and F_2.
- Sig. fig.'s are limited to 2 by the masses. If, for instance, I kept the full expression for F_1, I would get $F_2 = 54.6\ \text{N} = 55\ \text{N}$. As expected, the last digit sited is indeed questionable to some degree.
- The numbers seem reasonable.

EX. 4/5.25:

[MODEL] Standard N's 2nd law force analysis.

- Diagram:

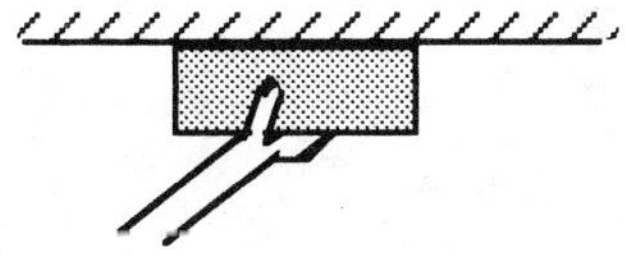

[SETUP]

- Data:

Given	Hidden	Want
$m_{\text{sponge}} = 0.045$ kg	$a_y = 0$	$F_{\text{min to get sponge moving}} = ?$
$\theta_{\text{hand}} = 25°$ from vert.		$a_{\text{x, once moving}} = ?$
$\mu_s = 0.55$		
$\mu_k = 0.21$		

- Receptor (object of interest): the sponge
- Sources of Forces: The Earth, your hand, the ceiling.
- Free-body Diagram: One will do for both situations (about to slide and sliding).

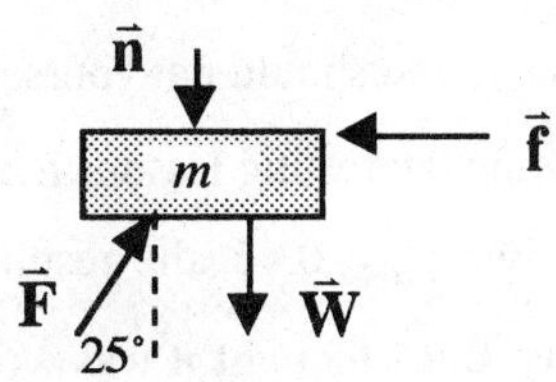

- Constraints: Sponge must slide along the ceiling.
- Accel.: (a) For $F_{\text{min to get sponge moving}}$, we use $a = 0$ to characterize the breaking point. Then F need only be a tiny bit more. (b) Once sliding, $\vec{a}$ is horiz. along the ceiling. $a_y = 0$ in both cases.
- Coordinate Axes: Might as well choose the vertical (↑) and horizontal (→) directions.
- Force Components: Start by quickly filling in the first column with the forces on your free-body diagram. Here we have $\vec{W}$, $\vec{n}$, $\vec{f}$, $\vec{F}$ (Do the easiest ones first to get rolling!)

Force	F_x (include the sign)	F_y (include the sign)
$\vec{W}$	0	$-W = -mg$
$\vec{n}$	0	$-n$
$\vec{f}$	(a) $-f = -f_{s,\max} = -\mu_s n$	0
	(b) $-f = -f_k = -\mu_k n$	0
$\vec{F}$	$+F\sin 25°$	$+F\cos 25°$

Components of F:

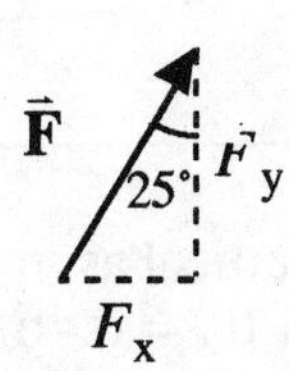

- N's 2nd At The Breaking Point: Just add the columns in the force component table and put in what you know about the acceleration components.

$$\underbrace{\sum F_x}_{F\sin 25^\circ - f_{s,max}} \quad \begin{matrix} = ma_x \\ = m\cdot 0 \end{matrix} \qquad\qquad \underbrace{\sum F_y}_{F\cos 25^\circ - n - mg} \quad \begin{matrix} = ma_y \\ = m\cdot 0 \end{matrix}$$

[SOLVE - part a] First, $f_{s,max} = \mu_s n$. From the vertical equation, $n = F\cos 25^\circ - mg$. So the horizontal equation becomes

$F\sin 25^\circ - (\mu_s n) = 0 \;\Rightarrow\; F\sin 25^\circ - \mu_s(F\cos 25^\circ - mg) = 0 \;\Rightarrow\; F(\sin 25^\circ - \mu_s\cos 25^\circ) + \mu_s mg = 0$

Thus,

$$F_{min} = \frac{-\mu_s mg}{\sin 25^\circ - \mu_s\cos 25^\circ} \xrightarrow{\text{or multiply top and bottom by}-1} \frac{\mu_s mg}{\mu_s\cos 25^\circ - \sin 25^\circ}$$

So,

$$F_{min} = -3.9\text{ N} \ldots \text{ Hmmm} \ldots$$

[ANALYSIS - part a] What?!! But you *are* pushing on the sponge in the direction shown, so the negative sign does NOT mean that the force is actually the other way. First, if you have managed to set up the problem correctly, you should pat yourself on the back and not worry *too* much about this little twist. But there are some interesting lessons here.

Had I chosen $\mu_s > 0.46$, the result would be positive and you probably wouldn't have thought twice about this. But I thought it was worth investigating a curious result that comes from an accurate analysis. Of course, the first question is whether or not the analysis *is* accurate. It is. Next, we can look at this mathematically. To do this, I'll put in the numbers and isolate the variable in question, namely μ_s.

$$F_{min} = \frac{\mu_s mg}{\mu_s\cos 25^\circ - \sin 25^\circ} = \frac{\mu_s(0.441)}{\mu_s(0.91) - 0.42}$$

Since the numerator is positive, the sign of F_{min} is determined by the denominator. For a positive result, we need $\mu_s(0.91) > 0.42$, i.e. $\mu_s > 0.46$. Then I graphed the above function on a graphing calculator and found the graph shown below. F_{min} vs μ_s:

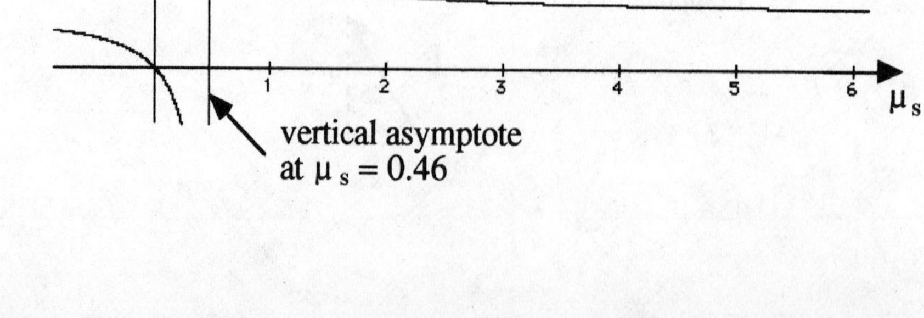

So what is the physical meaning of this result? It means that *any* force applied at 25° from the vertical that is large enough to satisfy the vertical condition that $a_y = 0$ will make the sponge accelerate along the ceiling. Let's try it out. Recall that the vertical 2nd law equation is $F\cos25^\circ - n - mg = 0$. Rewrite this as $F\cos25^\circ = n + mg$ to see that the y-component must minimally hold the sponge up. That makes physical sense. In that case $n = 0$ and $F = 0.49$ N. Now we rewrite the horizontal equation with this F, $n = 0$, and $a_x \neq 0$. This gives: $F\sin 25^\circ - (\mu_s n) = ma_x \Rightarrow F\sin 25^\circ = ma_x \Rightarrow a_x = 4.6\ ^{m}/_{s^2}$ ... wow, that's quite an acceleration. But I suppose I can get my hand from rest to 2 m/s in ½ s without much difficulty.

[SETUP & SOLVE - part b] As suggested in the hint section, let's take a look at what the acceleration is for $F = 4.0$ N (≈ 1 lb). We have all of the data organized, the free-body drawn, and the force components separated. The vertical 2nd law equation is the same: $F\cos25^\circ - n - mg = 0 \Rightarrow n = F\cos25^\circ - mg$. We can now use this in the horizontal 2nd law equation.

$$\underbrace{\sum F_x}_{} = ma_x$$

$$F\sin 25^\circ - f_k = ma_x$$

$$F\sin 25^\circ - \mu_k n = ma_x$$

$$F\sin 25^\circ - \mu_k(F\cos 25^\circ - mg) = ma_x \quad \Rightarrow \quad a_x = \frac{F\sin 25^\circ - \mu_k(F\cos 25^\circ - mg)}{m}$$

Thus, for a 4.0 N applied force, the acceleration across the ceiling is 8.6 m/s².

[ANALYZE] That's approaching the acceleration due to gravity! Is it really possible to accelerate your hand that much? Try it! Hold your pen in the air, drop it, and then move your hand below it to catch it. Hey! You can even give your hand an acceleration greater than 9.8 m/s²! Of course it has to stay attached to your arm, so this is only good for short strokes.

EX. 4/5.26:

[MODEL] Uniform circular motion. Use Newton's universal gravitational force equation for the strength of the pull on the Moon from the Earth. Use the method for applying N's 2nd law that we've been using all along.

- Diagram: See the problem statement and the free-body diagram below.

[SETUP]

- Data:

Given	Hidden	Want
$m_{pole} = 220$ kg	$a_x = 0$	$\vec{H}$
$T_1 = 440$ N	$a_y = 0$	
$T_2 = 330$ N		
$\theta = 31^\circ$ (angle shown)		

- Object of interest: the pole

- Free-body Diagram: I will guess the directions of H_x and H_y. It should be obvious that H_y is correct. If the direction of H_x is incorrect, I will get H_x = a negative number. (The negative tells me I guessed the wrong direction.)

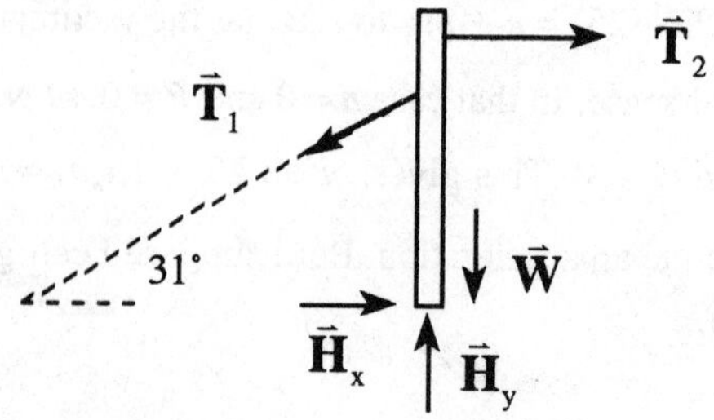

- Constraints: none really, just that it is in static equilibrium.
- Accel.: $a = 0$ (since static)
- Coordinate Axes: Might as well choose the vertical (↑) and horizontal (→) directions. $a_x = 0$, $a_y = 0$.
- Force Components: Start by quickly filling in the first column with the forces on your free-body diagram. Here we have $\vec{W}$, $\vec{H}$, $\vec{T}_2$, $\vec{T}_1$ (Do the easiest ones first to get rolling!)

Force	F_x (include the sign)	F_y (include the sign)
$\vec{W}$	0	$-W = -mg$
$\vec{H}$	$+H_x$	$+H_y$
$\vec{T}_2$	$+T_2$	0
$\vec{T}_1$	$-T_1\cos 31°$	$-T_1\sin 31°$

NOTE: In the component table of any force analysis, there are usually some vectors that only have one component and one or more that need to be broken down into two components. For vectors that need to be broken down, as $\vec{T}_1$ does above, it is usually a good idea to make a separate sketch so that you can make sure that you get the angles correct. I've skipped this step because the components of $\vec{T}_1$ are pretty easy to see in this case.

- Construct Equations With N's 2nd: Just add the columns in the force component table and put in what you know about the acceleration components.

$$\underbrace{H_x + T_2 - T_1\cos 31°}_{\sum F_x} = m\cdot 0 \quad (= ma_x) \qquad\qquad \underbrace{H_y - W - T_1\sin 31°}_{\sum F_y} = m\cdot 0 \quad (= ma_y)$$

[SOLVE]

$$\Rightarrow\ H_x = T_1\cos 31° + T_2 = +710\ \text{N} \qquad\qquad \Rightarrow\ H_y = mg + T_1\sin 31° = +2400\ \text{N}$$

Sketch $\vec{H}$: Since you have the components of $\vec{H}$, you can make a quick sketch to see the component triangle and the direction of $\vec{H}$. This sketch is NOT to scale, but it does show that $H_y > H_x$.

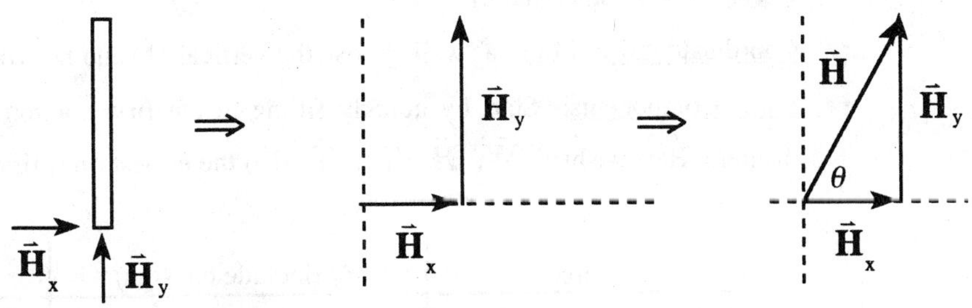

So, $H = \sqrt{H_x^2 + H_y^2} = 2500\ \text{N}$, and $\theta = \tan^{-1}\left(\frac{H_y}{H_x}\right) = 74°$

[ANALYZE] Let's see. If $T_1 = 0$, it makes both conceptual and mathematical sense that $H_x = T_2$ and $H_y = W$. The result we found is $H = 2500\ \text{N} \approx 625$ lb. The weight of the pole is mg = (220 kg) (9.8 m/s²) ≈ 2200 N, so we expect H to be a bit more, and it is. Looks OK to me.

EX. 4/5.22:

[MODEL] "Equilibrium" ⇒ $a = 0$, so $F_{net} = 0$, $F_{net,x} = 0$, and $F_{net,y} = 0$. "Static" ⇒ $v = 0$.

The ground is assumed to be able to push in any direction., thus we write the ground's force as the vertical and horizontal components H_y and H_x, respectively. (I used "H" for lack of a more compelling choice.)

- Diagram: See the problem statement and the free-body diagram below.

[SETUP]

- Data:

	Given	Hidden	Want
	$m_{pole} = 220$ kg	$a_x = 0$	$\vec{H}$
	$T_1 = 440$ N	$a_y = 0$	
	$T_2 = 330$ N		
	$\theta = 31°$ (angle shown)		

- Object of interest: the pole
- Free-body Diagram: I will guess the directions of H_x and H_y. It should be obvious that H_y is correct. If the direction of H_x is incorrect, I will get H_x = a negative number. (The negative tells me I guessed the wrong direction.)

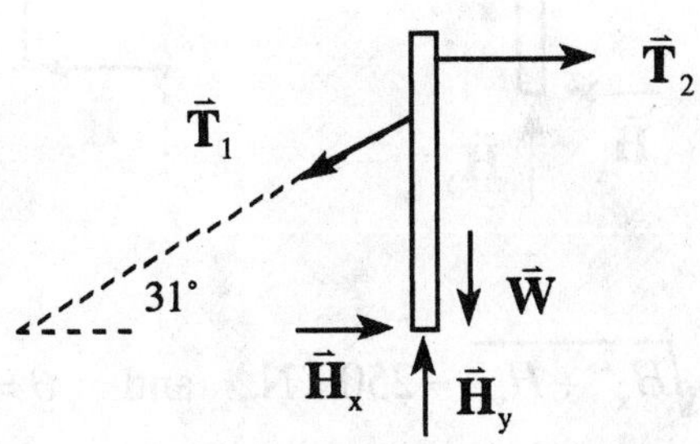

- Constraints: none really, just that it is in static equilibrium.
- Accel.: $a = 0$ (since static)
- Coordinate Axes: Might as well choose the vertical (↑) and horizontal (→) directions. $a_x = 0$, $a_y = 0$.
- Force Components: Start by quickly filling in the first column with the forces on your free-body diagram. Here we have $\vec{\mathbf{W}}$, $\vec{\mathbf{H}}$, $\vec{\mathbf{T}}_2$, $\vec{\mathbf{T}}_1$ (Do the easiest ones first to get rolling!)

Force	F_x (include the sign)	F_y (include the sign)
$\vec{\mathbf{W}}$	0	$-W = -mg$
$\vec{\mathbf{H}}$	$+H_x$	$+H_y$
$\vec{\mathbf{T}}_2$	$+T_2$	0
$\vec{\mathbf{T}}_1$	$-T_1\cos 31^\circ$	$-T_1\sin 31^\circ$

NOTE: In the component table of any force analysis, there are usually some vectors that only have one component and one or more that need to be broken down into two components. For vectors that need to be broken down, as $\vec{\mathbf{T}}_1$ does above, it is usually a good idea to make a separate sketch so that you can make sure that you get the angles correct. I've skipped this step because the components of $\vec{\mathbf{T}}_1$ are pretty easy to see in this case.

- Construct Equations With N's 2nd: Just add the columns in the force component table and put in what you know about the acceleration components.

$$\underbrace{H_x + T_2 - T_1\cos 31^\circ}_{\sum F_x} = \underbrace{m\cdot 0}_{ma_x} \qquad \underbrace{H_y - W - T_1\sin 31^\circ}_{\sum F_y} = \underbrace{m\cdot 0}_{may}$$

[SOLVE]

$$\Rightarrow \; H_x = T_1\cos 31^\circ + T_2 = +710 \text{ N} \qquad \Rightarrow \; H_y = mg + T_1\sin 31^\circ = +2400 \text{ N}$$

Sketch $\vec{\mathbf{H}}$: Since you have the components of $\vec{\mathbf{H}}$, you can make a quick sketch to see the component triangle and the direction of $\vec{\mathbf{H}}$. This sketch is NOT to scale, but it does show that $H_y > H_x$.

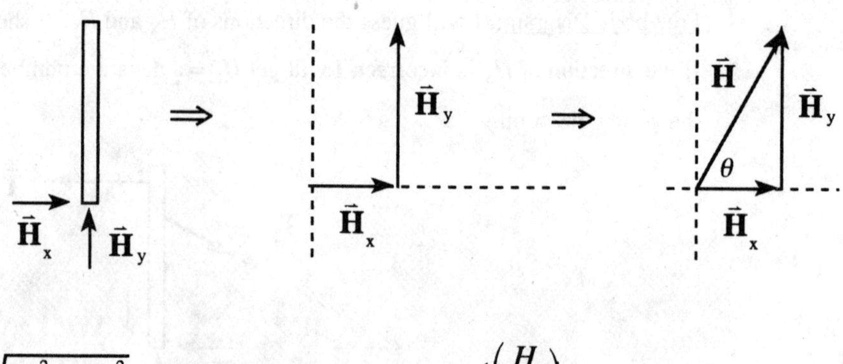

So, $H = \sqrt{H_x^{\,2} + H_y^{\,2}} = 2500 \text{ N}$, and $\theta = \tan^{-1}\left(\dfrac{H_y}{H_x}\right) = 74^\circ$

[ANALYZE] Let's see. If $T_1 = 0$, it makes both conceptual and mathematical sense that $H_x = T_2$ and $H_y = W$. The result we found is $H = 2500$ N ≈ 625 lb. The weight of the pole is mg = (220 kg) (9.8 m/s²) ≈ 2200 N, so we expect H to be a bit more, and it is. Looks OK to me.

EX. 4/5.22:

[MODEL] "Equilibrium" ⇒ $a = 0$, so $F_{net} = 0$, $F_{net,x} = 0$, and $F_{net,y} = 0$. "Static" ⇒ $v = 0$.

The ground is assumed to be able to push in any direction., thus we write the ground's force as the vertical and horizontal components H_y and H_x, respectively. (I used "*H*" for lack of a more compelling choice.)

- Diagram: See the problem statement and the free-body diagram below.

[SETUP]

- Data:

Given	Hidden	Want
$m_{pole} = 220$ kg	$a_x = 0$	$\vec{\mathbf{H}}$
$T_1 = 440$ N	$a_y = 0$	
$T_2 = 330$ N		
$\theta = 31°$ (angle shown)		

- Object of interest: the pole
- Free-body Diagram: I will guess the directions of H_x and H_y. It should be obvious that H_y is correct. If the direction of H_x is incorrect, I will get H_x = a negative number. (The negative tells me I guessed the wrong direction.)

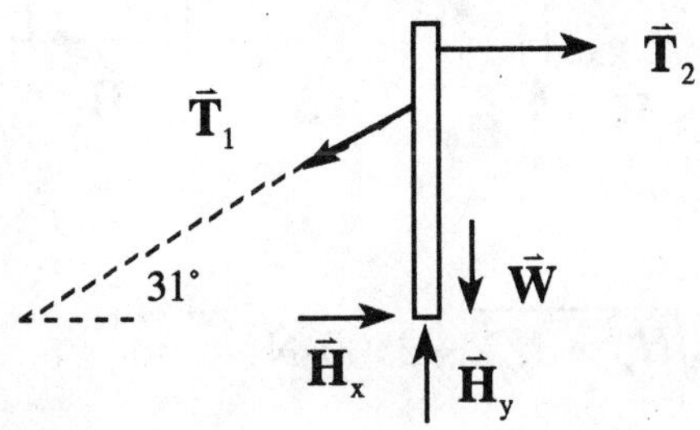

- Constraints: none really, just that it is in static equilibrium.
- Accel.: $a = 0$ (since static)
- Coordinate Axes: Might as well choose the vertical (↑) and horizontal (→) directions. $a_x = 0$, $a_y = 0$.
- Force Components: Start by quickly filling in the first column with the forces on your free-body diagram. Here we have $\vec{\mathbf{W}}$, $\vec{\mathbf{H}}$, $\vec{\mathbf{T}}_2$, $\vec{\mathbf{T}}_1$ (Do the easiest ones first to get rolling!)

Force	F_x (include the sign)	F_y (include the sign)
$\vec{\mathbf{W}}$	0	$-W = -mg$
$\vec{\mathbf{H}}$	$+H_x$	$+H_y$
$\vec{\mathbf{T}}_2$	$+T_2$	0
$\vec{\mathbf{T}}_1$	$-T_1\cos 31°$	$-T_1\sin 31°$

NOTE: In the component table of any force analysis, there are usually some vectors that only have one component and one or more that need to be broken down into two components. For vectors that need to be broken down, as $\vec{\mathbf{T}}_1$ does above, it is usually a good idea to make a separate sketch so that you can make sure that you get the angles correct. I've skipped this step because the components of $\vec{\mathbf{T}}_1$ are pretty easy to see in this case.

- Construct Equations With N's 2nd: Just add the columns in the force component table and put in what you know about the acceleration components.

$$\overbrace{H_x + T_2 - T_1\cos 31^\circ}^{\sum F_x} \overset{= ma_x}{= m\cdot 0} \qquad \overbrace{H_y - W - T_1\sin 31^\circ}^{\sum F_y} \overset{= may}{= m\cdot 0}$$

[SOLVE]

$$\Rightarrow\ H_x = T_1\cos 31^\circ + T_2 = +710\text{ N} \qquad \Rightarrow\ H_y = mg + T_1\sin 31^\circ = +2400\text{ N}$$

Sketch $\vec{\mathbf{H}}$: Since you have the components of $\vec{\mathbf{H}}$, you can make a quick sketch to see the component triangle and the direction of $\vec{\mathbf{H}}$. This sketch is NOT to scale, but it does show that $H_y > H_x$.

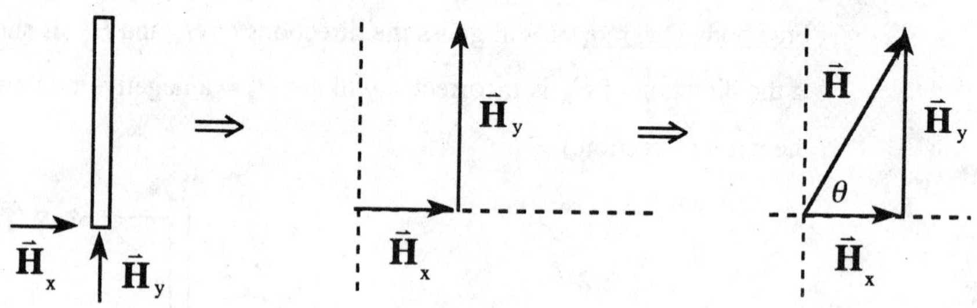

So, $H = \sqrt{H_x^{\,2} + H_y^{\,2}} = 2500\text{ N}$, and $\theta = \tan^{-1}\left(\dfrac{H_y}{H_x}\right) = 74^\circ$

[ANALYZE] Let's see. If $T_1 = 0$, it makes both conceptual and mathematical sense that $H_x = T_2$ and $H_y = W$. The result we found is $H = 2500\text{ N} \approx 625\text{ lb}$. The weight of the pole is mg = (220 kg) (9.8 $^{m}/_{s^2}$) ≈ 2200 N, so we expect H to be a bit more, and it is. Looks OK to me.